Building Gotham

To Tina Vidal —

From your record at FIU is it clear that we can expect great things from you.

KR

Building Gotham

Civic Culture and Public Policy
in New York City, 1898–1938

Keith D. Revell

The Johns Hopkins University Press
Baltimore and London

© 2003 The Johns Hopkins University Press
All rights reserved. Published 2003
Printed in the United States of America on acid-free paper
9 8 7 6 5 4 3 2 1

The Johns Hopkins University Press
2715 North Charles Street
Baltimore, Maryland 21218-4363
www.press.jhu.edu

Library of Congress Cataloging-in-Publication Data
Revell, Keith D., 1963–
  Building Gotham : civic culture and public policy in New York City, 1898–1938 / Keith D. Revell.
    p. cm.
"A catalog record for this book is available from the British Library."
  Includes bibliographical references and index.
  ISBN 0-8018-7073-9 (hard : alk. paper)
  1. City planning—New York (State)—New York. 2. Urban policy—New York (State)—New York. 3. City and town life—United States—New York. 4. New York (N.Y.)—History—19th century. I. Title.
  HT168.N5 R48 2003
  307.1'216'09747—dc21

2003000599

A catalog record for this book is available from the British Library.

*While all cities say the same thing,
New York says it first.*

O. HENRY, liberally paraphrased
by the *New York Times*

Contents

Preface and Acknowledgments ix

Introduction: Conceiving the New Metropolis: Expertise, Public Policy, and the Problem of Civic Culture in New York City — 1

PART 1: Private Infrastructure and Public Policy — 15

1 "The Public Be Pleased": Railroad Planning, Engineering Culture, and the Promise of Quasi-scientific Voluntarism — 17
2 Beyond Voluntarism: The Interstate Commerce Commission, the Railroads, and Freight Planning for New York Harbor — 58

PART 2: Public Infrastructure, Local Autonomy, and Private Wealth — 99

3 Buccaneer Bureaucrats, Physical Interdependence, and Free Riders: Building the Underground City — 101
4 Taxing, Spending, and Borrowing: Expanding Public Claims on Private Wealth — 143

PART 3: Urban Planning, Private Rights, and Public Power — 183

5 City Planning versus the Law: Zoning the New Metropolis — 185
6 "They shall splash at a ten-league canvas with brushes of comets' hair": Regional Planning and the Metropolitan Dilemma — 227

Conclusion: "An almost mystical unity": 268
Interdependence and the Public Interest in
the Modern Metropolis

Appendix 281
Notes 285
Index 321

Preface and Acknowledgments

I began this project as an inquiry into the ways that the concept of efficiency affected the regulation of skyscrapers and railroads in New York City during the Progressive Era. As I learned more about what the experts engaged in those projects were doing, I discovered that *efficiency* played a less important role in their worldview than *interdependence*—the latter a far more powerful concept with profound political implications. Interdependence, I concluded, provided the central ideal in a civic culture of expertise that conceptualized the city and the region in terms of common interests that could only be pursued adequately by invigorated public institutions. The book that has emerged thus explores the discovery (or perhaps, rediscovery) of the potential of government to solve collective problems. I do not hold up the solutions described here as models for our own time, but I must admit that I admire the commitment to the life of the city that they represent.

Among the many people who have aided me in this journey, I am most deeply indebted to Olivier Zunz—a scholar of truly remarkable insight, intellectual fearlessness, and generosity. Olivier taught me the craft of history and I am honored to have studied in his workshop. My ideas on railroad building and regulation have all been shaped by debates with Jameson W. Doig, whose unswerving focus on how government actually works kept me on my toes. Charles McCurdy, my mentor in legal history, made comments during my dissertation defense that guided my subsequent efforts; his critical reading of my research on the constitutional aspects of zoning has been invaluable.

The intellectual community at the University of Virginia nurtured this project from the beginning. I had the privilege of studying with William Taylor, Stephen Innes, and Robert Cross, each of whom helped me to understand what it means to explain social change. Dorothy Ross and William Holt both asked key questions early in the process that lingered with me throughout the writing of the book; I hope that I have finally answered them. I am very grateful to William Harbaugh, an inspiration as a historian and a teacher, whose words of advice—"to finish a book, you must be obsessed with it"—helped me through some difficult stretches. My friends Jack Brown and Bruce Fort provided encouragement when I needed it most.

The intellectual community at Davidson College likewise played a role in this effort. The late Ernest Finney Patterson's history of economic thought class set me on the path that led to this book, and Brian Shaw gave me a broader theoretical context for interpreting the ideas I picked up from Dr. Patterson. Hours of discussion with Murray Simpson were crucial in the formation of my early approach to this subject. And I want to extend a special thanks to David Shi, whose enthusiasm and breadth of knowledge convinced me to pursue my work in history.

Jeffrey Stine, of the National Museum of American History, offered insights into engineering culture in the formative stages of my research. My colleague at Florida International University, Howard Frank, helped me with the finer points of public debt management. Bob Brugger, history editor at the Johns Hopkins University Press, made illuminating suggestions for revising the manuscript and waited very patiently for them to be realized. I am also indebted to the members of the Social Science Research Council's Working Group on the Built Environment who allowed me to participate in their workshop on New York City, for it was there that I resolved to write this book.

This list would not be complete without acknowledging my family. My parents, Walter and Sheila Revell, have shown me, among many other things, what civic life is all about. Above all, I am grateful to my wife, Katherine, without whom I would not have completed this project. I dedicate the book to her, as a small gesture of my love and appreciation.

In spite of the good judgment of all of those from whom I have received counsel, there are, undoubtedly, errors, omissions, and flaws of style and interpretation in this book, for which I alone am responsible.

Building Gotham

INTRODUCTION

Conceiving the New Metropolis
Expertise, Public Policy, and the Problem of Civic Culture in New York City

> Invariably the newcomer to New York is puzzled to comprehend the city. From the bridge of an ocean liner coming up the harbor, the distant sky line of Manhattan is undoubtedly one of the greatest spectacles in the world, whether seen through the smoky blue mist of morning, or at night when the thousand windows of Wall Street glow through the darkness as though lit up by inner fires. No more impressive sight has ever been created solely by the industry and imagination of man. The superb majesty of the distant city at once gives the impression of possessing some inner unity and consistency. Yet when the vessel docks and we come to tread the crowded streets of the metropolis, this unity of New York City constantly eludes our search. So great has been [the city's] growth that year after year it has tended more and more to split up into separate communities, each with its own particular headquarters in the great city and its own particular interests and problems. Because of this fact we all tend, I feel, to lose our sense of citizenship in the metropolis itself.
> —EDWARD H. H. SIMMONS, president,
> New York Stock Exchange, 1927

In 1898, the residents of New York City embarked on an ambitious project in collective living. That year, the state legislature created Greater New York, uniting Manhattan, Brooklyn, the communities of Queens County, Staten Island, and the Bronx—a total of ninety-six governmental units in all—into a new metropolis of nearly three and a half million people—twice as large as Chicago, its closest domestic rival. By the time of consolidation, New York had secured its position as the nation's busiest port, its corporate and financial headquarters, its undisputed intellectual capital, and the most ethnically diverse of American cities. But this city of superlatives, a testament to the vitality and heterogeneity of nineteenth-century America, struggled with the new dimensions of urban life it comprised. As Paul Bourget, a member of the

French Academy, exclaimed in 1893, "This is not even a city in the sense which we [Europeans] understand the word. This is a table of contents of unique character. It is so colossal, it encloses so formidable an accumulation of human efforts, as to overpass the bounds of the imagination."[1] This profusion of activity resulted in an array of unprecedented problems that seemed beyond the capacity of the city's institutions of collective decision making—courts, parties, corporations, markets, and local governments—and threatened New York's future as the great American metropolis.

No one had ever tried to manage a city of this magnitude, diversity, and complexity, at least not in the United States. Only London surpassed New York in population, but the capital of Great Britain did not exhibit the hallmarks of modernity—skyscrapers, immigration, concentrated corporate power—in the same degree as Gotham, the capital of capitalism.[2] New York led the way, as usual, creating and striving to resolve the problems of size, concentration, and proximity characteristic of megacities in the twentieth century. Even today, only a handful of places around the globe have attempted such a feat, and those that have, especially in the developing world, continue to wrestle with many of the questions posed by the consolidated city a century ago. New Yorkers were the first to address the consequences of collective living on this new scale and thus to determine whether American culture could be adapted to establish the bonds of community—the terms of interconnection and mutual obligation—for a city of four, six, eight million.

Consolidation itself emerged from a perception of shared problems facing the communities that would make up Greater New York. Gotham's merchants had long clamored for metropolitan government to reorganize New York harbor, which illustrated on a vast scale the shortcomings of large corporations and local officials as economic managers.[3] Between the 1870s and 1900, New York's share of the nation's foreign commerce declined (although the volume and value increased), igniting a heated public debate over how to insure the long-term prosperity of the port district. The competition for space and fragmented ownership of the waterfront created a chaotic, inefficient, and costly system of freight transportation between Manhattan piers and New Jersey rail lines, prompting shippers, manufacturers, politicians, and boosters on both sides of the Hudson River to question the railroads' ability to manage the harbor for their mutual benefit. When the port became paralyzed during World War I—with railcars backed up as far as Chicago—evidence of the need for new institutions to oversee the most important economic asset in the region became even clearer.

This crisis of overcrowding and uncoordinated growth did not stop at the waterfront. A downtown packed with skyscrapers and sweatshops, department stores and factories, high-class hotels and squalid tenements led to the uncomfortable mixing of native and immigrant cultures and competition for space between small entrepreneurs and corporate giants. Population in sections of Manhattan's Lower East Side (just blocks from the national nerve center of high finance and corporate capitalism) increased to more than one thousand people per acre by 1905, producing some of the most densely inhabited areas in the world. Reformers blamed overcrowding alternately on the moral failings of immigrant tenants and rapacious landlords, but the skyscraper boom provided a larger context in which to interpret congestion. The tallest buildings of the prewar era—Cass Gilbert's Woolworth Building (the fifty-story "Cathedral of Commerce" that overshadowed City Hall), Ernest Flagg's Singer Building, and Napoleon LeBrun's Metropolitan Life Tower—served as but a prelude for the giant structures of the 1920s. By 1929, the city bulged with an astounding 2,479 buildings of more than ten stories—two thousand more than Chicago, the birthplace of the skyscraper.[4] New Yorkers thus confronted an urban landscape entirely new in the history of civilization—one that illustrated for many observers that private property, given undue protection by the courts, had run amok in the great city, requiring stringent regulations on individual rights and real estate markets to prevent Gotham from choking itself to death.

Population densities downtown and the sheer size of Greater New York placed extraordinary demands on infrastructure, highlighting the shortcomings of the political parties, private companies, and municipal technicians working to make the city livable. In 1898, the Brooklyn Bridge remained the only direct link for the 143 million commuters crossing each year between the region's central business district in Manhattan and growing suburban areas in Brooklyn and Queens. More than 70 million commuters a year from New Jersey, which functioned increasingly as both the residential and industrial periphery of lower Manhattan, still made their way across the busy Hudson by boat—the same method used by the Dutch explorers who had settled the island nearly three centuries earlier. More pressing still, neither the Croton waterworks, built in the 1840s, nor its subsequent additions could keep pace with the vertical and horizontal expansion of the city. The steel and concrete skyscrapers that housed armies of white-collar workers at the turn of the century required as much as twenty times more water than the low-rise brick structures they replaced, and the buildings only got bigger. In many places in the city, pressure was so low that water did not rise into pipes above basement level without supplemental pumping, causing considerable anxiety among insurance executives who knew just how much flammable wealth sat in midtown warehouses and factories. The citizens of Brooklyn, who relied on Long Island reservoirs, lived under an annual threat of water famine. Despite the shortage of water, the city dumped prodigious amounts of raw sewage into surrounding rivers—roughly 300 million gallons per day by the time of consolidation. With local officials unwilling and unable to remedy the problem—financially, technically, or politically—the pollution oozed its way beyond the harbor, forcing health authorities to close beaches on the outskirts of the city and prompting Robert Moses to build swimming pools to prevent outbreaks of typhoid and cholera. To clean up the pungent mess lapping at the city's waterfront, to provide enough water for domestic and industrial use (and public safety) in the five boroughs, and to cope with the ebb and flow of economic activity linking Manhattan to Long Island and New Jersey, municipal officials would have to think beyond electoral districts and significantly expand public claims on private wealth in spite of the lingering shadow of Tweedism.

Responding to these challenges forced New Yorkers to face the limitations of their governing institutions and approaches to collective decision making. The failure of railroad executives, bankers, machine politicians, judges, elected officials, and real estate developers to resolve the problems of the modern city precipitated a crisis of legitimacy that opened the door for new actors—experts in engineering, law, finance, public health, and architecture, armed with specialized knowledge, technical skills, and a new perspective on the metropolis—who aspired to remake the institutional relationships necessary to build and manage Gotham. This book examines how those experts, working in conjunction and often at odds with powerful economic and political groups, fought to establish new terms of togetherness for New York City between the 1890s and 1940s and set an example for the rest of the world.

Civic Culture, Public Policy, and Institutional Change

New Yorkers had few cultural tools for conceptualizing community on this new scale. Like most Americans, they inherited from the nineteenth century a civic culture of privatism embedded in their institutions of collective decision making.[5] The complex of courts, parties, corporations, markets, and local governments that New Yorkers used to respond to common problems favored voluntary over compulsory commitments to the public welfare, individual rights over collective needs, and parochial orientations to policy issues over more comprehensive perspectives. Even the greater city itself, the product of a compromise between machine politicians and reformers, strained to accommodate the broader conception of public good envisioned by Andrew Green, the father of consolidation; although the charter of 1898 did set up a centralized government, it gave way to a new charter in 1902, much to Green's dismay, which reestablished a leading role for borough officials, shifting power over city-building policies back to a more local level. New Yorkers thus approached the shared circumstances of modernity (problems that cut across electoral boundaries, class lines, economic interests, and ethnic ties) with a cultural inheritance that eschewed the institutionalized collective commitments that we now associate with urban life (at least in many places around the world), making it difficult for them to articulate and pursue general interests for so large and diverse a community.

By analyzing debates over commuter and freight transportation, municipal infrastructure and fiscal management, zoning and regional planning, this book shows how a civic culture of expertise developed as a response to the crisis of legitimacy of the city's institutions of collective decision making in the early years of the twentieth century. By virtue of living and working in a city of unique scale, density, and diversity, experts in business, government, academic, philanthropic, and consulting roles together formulated a distinctive *civic* culture—a shared way of seeing the interconnections among the multitudinous communities of the greater city. Their cultural outlook provided the context for discussion of the burdens and restrictions that an invigorated government could impose on private property, individual liberty, group prerogatives, and local autonomy in pursuit of collective interests.[6] In place of voluntarism, individualism, and localism, the civic culture of expertise offered centralization (the expansion and coordination of regulatory and administrative powers), interdependence, and a citywide perspective on urban problems.

The historical problem posed by this way of conceptualizing the new metropolis arose because cities, like nations, contain multiple cultures and thus multiple approaches to social integration. "American history has been in considerable measure a struggle between rival ways of getting together," historian John Higham noted in one of his most insightful arguments. "In actual experience the alternatives have overlapped very greatly. Instead of facing a clear choice between commensurate loyalties, Americans have commonly been enmeshed in divergent systems of integration."⁷ Metropolitan New York City exhibited many such forms of social integration at the turn of the century: ethnic, religious, and fraternal communities; communities of economic interest, such as labor unions, taxpayer associations, and producer groups; communities organized along geographical lines such as neighborhoods, wards, and boroughs; and of course political communities like Tammany Hall (the regular Democratic Party organization in Manhattan) and the Fusion (a perennial coalition of anti-Tammany forces, including Republicans, Independents, and reform Democrats). But New York also operated at more extensive levels of connectedness—at the level of the city as a whole and sometimes at a regional level—and the civic culture of expertise evolved around the notion (inherent in consolidation) that the city required a more comprehensive level of social integration to deal with problems that spanned carefully constructed political, ethnic, economic, and geographical boundaries. Citywide problems required citywide solutions; regional problems required regional solutions; and New York suffered from the limitations of institutions that could not quite manage to operate at those levels. Connecting the whole with the parts—finding ways to bring the larger-scale organization of urban life into existence while negotiating terms of coexistence with the less comprehensive forms of community—constituted the basic difficulty of civic culture and institutional change in early-twentieth-century New York City.

This process of cultural conflict and institutional innovation included, but was not limited to, the classic patterns of urban coalition building. Corporate executives, small businessmen, machine and reform politicians, neighborhood associations, real estate developers, civic leaders, and a variety of technically trained professionals contributed to the decisions necessary to construct railroad tunnels underneath the Hudson River, expand the city's borrowing capacity, create a regional sewage-treatment system, and regulate skyscrapers. But these projects involved more than power struggles and political alliances. Although experts often took the lead building coalitions in these policy debates, they also pushed for more fundamental change, redefining the values embedded in decision-making institutions and expanding the roles and perspectives they brought to bear on common problems. The institutions pushed back, of course, resisting and redirecting innovation by reasserting the values and prerogatives of private property, individual liberty, corporate and electoral power, local control, and fiscal conservatism that animated them. Institutional change and cultural conflict thus proceeded hand in hand, with a cycle of punch/counterpunch driving the state-building process even more so than pluralistic bargaining.[8]

By shifting the focus of inquiry to the technical actors involved in this dual process, this book demonstrates how engineers, economists, public health specialists, architects, and lawyers attempted to change the terms of public policy discourse by reconceptualizing the metropolitan community. Through their efforts to solve the problems associated with railroad planning, public works construction, and land-use regulation, they fought existing approaches to decision making and the groups invested in them in order to establish alternative standards of governmental legitimacy, expand existing public powers, and create new municipal and regional institutions to implement their vision of a planned metropolis.[9] That process, riddled with conflicts and contradictions, produced neither a clear centralization of power nor an obvious triumph of any one political or economic group; instead, it generated a diffuse centralization of governmental authority—a multiplication of partially centralized, fragmented, and overlapping centers of public and quasi-public power governing a metropolitan community that struggled to achieve a marriage among divergent systems of integration. By the 1960s, the resulting intergovernmental problem had grown to breathtaking proportions, with 1,467 distinct public entities exercising political authority in the New York region—a thousand more than forty years earlier.[10]

Problematizing the Civic Culture of Expertise: Beyond Efficiency

This book explores the institutional and cultural dynamics of organizing the modern city along the lines proposed by the experts who became such a force in late-nineteenth-century America by virtue of the central roles they played in big business, government, academia, and philanthropy. While myriad other actors and approaches to social integration shaped New York City, the civic culture of expertise provided the decisive linkages between the technical discourses of problem solving and the larger public life of the city that explain the diffuse centralization of authority that resulted from Progressive Era state building. Historians have typically relied on the concept of efficiency to demonstrate that connection, and the transformation of this powerful industrial principle into an influential social ideal certainly affected the governance of cities.[11] Although efficiency played an important role in the policy debates explored here, it does not fully comprehend the alternative approach to urban problems offered by the civic culture of expertise. In addition to privileging an abstraction from engineering and economics over those from law, architecture, and public health, the idea of efficiency has lent itself to narratives of exploitation, which tend to collapse the complex problem of "divergent systems of integration" in large cities into a simplistic battle between democratic and undemocratic approaches to decision making, especially when coupled to tensions between native and immigrant cultures: efficiency versus democracy, rationality versus participation, top-down versus bottom-up.[12]

As compelling and illuminating as that formulation seems, the presumed conflict between efficiency and participatory democracy played at best a peripheral role in efforts to centralize public authority in New York. Most policies made in American cities, especially with regard to public monies and regulatory power, did not involve mass politics, either in the nineteenth century or in the twentieth, in spite of the successful movements for inclusion that reshaped part of our political landscape during this period.[13] Government with the consent of the governed, rather than government by the people directly, has generally characterized urban decision making in this country. The experiments in railroad regulation, infrastructure development, and land-use planning undertaken by experts in New York City did not rely on the suppression of mass political movements, although electoral changes affected state building (profoundly, but usually indirectly) and elected officials influenced every area of policy. Instead, the institutional changes examined in this book resulted in a transfer of decision-making power between intermediate forms of centralized authority—from the invisible government of corporations, party bosses, judges, and property owners to municipal bureaucrats—or established public controls over previously unregulated private choices or market processes.[14] To be sure, property owners did object to the use of newly centralized regulatory authority as "undemocratic," but they defended "democracy" as individual economic rights and the prerogatives of ownership rather than "democracy" as group rights of cultural self-determination and political participation.

To understand the nature of the conflicts and changes generated by the entrance of technical discourses to the public sphere, therefore, this book looks beyond efficiency to other concepts that experts brought to city building and that threatened the "ways of getting together" represented by courts, parties, corporations, markets, and local governments. Rather than producing a split between democratic and undemocratic forms of political interaction, the civic culture of expertise caused fractures along three interrelated fault lines of public discourse affecting the creation, operation, and structure of new governing institutions: the legacy of active government as a foundation for state building, the implications of interdependence for policy making, and the constitution of general interests vis-à-vis the structure and scope of government power.

The civic culture of expertise built on an ambiguous legacy of active government that served as both a starting point for state-building coalitions and a potential barrier to fundamental institutional change: The experts who wanted to expand the role of government to solve urban problems in the New York region did not start from scratch. Although it had long been something of a national conceit that Americans considered government a "necessary evil" and strove for minimal public interference in private affairs, cities generally, and New York especially, had an almost unbroken heritage of public activism—as evident in three of the most significant infrastructure achievements of nineteenth-century government: the Croton waterworks, Central Park, and the Brooklyn Bridge.[15] British observer James Bryce captured this contradiction in *The American Commonwealth* (1888): "Though the Americans have no theory of the State and take a narrow view of its functions, though they conceive themselves to be devoted to *laissez faire* in principle, and to be in practice the most self-reliant of peoples, they have grown no less accustomed than the English to carry the action of government into ever-widening fields."[16] This vernacular political philosophy of limited government had such a pervasive influence on the American mind that each generation had to discover the governmental habit anew. "Land grants, franchise steals, favorable court decisions, and supple politicians appeared in a bewildering array" during the nineteenth century, observed Queens-native Benjamin Parke De Witt in 1915; "long before the country realized it, the government *was* being used—not in the interests of the many, but in the interests of the few." Although muckrakers focused on the deals between political bosses and corporate moguls, businesses from transcontinental railroads to local saloons benefitted from some form of public assistance. "What these big and little businesses all had in common," journalist Lincoln Steffens concluded, "was not size but the need of privileges," from franchises and protective tariffs to lenient enforcement of blue laws. New

York's well-deserved reputation for political corruption—due largely to the brief but spectacular reign of boss Tweed—did result in a significant reduction of direct municipal spending, but businessmen continued to ask for public favors, politicians continued to grant them, and municipal government continued to pave streets, build parks, furnish street-car franchises, and construct and maintain waterfront facilities, among its many promotional activities.[17]

This division between the theory and practice of active government shaped the process of institutional innovation in New York City at every turn. On the one hand, assisting local businesses and promoting economic growth provided ready-to-hand strategies to garner support for expanded municipal powers, and experts interested in city building often enlarged the role of local government by linking their efforts to promotional purposes. On the other hand, this rather opportunistic approach to state building meant confronting the differences between pursuing some group's particular interests and creating a durable intellectual foundation and widely accepted political rationale for the regulatory state. While accessibility to public economic and legal assistance provided a consistent justification for governmental activism, more often than not it resulted in uncertain standards of bureaucratic legitimacy. Since they could rarely find allies pure in motive, the experts who tackled the problems of freight planning, water supply, and land-use regulation in New York City had to cut deals where they could and attempt to induce their fellow citizens to embrace a new understanding of the relationships between whole and parts in the modern metropolis. To what extent, they were forced to ask, could they rely on the parochial motives of collaborators to buttress their vision of the collective good before subverting that larger goal altogether?

The civic culture of expertise emphasized the physical and economic interdependence of the modern city—a concept with radical implications for policy making when employed by public sector institutions: The institutional implications of interdependence influenced every phase of city building in Greater New York as experts learned to think in terms of the larger trends and processes necessary to sustain the shared working and living environments of the regional metropolis. Engineers studied traffic flows between commercial centers and bedroom communities and responded with bridges, tunnels, rail lines, and highways. Sanitarians analyzed the combined effects of uncoordinated sewage-disposal programs and then devised the subterranean structures that linked tenements to skyscrapers and suburbs to factories in order to clean up the harbor. Comptrollers and bankers discerned and then tried to manage the reciprocal relationships between private real estate values and the city's capacity to fund subways and schools. Economists uncovered the developmental processes that linked downtown congestion to the diversification of cities on the metropolitan periphery in an effort to make the region—that "new dynamic Something" explored by the authors of the landmark *Regional Plan of New York and Its Environs*—a more concrete concept and thus an imperative focus of public policy.[18] As their perception of interdependence blurred the boundaries between city and suburb, private property and public welfare, individual wealth and collective resources, the experts saw the political and legal distinctions that divided the metropolis as less and less meaningful. Instead, they thought in terms of the city as a whole and the city as a system, and other influential actors increasingly borrowed those concepts to describe urban problems and their solutions.[19] By the 1920s, even a career party politician like Mayor Jimmy Walker adopted the experts' language, urging his fellow citizens to "look upon the city as a whole" and "to plan our improvements with a view to the best development of the city as a whole."[20]

Armed with mounting evidence of metropolitan interdependence and its acceptance by powerful decision makers, technicians who took an interest in the broader issues of city building and urban management attempted to make "the city as a system" the operating principle of public sector institutions. They argued that public officials should treat building regulations demanded by Fifth Avenue merchants, sewer construction undertaken by borough officials in Queens, and land-use decisions made by real estate developers in Hoboken as opportunities to solve larger problems affecting metropolitan growth. Planning, as the coordination of city-building policies, became their primary goal. Toward that end, they pushed for even more explicit control over policy making to implement their vision removed from the parochialism built into the market decisions, interest-group lobbying, and urban political structures that seemed to undermine the very notion of citywide or regional interconnection; in so doing, they threatened established patterns of distributive politics carried on by business interests, party bosses, and property owners. Particular groups, localities, and institutions did not always fit into or want to contribute to the citywide and regional systems of freight and commuter transportation, underground infrastructure, or land use envisioned by experts. Who would compose those systems, at what level of organization (borough, city, region, nation), and at what cost put the city's emerging sense of togetherness to the test.

As they stepped from conceptualizing to institutionalizing, therefore, experts ran the risk of pitting the durability of interdependence against claims of legal and political autonomy, and urban planning against distributive policy making at multiple levels of aggregation. To what extent, they were forced to ask, could interdependence—as the necessary starting point for addressing the problems of the modern city—coexist with political and legal values that promoted independent, local, voluntary action (or inaction)? And to what extent would accommodating that independence interfere with the necessary coherence of the urban system?

The civic culture of expertise vacillated between aggregative and disaggregative approaches to the general interest: Because they saw New York in terms of economic and physical linkages among groups and localities, civil engineers, economists, and other professionals believed that the city possessed "some inner unity and consistency" in the form of identifiable general interests. Everyone in the greater metropolitan area relied on transportation, water, sewage, and land-use systems and everyone would benefit from better coordination of public and private resources. Whole and parts would fit together if properly configured, they imagined, which endowed the civic culture of expertise with a fundamental optimism about the potential coincidence of individual and collective welfare.[21]

This did not mean that experts tried to force urban diversity into the Procrustean bed of a unitary public interest. They knew too much about the city to think in such one-dimensional terms; they recognized that New York was a community of communities—subdivided down to the street level, with sense of place and group loyalties stronger for smaller units than for the city or region as a whole. Instead, their understanding of the city's essential interdependence contributed to a persistent faith that enlightened policy making could reconcile profit seeking and the public interest and permit distinctive local communities to coexist with assertive citywide and regional institutions. Experts in New York City tried to avoid casting policy battles in terms of stark choices between business and community, self-interest and public interest, individual and collective pursuits, by emphasizing the need for new policies and institutional tools to act on general interests when courts, parties, corporations, markets, and local governments—because of their limited perspectives on metropolitan life—could not.

These conflicts existed nonetheless, and state builders struggled to configure new institutions to reflect the illusive nature of the general interests they perceived; and this difficulty expressed itself as a constant tension between aggregative and disaggregative approaches to the general interest: that is, whether to treat the whole (city or region) merely as the sum of its parts, or as something more. In practice, the relationships between individual railroad facilities and the port system, local sewage networks and regional pollution problems, and neighborhood land-use preferences and metropolitan development needs did not mesh seamlessly, even when aided by the broader perspectives of enlightened business leaders and politicians; the parts added up to a collection of parts, not a coherent whole that resolved shared problems or improved standards of living or made the city more habitable. A more rational freight system, a cleaner harbor, and a less congested downtown all seemed like clear, widely supported general interests, but organizing railroads, municipalities, and real estate owners to pursue them involved changes in behavior that many did not desire or could not accommodate. Creating institutions predicated on a unity of interests only exposed the differences between individual and collective.

The time had come to move beyond existing institutions of collective decision making that could not adequately pursue general interests, in other words, but configuring new institutions to act on such interests raised difficult philosophical and practical questions. Was there a difference between recognizing and executing general interests where they existed by virtue of clear public agreement, and creating and enforcing general interests where that agreement remained weak or uncertain? Should government merely respond to conceptions of general interests as voiced by particular groups, or could it formulate its own? To what extent, the state builders had to determine, could they implement general interests against the wishes of groups whose interests they presumed to pursue? And to what extent did a timid approach to the general interest undermine the very notion of its existence? Did general interests cease to exist in the face of disagreement among the constituent groups of the city or did they exist permanently, waiting for sufficiently empowered institutions to act on them? If general interests did disappear into disagreement, what sort of institutions could capture that evanescence? In this sense, the protagonists of this book wrestled with the same vexing issues of political theory that concerned Rousseau in *The Social Contract* (how to distinguish between the General Will and the Will of All) and Madison in *Federalist 10* (how to structure government to compel social and economic factions to act in the general interest).[22] As such, they confronted one of the great institutional problems of the twentieth century: are bureaucratic organizations the proper instruments for determining and carrying out the public interest in a democratic society?

This book thus captures the moment when the engineers, sanitarians, and other professionals who played important technical roles in many American cities in the nineteenth century transformed urban political culture. They did that even as traditional reform and muckraking generated more sound and fury. As Steffens, the greatest muckraker of them all, acknowledged in his autobiography, the argument that bad government was attributable to bad men and effective government a matter of moral character lost much of its explanatory power and political impact during the perennial good-government crusades that followed Tweed's downfall. Steffens saw most reformers as "merely destructive" since they lacked any credible plan for replacing the bosses, who often got things done in spite of rampant corruption, with constructive policymakers. "I did not find anybody with any intelligent plan for the reform of a city," he lamented; "facts we had, but no generalizations and no capacity to generalize." Steffens focused so intently on the battle between bosses and reformers that he did not see the impact that the new breed of experts had on the city.[23] But away from the limelight on the more sensational misdeeds of government and business, a new approach to urban problems had already emerged that did provide the generalizations Steffens knew American cities needed to survive.

The chapters that follow show how experts from various disciplines provided those generalizations in key policy areas, each with its own history, its own timing regarding the creation of administrative mechanisms, and its own institutional contexts within the city, state, region, and nation. Viewed together, they demonstrate how similarities between apparently unrelated debates emerged from a flawed but influential civic culture that explains the uneven, incomplete, and in many ways unsuccessful centralization of authority in the nation's leading metropolis in the early twentieth century.

Part 1 / Private Infrastructure and Public Policy

CHAPTER ONE

"The Public Be Pleased"

Railroad Planning, Engineering Culture, and the Promise of Quasi-scientific Voluntarism

"There is obviously a great difference in outlook between the Vanderbilt policy of 'the public be damned' and the McAdoo policy of 'the public be pleased,'" observed Walter Lippmann in *Drift and Mastery* (1914). A radical alteration in the "cultural basis of private property" had occurred with the emergence of the modern corporation; control of big businesses had shifted to salaried executives whose authority rested on expertise and competence rather than on legal claims to ownership. Because they rose to positions of power by virtue of their technical abilities, new incentives inspired this generation of corporate managers: a commitment to a professional community, a sense of craftsmanship, and a concern for the public interest, not merely the profit motive. "The men who are connected with these essential properties cannot escape the fact that they are expected to act increasingly as public officials," Lippmann remarked, and nowhere did they fulfill that expectation more successfully than in New York City, where railroads provided the vital linkages between the water-bound central business district in Manhattan and emerging suburban areas in Brooklyn, Queens, and New Jersey.[1]

This chapter shows how the companies that so impressed Lippmann—the

Hudson and Manhattan Railroad (H&M) and the Pennsylvania Railroad (PRR), in particular—provided an institutional context for civic discourse in the new metropolis, with important implications for the debate over public franchises.[2] The tunnels and stations built by the H&M and the PRR, when combined with the city's subway system, created a commuter network far beyond the capacity of any single company or city government acting alone, thus solving one of New York's most pressing transportation problems. In the process, the railroads sustained a network of engineers inclined to think in terms of the comprehensive planning issues confronting the greater city. The interaction of railroad corporations and the engineering profession—both translocal institutions—thus allowed a community of discourse to flourish around the problem of how best to move passengers around the region, across the Hudson River, and through the densely inhabited city.[3] As a result, an emerging civic culture of expertise held the promise of making centralized private economic power genuinely responsive to public needs.

Commuter Geography

Gotham's need for commuter facilities to link center and periphery arose from the formidable natural obstacles around it: the Harlem, East, and Hudson Rivers encircled Manhattan; and the East River and Long Island Sound separated Queens and Brooklyn from Manhattan, the Bronx, and eastern New York state. In spite of these physical barriers, lower Manhattan served as both the center of the city and the region, with financial, corporate, manufacturing, and retail activity concentrated below Central Park. Though Manhattan's extensive waterfront would remain the principal point of exchange for seaborne commerce until the 1950s, the transportation requirements of a corporate metropolis differed from those of a nineteenth-century trading city.

For so important an economic center, Manhattan offered commuters very few options for entering the bustling downtown area. In 1897, 143 million people crossed the East River; of those, about 50 million used the Brooklyn Bridge (completed 1883). By 1906, commuter traffic between Long Island and Manhattan more than doubled, to 295 million, with 100 million taking ferries and 195 million using the railways across the Brooklyn and Williamsburg Bridge (completed 1903). Millions of others simply walked across the bridges to work.[4]

From the New Jersey side of the Hudson, where the residential population within twenty miles of Manhattan doubled between 1890 and 1910, commuters entered the city by boat. Eleven major rail lines terminated on the west bank of the Hudson, where passengers boarded ferries for the journey across the river. Six ferry companies transported more than 72 million commuters in 1890, and by 1896 the PRR's ferries alone carried 34 million passengers.[5] The crossing usually took about fifteen minutes, but ice floes or fog could extend the trip to an hour. Delays abounded on the Manhattan side as well since competition for space forced ferries to dock between the busy freight piers. Passengers then walked to the business district through narrow, cross-town streets, already congested with carts and heavy wagons moving goods around the waterfront.[6]

Manhattan had only one direct rail link to the mainland—the tracks of the New York Central Railroad, which ran along the Hudson from the Bronx, across the Harlem River, and down the west side of Manhattan. But the Central did not serve New Jersey commuters; in fact, it only made the area more congested, adding to the delays experienced by passengers and freight handlers moving to and from their river slips.

Railroad executives fully recognized the economic opportunities created by Manhattan's water-bound condition. As Samuel Rea, a Pennsylvania Railroad vice president, observed in 1892, "The railroad situation at New York is unique: A parallel does not exist. Here is a great seaport, with an aggregate population and of commercial importance second to none in the world, separated by the navigable waters of a river from all the rail transportation systems of its country, with but a single exception, namely, the New York Central system."[7] Between 1900 and 1910, the H&M and PRR completed tunnel projects designed to challenge the privileged position of the New York Central and, as one observer put it, to "emancipate New York City from the limitations of its insular position."[8]

William Gibbs McAdoo and Entrepreneurial Planning

Credit for the first successful crossing of the Hudson by rail goes to entrepreneur-turned-statesman William Gibbs McAdoo. McAdoo began his eventful career in Chattanooga, Tennessee. In 1889, he purchased a controlling interest in the Knoxville Street Railway, which he conceived as the seed of a citywide transit network, but technical and financial difficulties forced the company into receivership in 1892. McAdoo then decided to seek his fortune in New York City, where he channeled his extraordinary vision and tireless self-promotion into the Hudson and Manhattan Railroad, which transformed him into one of the nation's foremost representatives of enlightened business management. McAdoo then lost no time translating his business success into a major political career. He met Woodrow Wilson in 1909 (when Wilson was president of Princeton University) and cultivated this relationship socially and politically, courting Wilson's daughter, Eleanor, whom he married in 1914. McAdoo became vice chairman of the Democratic National Committee in 1912 and naturally supported Woodrow Wilson for president. Wilson returned the favor by appointing McAdoo secretary of the treasury in 1913, where he helped create the Federal Reserve system, and later selected him as director-general of the Railroad Administration, the federal agency that ran the railroads during World War I. McAdoo made two unsuccessful runs for the Democratic presidential nomination, in 1920 and 1924, and played a central role in nominating Franklin Roosevelt during the deadlocked party convention in 1932.[9]

That a failed traction mogul from Tennessee could catapult himself to this level of political success stemmed in large measure from the reputation McAdoo developed for pulling off the H&M tunnels, which had a daunting history of catastrophe and disappointment. De Witt Haskins had first proposed tunneling under the river in 1871, but that effort was abandoned in 1880 after the tunnel reached a distance of twelve hundred feet under the river and a blowout occurred, killing twenty men and bringing construction to a halt. A British company took over in 1888 and extended the tunnel another eighteen hundred feet before a shortage of capital interrupted construction four years later.[10] Subsequent efforts to raise money had proven quite difficult because the stockholders of the defunct tunnel company thought of New Jersey trolley companies as the principal source of commuter traffic—a very small market for so dangerous and expensive an engineering endeavor.[11] McAdoo recognized that the tunnel should connect the steam railroads on the Jersey shore with the major transportation points in New York; therein lay a real source of commuter traffic and thus the only sure way to guarantee investors a return on their money.[12]

With this idea in hand, McAdoo revived the Hudson Tunnel Company by pulling together the leading financial, technical, and transportation networks in New York City to address the needs of commuters to the nation's corporate headquarters. McAdoo's real skill as a promoter, therefore, lay in his ability to link economic and technical communities in the region and focus their resources on one of the city's most challenging planning and engineering problems.

Finding the financial muscle to resuscitate the tunnel plan proved to be McAdoo's greatest achievement. During the 1890s, McAdoo sold bonds on commission, and that experience brought him into contact with the few players in the city's financial community who could muster the resources for so large and dangerous an undertaking: Walter G. Oakman, president of the Guaranty Trust Company, then involved with the Brooklyn Rapid Transit (BRT) Company; Anthony Brady, who controlled the BRT and aspired to enter the Manhattan subway game; E. C. Converse and Judge Elbridge Gary, both directors of U.S. Steel, the Pennsylvania Railroad's biggest customer.[13] By establishing relationships with these key figures, McAdoo insured that the H&M, unlike its predecessors, would not fail due to lack of capital. There would, however, be many close calls. During the panic of 1903, when new securities glutted the stock market, McAdoo's investors stood by him. When the company needed fresh capital in 1905, Oakman turned to Pliny Fisk, head of the Wall Street banking firm of Harvey Fisk & Sons, who began raising capital for the H&M at a time when the vast scope of the tunnel plan demanded secure long-term financing: by 1908, McAdoo had spent nearly $60 million. Fisk also helped the company weather the devastating panic of 1907, an achievement that McAdoo gratefully acknowledged as a feat of genius.[14]

McAdoo also had the good fortune to happen upon a project already staffed by some of the best engineers in New York. Charles M. Jacobs and John Vipond Davies were two of the most experienced railroad and tunnel engineers in a city full of engineering talent. Born and trained in Britain, Jacobs came to New York at the behest of Austin Corbin, president of the Long Island Railroad (LIRR). In 1891, Jacobs opened a consulting firm in New York City, with Davies, a Welsh engineer who had emigrated to the United States in 1889 to work for the Philadelphia and Reading Railroad, as his chief assistant. In 1894, Jacobs and Davies secured their reputations by completing the stalled East River Gas tunnel, a success that made them the obvious choice to salvage the failed Hudson tunnel project, which presented an even greater challenge.[15] The original project consisted of a tunnel between Hoboken, New Jersey, and Christopher Street in Manhattan, but the ambitious McAdoo envisioned two sets of parallel tunnels connecting the major traffic points of New Jersey and Manhattan (figure 1). These routes required engineers to dig through silt, solid rock on both sides of the river, and an alternating array of rock, sand, and clay—work that John Davies described as "more varied in character than any underground project ever executed," involving "every known type of tunnel construction."[16]

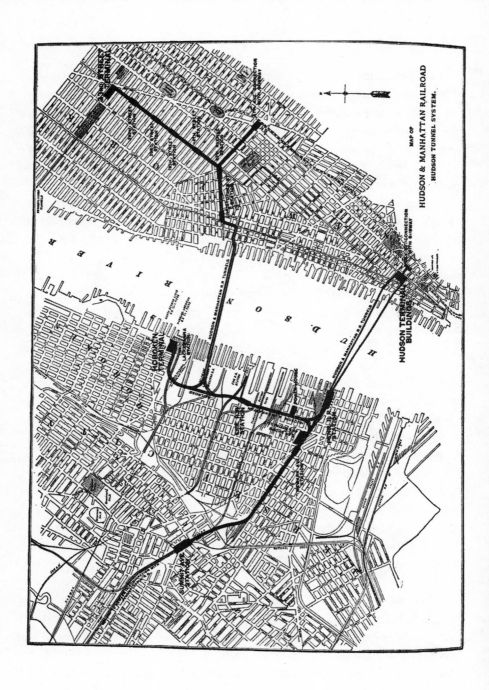

The cooperative arrangements McAdoo made with the PRR insured that all of this expensive engineering work served a real need. Fortunately for McAdoo, PRR president Alexander J. Cassatt understood the role of transportation in a regional metropolis. In a 1903 meeting with McAdoo, Cassatt readily admitted that the H&M would "destroy" the profitable Jersey City–Cortlandt Street ferry business; in spite of the loss, however, he agreed to a connection between the H&M and the PRR at Jersey City.[17] The Erie and the Delaware, Lackawanna & Western (DL&W), which had resisted the H&M project with lawsuits, followed the PRR lead. When the H&M opened for business, it drew on a wealth of commuter traffic from the steam railroads.[18]

The completion of the H&M's Hoboken–Christopher Street tubes in 1908 and the Jersey City–Cortlandt Street Terminal tubes in 1909 offered PRR commuters a vastly superior way to cross the Hudson. In its first week of operation, the Cortlandt Street tunnel carried twelve thousand passengers who would have otherwise taken the railroad ferries: a PRR executive noted that business on the company's ferry between Jersey City and Cortlandt Street fell by eleven thousand that week.[19] Samuel Rea took one of the very first H&M trains to work and was suitably impressed by the results: "My experience this morning verifies what we have anticipated since Hudson Tunnels were projected," he reported to his colleagues; "namely, that by their use a passenger would arrive in New York by the time he was due to leave Jersey City. I alighted from [my commuter] train at 9:30, left by tunnel at 9:32½, and was in our office just 12 minutes after leaving the train in Jersey City Station. We have repeatedly timed this journey [by ferry] and never made it in less than 25 minutes."[20] Like Cassatt, Rea believed the benefits of the H&M connection would more than compensate for the loss of ferry traffic. The PRR station in Jersey City was located directly above the H&M station thanks to the agreement between Cassatt and McAdoo, so commuters had only to take an elevator from one level to another to travel into Manhattan completely by rail. When the H&M tunnels opened, conductors on the PRR lines reported seeing "many new faces."[21]

A brilliant public relations pitch provided the final ingredient in the H&M success story. McAdoo coined the slogan "the public be pleased" to put his

Figure 1. The Hudson and Manhattan Tunnel System, 1910. Conceived by promoter William Gibbs McAdoo, the H&M provided commuters with the first all-rail route between New Jersey and Manhattan. *Source:* Gilbert H. Gilbert, Lucius I. Wightman, and W. L. Saunders, *The Subways and Tunnels of New York* (New York: Wiley, 1912), 154.

railroad squarely in the minds of potential commuters and to differentiate himself clearly from Cornelius Vanderbilt. After all, New York City had provided the stage for some of the most memorable episodes in the history of railroad corruption: the robber barons of Charles Francis Adams's classic "A Chapter of Erie" directed their railroad wars from there; during the 1870s, the Hepburn Committee, composed largely of city merchants, revealed that the New York Central and other railroads used their monopolistic positions to water stock, overcharge customers, and buy off legislators. The Hepburn investigations supplied the setting for Vanderbilt's famous pronouncement on the nascent regulatory movement, "the public be damned." The memories of predatory railroad management remained strong in New York City, and McAdoo played upon that history to drum up business.[22]

But McAdoo intended his slogan to represent much more than a successful advertising gimmick. He set out to illustrate that a successful entrepreneur could take his public responsibilities seriously.[23] The H&M had two clear tasks: to facilitate commuting across the Hudson and to produce a return on investment. It did both.[24] McAdoo achieved the twin goals of private gain and public benefit by building where the volume of traffic warranted investment, by tapping into the right financial networks, by enlisting the expertise of engineers familiar with local conditions, and by securing the cooperation of the PRR, the Erie, and the DL&W. The combination of voluntarism and expertise necessary to complete the H&M tunnels transformed opportunism into enlightened self-interest, thus allowing concentrated private economic power to serve public purposes.

The Civic Vision of the Pennsylvania Railroad

The work of the PRR demonstrates that a corporation whose main interests lay outside of the region could serve the metropolis in the same manner as a local company like the H&M. When the PRR finally reached into New York (culminating nearly forty years of corporate planning with the opening of Penn Station in midtown Manhattan in 1910), it brought the city into *its* transportation system—a network that ranged from Saint Louis to the Hudson River, from Chicago to Philadelphia, and from the Great Lakes to the eastern shore of Maryland. Despite the company's focus on its national rail network, however, it became one of the most important private planning agencies in New York City. To an even greater extent than the H&M, therefore, the PRR, as translocal economic institution, contributed to an emerging civic culture of expertise through its planning and construction efforts in, under, and around the city.

The PRR network of tunnels, stations, bridges, and rail lines through New Jersey, Manhattan, Brooklyn, Queens, and the Bronx resulted from a civic vision crafted by several generations of corporate managers who began thinking about a rail connection to Manhattan as early as 1871. Although company executives looked enviously at De Witt Haskins's tunnel project, the panic of 1873 and the railroad strikes of the 1870s prevented them from making any immediate investment in a river crossing. Over the next thirty years, technical and economic obstacles hindered the company's subsequent efforts to reach New York City by rail. During this extended period, company engineers and their consultants in the city conducted a crucial transportation planning debate over whether to cross the Hudson by tunnel or bridge—a debate continued by government agencies through the early twentieth century.[25]

From the 1890s through the 1920s, engineer Rea served as the PRR's point man in this debate. Rea had an extraordinary career with the company, developing a broad understanding of the potential of railroad work in New York City. Born in Hollidaysburg, Pennsylvania, in 1855, Rea joined the PRR as a chainman and rodman in 1871. Although he had no formal training as an engineer, Rea worked his way through various engineering posts at the PRR, specializing in construction projects. During a brief tenure as vice president of the Maryland Central Railway, he oversaw the construction of the Baltimore Belt Railroad. He returned to the PRR in 1892 as special assistant to the company president, George B. Roberts, and began planning the PRR's New York projects. Rea moved steadily up through vice-presidential positions and became president of the company in 1913—a remarkable twenty-year stretch during which the PRR planned and completed Penn Station and its own Hudson tunnels.[26] Thus, the president of one of the nation's most important railroads ascended to his office as an expert in planning and building in New York City. Like Jacobs and Davies, Rea moved from construction to planning, expanding his expertise from a narrowly technical field to the larger domain of urban infrastructure development. His career illustrates how corporations operating in the New York region could encourage the development of a civic vision in their ranks, linking debates within companies to a larger professional discourse about city planning.

Beginning in the early 1890s, Rea considered a variety of approaches for bringing the PRR into New York City. He preferred a plan developed by civil engineer Gustav Lindenthal in 1884, which contemplated a multispan bridge from Jersey City to a terminal between Canal and Desbrosses Streets, since this solved all the PRR's problems by allowing freight and passenger traffic into the heart of the city using standard steam engines. However, Rea estimated that the bridge would cost $100 million, necessitating a joint venture with other railroads in the region. Although he initiated discussions with these potential partners, the panic of 1893 postponed further action on the bridge until 1900, when lack of cooperation by the other companies forced the PRR to abandon the Lindenthal project. Rea then realized that "if we were ever to build into New York we must do it alone."[27] (Although the PRR would eventually cross the Hudson by other means, Lindenthal continued to advocate for a bridge, and other engineers working through other institutions took up his dream. The bridge plan continued to be discussed during the ongoing debate over the need for better transportation for the other railroad companies who moved goods and passengers between New York and New Jersey. During the 1920s, engineer O. H. Ammann transformed the idea of a joint rail bridge to lower Manhattan into a project for an automobile bridge much further north on the River—the George Washington Bridge—which he completed in 1931 under the auspices of the Port of New York Authority.)[28]

Rea also contemplated constructing a tunnel to reach Manhattan by rail. He recognized the feasibility of tunnels after his 1892 investigation of the construction of the rapid transit tunnels under the Thames. Rea knew the chief engineers on the London project, Benjamin Baker and James Greathead, who allowed him to inspect the tunnels closely. He discovered that the success of the British effort lay in their use of electric locomotives. Steam engines required elaborate ventilation facilities and made tunnels "grimy and disagreeable," whereas electric engines ran much cleaner; but electric trains simply could not compete with steam when it came to pulling heavy loads. By 1900, however, electrical technology had improved considerably. When Cassatt, who became president of the PRR in 1899, saw electric engines used successfully by the Orléans Railway in Paris in 1901, he ordered a reevaluation of the tunnel scheme. Cassatt, Rea, and engineers Jacobs and Davies (who worked for the LIRR, then controlled by the PRR) concluded that electric traction made the use of full-sized tunnels a financial possibility.[29]

Thanks to electric traction, the completed PRR tunnel project became a great engineering and planning achievement (figure 2). From portal to portal,

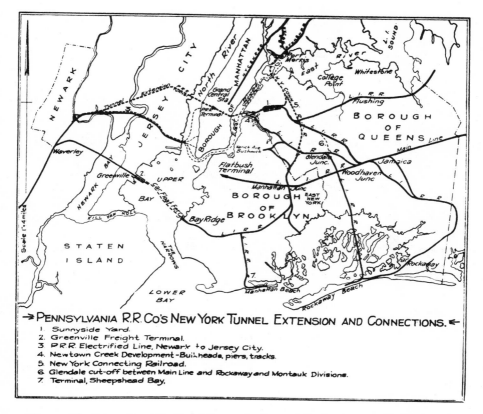

Figure 2. The Pennsylvania Railroad System in Greater New York, 1907. The PRR tunnels under the Hudson River into midtown Manhattan and under the East River to Queens ended the isolation of Long Island from continental rail systems. *Source:* A. J. County, "The Economic Necessity for the Pennsylvania Railroad Tunnel Extension into New York City," *Annals of the American Academy of Political and Social Science* (March 1907), reprint in James Forgie Collection, box 46, Smithsonian Institution, Washington, D.C.

the tunnels extended five and one-half miles, through solid rock at Bergen Hill, through mixed sand, rock, and gravel in New Jersey, and through Hudson River silt, underneath Manhattan Island and the East River, to Queens. These massive tunnels could move up to 144 trains per hour from the mainline of the PRR in New Jersey through Penn Station in midtown to Long Island.[30] The PRR's new facilities also provided for the movement of freight from Newark, New Jersey, to its piers in Greenville, where cars were loaded on barges bound for Bay Ridge in Brooklyn. In a joint venture with the New York, New Haven &

Figure 3. Pennsylvania Station, c. 1910. This monumental terminal, designed by McKim, Mead, and White, became the jewel of the PRR system and one of the most important civic structures ever constructed by a private company. *Source: The New York Improvement and Tunnel Extension of the Pennsylvania Railroad* (July 1910), promotional pamphlet, James Forgie Collection, box 46, Smithsonian Institution, Washington, D.C.

Hartford Railroad, the PRR bought and upgraded the New York Connecting Railroad, which linked Bay Ridge to Sunnyside Yard, a 153-acre switching area in Queens, and removed "the disadvantages of the insular isolation of Brooklyn and Queens," thus clearing the way for increased commercial development.[31] The company hired Lindenthal to build the Hell Gate Bridge over the East River to link the Connecting Railroad to the New Haven lines. When completed in 1916, the Hell Gate was the longest arch span in the world (surpassed by the George Washington and Bayonne Bridges in 1931) and remains the heaviest arch bridge in the world.[32] By building tracks from New Jersey, under Manhattan, and over Hell Gate, the PRR created the final pieces in an unbroken rail line that extended from Key West, Florida, to Halifax, Nova Scotia.

Perhaps the most decisive statement of the PRR's commitment to the civic life of the metropolis was Penn Station—a structure that was "built for all time," in Rea's words (figure 3).[33] McKim, Mead & White, the city's most prestigious architectural firm, designed the building to be much more than a train station. They intended its Doric style to suggest Roman temples or basilicas, and to authenticate the effect they imported the same sort of Italian marble used in imperial Roman buildings.[34] Their efforts paid off—so much so, in fact, that architectural historian Fiske Kimball would later describe the practical functions of the "soaring, musical spaces" of the station's waiting room as "insignificant": "It is conceived, rather," he affirmed, "in accordance with its higher, ideal function—as a civic vestibule to the world metropolis."[35]

Unlike most large-scale private buildings in New York City, Penn Station was in fact designed as a civic structure. Other companies built vertically, both to create important corporate symbols and to recoup the massive investments incurred by building in downtown Manhattan. The Woolworth and Equitable Buildings, for instance, were both speculative skyscrapers.[36] So, too, was the H&M station at Cortlandt Street (figure 4); with twenty-two stories above ground and four below, the Cortlandt Street Station had nine hundred leaseholders—including Armour & Company, the Erie Railroad, General Electric, and U.S. Steel.[37] Along these lines, PRR President Cassatt had originally envisioned a skyscraper hotel attached to Penn Station, similar to British and Continental railway terminals, which would have provided a source of income for the massive structure. But architect Charles McKim convinced Cassatt that a skyscraper could never serve as a grand civic structure. As a result, Penn Station included almost no rental space to generate revenue directly. Instead, Cassatt believed that the magnificent station would improve business for the company generally, serving as an appropriate monument to the greatness of both the city and the railroad.[38]

Technical considerations forced the company to build the station away from the center of activity in midtown. Cassatt originally wanted the massive terminal on the east side of Fourth Avenue in the heart of the city, right on the first subway route—for what more suitable place could there be for "a civic vestibule to the world metropolis"? But when Cassatt sought advice on the location of the station from local experts Jacobs and Davies (the two had previously investigated the possibility of an LIRR tunnel under the East River into Manhattan), he discovered that grades for tunnel approaches to Manhattan would not permit a station anywhere except west of Broadway. This meant that the company would have to locate the station in the tenderloin district around Seventh Avenue—thus removing the PRR's great public building from the heart of Manhattan—where it awaited the construction of a new subway to

Figure 4. Cityscape with the H&M's Cortlandt Street Station on the right, c. 1915. The giant Hudson Terminal Building was a speculative skyscraper as well as a commuter rail station. With twenty-two stories above ground and four below, it had nine hundred leaseholders, among them Armour and Co., the Erie Railroad, General Electric, and U.S. Steel. The Port Authority's ill-fated World Trade Center buildings were built on the site in the 1960s. Note the Singer Tower at extreme left. *Source:* Detroit Publishing Co. Collection, Library of Congress, Washington, D.C.

link it with the rest of the city.[39] But such was Cassatt's commitment to a truly citywide transportation system that the company proceeded with its plans, going so far as to build Penn Station some forty feet deeper than necessary in order to provide the clearances for the future subway, even though the city could offer no firm date for the completion of the Seventh Avenue line.[40]

Financing these ambitious (and imperfect) plans proved to be the most daunting obstacle faced by the PRR, but one that it overcame with considerable flair. The company could not depend on financial support from other railroads, city, state, or federal governments—the failure of Lindenthal's North River bridge plan proved that much—and it had enormous needs. Instead, it began raising capital through "spectacular equity financing." Between 1898 and 1906, the PRR invested nearly $400 million in its various projects, including

$180 million in new stocks, $104 million in long-term borrowing, and $92 million of retained earnings (capital remaining after paying fixed charges and stock dividends).[41] By drawing on the superior financial strength of the entire PRR system, company officials carried out their mammoth plans for Greater New York.

The PRR building program in New York was both city boosting and city planning; because those activities were interconnected, corporations engaged in city building became incubators for an important dimension of civic culture through the work of the engineers they employed. PRR managers recognized New York as the country's commercial capital and they spent lavishly to make the city an integral part of the PRR system by tying its improvements to the other pieces of the city's emerging commuter system—the subways, the els, and the H&M tunnels. The resulting rail network met the needs of both the city and the company since public and private institutions wrestled with aspects of the same challenge: how to move people to and from lower Manhattan. Thanks to a combination of ambition and expertise, private companies could contribute to the resolution of public problems; in the process, they produced a generation of engineers—Rea, Jacobs, and Davies chief among them—who were deeply engaged in civic discourse.

Civic Discourse and Engineering Culture

The salience of Lippmann's observations about the cultural transformation of private property becomes clearer when we trace the discourse of railroad engineers beyond the corporate environment that sustained them. Engineers found practical solutions to the problems of how to tunnel under rivers, where to locate rail lines, and how to move commuters into the city center. These were technical matters, of course—the stuff of narrow discussions among specialists in excavation methods, caisson construction, and electric traction. But they were also intimately related to the planning challenges facing New York City. Thus, instead of removing experts from public discourse, the opportunities offered by the H&M and PRR made the city, rather than the corporation or the profession, the center of an engineer's concern.

There can be little doubt that corporate construction offered engineers unique career opportunities. Charles Jacobs, for instance, was the first man to walk dry-shod underneath the Hudson River, through the completed H&M tunnel—a moment he considered to be the greatest of all his professional successes. Such "firsts" were a source of both amusement and prestige in the engineering community; engineers at the PRR even went to the trouble of lowering an automobile into their tunnels in order to be the first group to motor under the Hudson, nearly twenty years before the Holland Tunnel opened.[42] These engineers knew that they were engaged in building projects on a heroic scale and they reveled in their work.

Engineers also recognized the significance of the work completed by corporations. They knew the long history of failure of Hudson tunnels and understood the legal, technical, and financial obstacles that blocked tunneling efforts. They personally experienced the difficulties of building in and under New York City and they held the corporations that succeeded on a spectacular scale in high esteem. By the same token, engineers who knew how to get things done could build their résumés by carrying out the plans of their corporate clients.[43]

Like most American engineers, those who worked on the Hudson tunnel projects relied heavily on corporations for employment. Even though engineers continued to emphasize their entrepreneurial roots well into the early twentieth century, they also moved in startling numbers into the corporate world. By 1925, only 25 percent of engineers described themselves as either self-employed or consultants; 75 percent worked in corporate or government jobs.[44]

In spite of the importance of corporate employment, however, many railroad engineers in New York City maintained a strong sense of professional independence. Engineers who contributed to corporate projects in the city were frequently itinerant, moving from company to company, from construction job to construction job. Many of the engineers who worked on the Hudson tunnels spent only a fraction of their careers with the PRR. Eugene Allen, for example, received a civil-engineering degree from Princeton and went to work as a construction engineer for the Essex County, New Jersey, park system. He conducted surveys for the LIRR in Brooklyn and for the PRR tunnels and then helped build the H&M tunnels. He spent two years selling railroad and mill supplies before moving into railroad valuation work for the New Jersey State Board of Assessors. Allen then took his skills back into the private sector as a valuation engineer for the Central Railroad of New Jersey and later for the Philadelphia and Reading Railroad. Daniel B. McAllister studied engineering at the Lawrence Scientific School at Harvard before becoming a transitman on the PRR tunnels. He worked for mining companies in West Virginia and Michigan and then moved to Mexico, where he managed several mineral mines. Herbert Wild became an assistant construction engineer on the PRR tunnels after working his way up from transitman on the Maine Central Railroad. He became a professor of engineering at the Pennsylvania Military College in 1912 and entered the Army Corps of Engineers as a captain in 1917, where he rose to the rank of major and had a successful career in the public sector. Thus, while Allen, McAllister, and Wild benefitted from their corporate employment, they were not tied to particular businesses and moved easily from railroads to construction companies, to mining operations, to government positions.[45]

Many of the engineers who worked on the New York tunnels were also generalists. Even though the engineers who created the H&M and PRR systems often performed narrowly technical roles within these corporations, they developed a more comprehensive understanding of the connection between railroad building and city planning. Few could match Samuel Rea's record of advancement in one company, but engineers at all levels of the PRR project used their construction experience to gain a broader perspective on infrastructure development, making the transition from technical specialist to planning generalist.

An important group of itinerant generalists moved primarily from project to project in New York City and helped create a community of discourse that extended from the corporation to the city. Scottish engineer James Forgie, for instance, followed the same career path as Jacobs and Davies. Forgie had worked on the London subways for James Greathead, inventor of a widely used subaqueous tunneling shield that had impressed Samuel Rea. In 1902, Forgie became chief assistant engineer on the PRR's North Hudson River project, where he designed the shields used to cut the tunnels and oversaw their construction. There, Forgie met Jacobs and Davies, and in 1909 he joined their consulting firm. Jacobs and Davies connected Forgie with the H&M, where he oversaw its Sixth Avenue tunnel extension.[46] Forgie's expertise then took him into the public sector, as one of the chief engineers on the Holland Tunnel, the first automobile tunnel under the Hudson.

Benjamin Franklin Cresson Jr. also used his experience at the PRR to move into the public sector, where he occasionally came into conflict with his old employer. Cresson studied engineering at Lehigh and the University of Pennsylvania before taking a job on the Lehigh Valley Railroad. Between 1897 and 1899, he did tunnel and subway work for the Reading Railroad in Philadelphia. He then joined Davies in the survey and inspection work for the Hudson River tunnels. After a brief tour with the West Virginia Short Line Railroad, Cresson returned to New York as an engineer for the LIRR's Atlantic Avenue improvements in Brooklyn. He subsequently became the first engineer on the PRR tunnel surveying work and was promoted to resident engineer for terminal excavation. He left the PRR in 1910 to be deputy commissioner of the New York City Department of Docks, under Calvin Tomkins, where he oversaw their engineering work until 1913. At the docks department, Cresson joined the heated debate over the need for public management of the port, becoming a leading expert on freight planning in the region. He consulted on several large studies of the port problem that identified mismanagement of the waterfront by the railroads as the principal obstacle to efficient freight movement. In 1921, Cresson became chief engineer for the Port of New York Authority, where he argued for cooperative planning of freight transportation facilities.[47] Cresson thus translated his construction experience into planning expertise in much the same way as Samuel Rea, but he chose a career in government rather than corporate institutions.

Like Cresson, Davies expanded his professional competence from construction to planning. As he worked closely with McAdoo to plan the extension of the original tunnel, Davies developed an understanding of the social and economic contexts of large-scale engineering works. By the time the H&M tunnels were completed in 1910, he saw them as part of a private transportation system linking distant residential areas with the region's economic center. In the process, Davies became critical of unplanned regional development. The street patterns created in the early nineteenth century suited a compact walking city, with traffic moving primarily between the Hudson and East Rivers. But the city had expanded haphazardly, and in Davies's view, private companies like the H&M stepped in to provide subsurface transportation to accommodate these new traffic patterns. Working alongside municipal government, which planned the subways, the H&M, the PRR, and even trolley companies helped to create a commuter network to integrate the sprawling region—a vision shared by many of Davies's fellow engineers.[48] Thanks to his experiences with the H&M, Davies grew into an authority on transportation and city planning, publishing his work alongside such well-known planners as Frederick Law Olmsted Jr., John Nolen, Edward Bennett, and Charles Mulford Robinson.[49]

The PRR and H&M were not the only railroads producing engineers with expertise in city planning. William Wilgus's career at the New York Central rivaled Rea's at the PRR. Wilgus oversaw the electrification of the Central's tracks in New York and the construction of Grand Central Station—another extraordinary civic structure created by a private company. Wilgus left the company in 1907 to become a consultant and, for the next twenty years, produced some of the most innovative freight transportation plans for the port of New York, including an elaborate subway system to move freight between New Jersey and Manhattan. Although the subway was never built, it became a centerpiece of regional freight planning debates conducted by the Port of New York Authority and the Committee on the Regional Plan during the 1920s, where it remained a nagging alternative to the railroads' chaotic approach to freight management.[50]

The career of Francis Lee Stuart, who started at the Baltimore and Ohio Railroad (B&O) in 1884, likewise illustrates just how varied the professional lives of the city's itinerant generalists could be. After gaining experience with other railroads, in coal-mining operations, and in public works posts, he joined the Nicaraguan canal surveying team in 1897. In 1905, he was appointed chief engineer of the Erie Railroad when that company conducted its most extensive improvements campaign, including a terminal and station expansion project in Jersey City. Stuart returned to the B&O in 1910 as chief engineer, then in 1915 launched a career as a consultant. He served as an adviser for the War Board of the Red Cross, assisted in the layout of army bases, and chaired the budget committee at the Railroad Administration, later consulting on port projects in Baltimore and Los Angeles and on rapid transit matters in Philadelphia. Stuart also advised the Cunard Steamship Company on its Weehawken improvements and on the North River Bridge project during the 1920s. He later became the vice chair of the Technical Advisory Committee to the Port Authority.[51]

Jacobs, Davies, Forgie, Cresson, Wilgus, and Stuart were part of a network of engineers in New York City that grew within the institutional structures provided by companies like the PRR, the H&M, the New York Central, the B&O, and the Erie. Railroad work gave these engineers invaluable opportunities to expand their expertise, allowing them to make the transition from construction to planning. The magnitude of private building in the early twentieth century thus nurtured this community of discourse concerned with the problems of planning the city and the region. These engineers benefitted from corporate employment, of course, but the metropolis rather than the corporation became their domain.

In the process, itinerant generalist engineers created a new type of civic culture—a way of seeing the connections among apparently unrelated components of the city—that highlighted the interdependence of the emerging metropolitan environment. Their shared experiences building tunnels under rivers, through the city's subsurface, and around the expanding region gave them a unique view of urban life. They experienced firsthand how competition for space in the skyline and on the streets extended underneath the city. Just as skyscrapers pushed the city upward, infrastructure extended the city downward. Supporting growth in both directions posed serious difficulties for engineers, emphasizing the need for new approaches to the challenges of collective living in the modern metropolis.[52] At Sixth Avenue and Thirty-third Street in Manhattan, for instance, there were six different layers of transportation infrastructure: at the highest level were elevated trolley lines; these ran above surface trolley lines; then there was the street; and beneath it, the Broadway subway; the H&M tunnel crossed just under the subway tunnel, and the PRR tubes ran below the H&M tunnel—all interlaced by city sewage lines. And the city planned another subway below the PRR.[53] The problems of subsurface crowding became only more complex as the city grew. Engineers for the new Waldorf-Astoria Hotel (1931) had to build the gigantic, forty-two-story structure's foundations over the tracks of the New York Central; they devised a system of 216 footings spaced between rail lines and composed of alternate layers of lead, asbestos, and iron to prevent vibrations passing between moving trains and the hotel superstructure.[54]

Overlapping layers of development and complex partnerships between companies inevitably led to technical, financial, and legal conflicts that made experts indispensable to the process of planning and building in the congested downtown. Consultants did a booming business mediating disputes between parties on construction projects, for example. Several companies employed James Forgie to assess the validity of cost estimates, work timetables, and construction plans. He examined materials costs, wage rates, and technical problems to determine whether contract bids were realistic and profit margins within accepted norms.[55] Expertise and experience were thus vital elements in the professionals' approach to conflict resolution.

Engineers also had to deal with problems of financing, another important part of planning in the new metropolitan environment. Building in the New York region involved enormous amounts of money, and projects almost inevitably consumed more than originally estimated. Companies also had to build for the very long term, which required solid financial backing from bankers with as broad a vision of the region's needs as the executives and engineers in charge of large projects. As one observer of the H&M experience described it, "It is an old saying that 'money makes the mare go,' and what is true of the 'mare' is true of tunnels. They cannot be built without money—plenty of money."[56]

In addition to this close collaboration with the financial community, engineers recognized that a de facto planning partnership existed between private business and city government. The PRR and the H&M designed their tunnel projects to connect with city subways that were planned but not yet built. Anticipated connections with the Seventh Avenue subway eventually saved Penn Station from nearly complete isolation. Although PRR officials worked out much of their New York improvements scheme privately, the success of their plans rested heavily on the reliability of city officials.[57] At the same time, the engineers who made the transition from construction to planning recognized just how important private companies were in the new scale of urban life represented by the New York region. Corporate engineers like Rea and consultants like Davies helped to link the greater city to the region when no governmental institutions addressed that scale of metropolitan life.[58]

The technical discourse of railroad engineers therefore had a vital civic dimension. That discourse was sustained by, but not confined to or beholden to, large corporations. Private planning in New York succeeded because it relied heavily on technical debates about the same issues that concerned public officials in the city and the region, even while those officials lacked the regional institutions to address those problems. Engineering discourse thus served as a bridge between private and public goals, allowing corporations to pursue both profit and public service. This enlightened voluntarism made it possible to believe that railroads might actually fulfill their role as *public service* corporations, just as intended in their franchise grants.

Business versus Politics: The PRR Franchise Battles

Although the PRR planned, financed, and constructed its bridge, station, and tunnel projects on its own, the company's ambitious infrastructure program had to meet with the approval of the state legislature and the city's rapid transit commission and board of aldermen. The PRR required two franchise grants to complete its New York projects: one for its tunnels under Manhattan and a second for the Connecting Railroad through Brooklyn and Queens. This requirement created the potential for a clash between company officials, who thought in terms of citywide and regional planning, and local politicians more concerned with winning elections in particular wards. At the same time, it brought the PRR into the ongoing partisan conflicts between Democrats and Republicans, between Manhattan and Brooklyn, and between city government and the state legislature. The PRR franchise battles thus played out as conflict between civic cultures—one rooted in electoral politics and party competition and the other sustained by corporations and professions.

In the end, this cultural conflict resulted in a centralization of franchise-granting authority, and at first blush it appears to represent the triumph of big business over local politics. It was in this tension—between interpersonal, localistic politics and impersonal, large-scale corporate decision making—that historian Samuel Hays saw the great political battle of the early twentieth century. Hays described this conflict between decentralization and centralization as essentially an effort "to substitute one system of decision-making, that inherent in the spirit of modern science and technology, for another, that inherent in the give-and-take among lesser groupings of influence freely competing within a larger system." Centralization thus represented the conquest of a closed system of decision making over a more open, democratic, participatory approach to politics.[59]

Although Hays aptly characterized the spirit of rational decision making, with its emphasis on system and expertise, his description of the battle between centralization and decentralization obscures the fact that both business and politics were centralized institutions in New York City, at least partially. Although the board of aldermen drew its membership from wards throughout the city, Tammany Hall had long acted as a centralizing institution, especially in Manhattan and the Bronx. Well before the legislature created the consolidated city, Tammany Hall had attempted to bring some order and coherence to the vigorous localism of ward representation. Like all party organizers (from ward heelers to national party bosses), Tammany took on the task of interest aggregation—creating a workable semblance of unity out of the multiplicity of interests represented within its ranks. Too often, however, Tammany's leaders focused on winning elections and lining their own pockets, rather than using centralized power to address the larger problems facing the city. Like big corporations, political machines could be narrowly self-serving.

The PRR franchise battles thus had more to do with creating an additional centralized perspective on the new problems of metropolitan governance (one that had been sorely missing) than it did with eliminating democratic politics. Indeed, the debate over the tunnel franchises had much more to do with competing approaches to the use of centralized power. The centralization of franchise-granting authority that occurred at the instigation of the PRR did not result in a simple victory for the railroad since the new approach to granting franchises never rose above the give-and-take of political conflict. Instead, the franchise battles raised the question of which centralized institutions—parties or corporations—were most capable of taking a citywide perspective on the transportation problems confronting the new metropolis. In short, these battles forced New Yorkers to ask whether it was possible to manage a regional metropolis with the perspective of a ward politician.

The first PRR franchise "hold-up" began as a partisan conflict between state legislators and city aldermen who had inherited an approach to public policy from nineteenth-century party politics that emphasized distribution as the primary function of government.[60] Railroads worked adeptly within this distributive political system. What was true of the New Haven Railroad (the New York, New Haven, and Hartford Railroad) was just as true of the PRR: as an Interstate Commerce Commission report put it, "the New Haven Railroad had no politics. It was Democratic in Democratic States and Republican in Republic States." The railroads always attempted, in the words of the New Haven president, Charles Mellen, to "get under the best umbrella," regardless of party.[61]

At Albany, in the New York state legislature, where the PRR first sought the necessary legislation for its tunnel plans, Republican Party boss Thomas C. Platt held the best umbrella. In 1902, the Republican Party controlled the legislature and Platt controlled the Republican Party, where he expanded government support of business into a regular system of exchange for campaign contributions. Not surprisingly, Platt became a compliant political ally when PRR President Alexander Cassatt called on him to amend the city's charter to allow the company to build its Hudson tunnel. Although Cassatt did not bribe Platt outright, he did promise to "acknowledge" the boss's assistance at some future date. Platt arranged for the passage of all the measures Cassatt requested, making changes in legislation as PRR managers modified their plans. Platt thus treated the PRR franchise request as a distributional problem. Although the boss probably realized the importance of rail connections between New Jersey and New York and the new level of transportation planning required for the greater city, he did not use the power of state government to plan, only to facilitate private planning.⁶²

Cassatt got everything he wanted from Platt, but the PRR could not entirely bypass city officials, and this injected the battle between Albany and the city into the company's plans. Prior to Platt's efforts to amend the city's charter, the board of aldermen had authority to negotiate railroad franchises, but the Republican legislature amended the charter to give the rapid transit commission (controlled at the time by anti-Tammany Democrats, Independents, and Republicans) authority to negotiate the PRR deal. The charter amendment further specified that the board of aldermen (where Tammany Democrats controlled enough votes to block legislation) could only approve or reject the franchise; they could not amend it. The scale of the PRR's plans made this political snub all the more insulting. As the conservative *New York Tribune* observed, "The Pennsylvania is preparing to pour out money like water for the privilege of placing a station in the heart of the nation's metropolis," and the arrangements made by the Republican legislature and the rapid transit commission effectively prevented the Democratic Party from participating in that bounty. So, while the Republicans had an opportunity to treat the franchise as a familiar distributive matter, the deal they struck with the company cut municipal politicians out of the process.⁶³ As a consequence, the board of aldermen did not approve the PRR deal with the same dispatch as the legislature, and on July 22, 1902, by a vote of fifty-six to ten, they refused to pass the tunnel franchise.⁶⁴

Initially, this hold-up probably resulted from widespread resentment against the imperious Republican legislature and its corporate ally. Local politicians decried the PRR's exploitation of the city during the franchise debate as further evidence of typical upstate disregard for local concerns: the company would import "foreign" labor to work on the tunnel; a perpetual franchise would invite abuses; the franchise fee was too low; the contract allowed the company to tear up city streets with impunity; the tunnel threatened property interests; and the benefits to the city were minimal compared with the gains for the railroad.⁶⁵ More telling were the issues not raised in the public debate. Creating the monumental Penn Station involved the same sort of urban redevelopment that drew such criticism in the post–World War II era. Construction began with "house-wrecking on a grand scale": "Tenements four and five stories high, densely inhabited with Italians" covered the midtown property chosen as the station site.⁶⁶ Between Ninth and Tenth Avenues, ninety-four buildings stood in the way of the proposed terminal. In the case of the Church of Saint Michael at Ninth and Thirty-first Street, the PRR agreed to build a new church, rectory, convent, and school at another location. Often times, however, the company encountered recalcitrant owners whom it did not treat as generously as Saint Michael's. In such cases, the PRR instituted "condemnation proceedings," and the city marshal forcibly removed the tenants.[67] But the public debate never focused on the issue of the location of the station or any other specific feature of the PRR's plans for building in the city.

Instead, it was the political battle between Tammany and anti-Tammany politicians in the city that drove the franchise hold-up. Tammany Democrats had fared poorly in the 1901 elections thanks to the efforts of the Fusion coalition and the Mazet Committee's investigations of party corruption, which left the Hall in disarray and prompted the departure of the rapacious boss Richard Croker, who, in the face of the resounding electoral defeat, retired to his estates in Palm Beach, England, and Ireland. Tammany lost each of the citywide races to Fusion candidates, with Columbia University president Seth Low taking the mayoralty.[68]

More important for the franchise question, anti-Tammany candidates won thirty-nine of the seventy-three seats on the board of aldermen. The board, which controlled the city's purse strings, was composed of biennially elected officials from assembly districts in the greater city, the five borough presidents (elected by borough), and the board president (elected at-large). The loss of so many aldermanic seats meant that Tammany was weak where it should have been the strongest, at the local level where candidates had close ties to urban constituencies. For those Tammany members who were elected, the board of aldermen often constituted the first step in a long political career. For example, Alderman Frederick Richter, elected from Manhattan's fifteenth district, subsequently became an examiner for the Department of Plant and Structures for thirteen years, rising to deputy commissioner in 1931. For John J. Dietz, the son of German immigrants, getting elected to the board of alderman in 1901 was the start of a thirty-two-year career as a city employee. Dietz, who served as Tammany leader in the Eighteenth Assembly District, South, from 1905 to 1935, moved from the board to equity clerk in the county clerk's office, then rose to chief clerk for the Eighth District Municipal Court, later to deputy commissioner of water supply, and then commissioner in 1928. Frank Dowling, born on West Twenty-seventh Street in the old Chelsea district of Manhattan, became a Tammany leader in the Twenty-fourth Assembly District in the 1890s and served on the board of aldermen for eighteen years before becoming Manhattan borough president. Tweed, Kelly, and Croker—all future Tammany bosses—had taken their turns as aldermen.[69]

The franchise debate gave Tammany aldermen the opportunity to cultivate the sort of grassroots political support that they had lost during the 1901 elections, especially from the city's largest labor organization, the Central Federated Union (CFU), which opposed the tunnel franchise. Although Democrats were generally on better terms with the CFU, Tammany Hall had a mixed record of support for labor interests. And while the Fusion administration of Seth Low had courted labor during the 1901 campaign, by the time of the franchise debate at least one labor leader had concluded that "one administration is as bad as another."[70] Especially in light of Mayor Low's solid support for the PRR project, Tammany aldermen used the tunnel issue to show their concern for working-class interests. Alderman Dowling, whose district bordered on the proposed tunnel route, suggested that the franchise require the company to recognize the American Federation of Labor and employ New York workmen, and other aldermen echoed a CFU demand by calling for an eight-hour clause.[71]

In spite of this show of concern, the aldermen's resistance to the tunnel franchise hardly represented the emergence of a class-based or interest-group political culture. Although Tammany aldermen jumped on the eight-hour clause as though they represented labor, Democratic concern was fleeting at best, and it melted away when the political utility of the hold-up dwindled. Drumming up support was just another tactic to reestablish political control over the board of aldermen, rather than an attempt to use the party to translate local concerns into public policy. The franchise battle thus confirms what other historians have shown about the relationship between unions and politicians—that labor politics and party politics were different things.[72]

The subsequent fate of the franchise illustrated that point when the aldermen approved the deal after several months of wrangling. The first break came when several local labor organizations passed resolutions calling for the passage of the franchise with or without an eight-hour clause. Merchants and property owners with businesses near the station also called for an end to the delay, while the Merchants' Association and the chamber of commerce hailed the benefits of the project. The company made much of this support from "the public" (figure 5), and on December 16, 1902, the board of aldermen finally granted the franchise by a margin of forty-one to thirty-six, just one more vote

CLEAR THE TRACK.

Figure 5. "Clear the Track," *New York Herald,* c. 1902. Silent Charlie Murphy, the new boss of Tammany Hall, engineered the tunnel-franchise "hold-up" by the board of aldermen to secure a share of the construction work on the PRR tunnel and station project, but this only made the company look like it was doing a better job of serving the city than the political machine. *Source:* PRR chief engineer's review, book 2 of vol. 1, James Forgie Collection, Smithsonian Institution, Washington, D.C.

than was necessary.[73] The vote also showed just how much influence the boroughs of Manhattan and the Bronx—Tammany strongholds—exercised over the board of aldermen. Tammany had a total of forty-four of the seventy-seven votes cast, and twenty-six of the thirty-six votes against the franchise. Had it not been for yea votes by the president of the board and the presidents of the boroughs of Brooklyn, Queens, Richmond, and the Bronx, the franchise would not have passed.[74]

But the PRR triumph had more to do with machine politics than public opinion. Six months after the franchise passed, the PRR awarded a $2 million excavation contract for its tunnel project to the New York Contracting and Trucking Company, headed by Charles Francis Murphy, who had recently succeeded Croker as Tammany boss. Although later in his career Murphy became something of a reformer within Tammany, reducing police corruption and supporting progressive social legislation, he certainly had his share of the sort of contract graft typically associated with party politicians. Born to Irish immigrant parents in 1858 in the "Gas House" district bordering the East River between Fourteenth and Twenty-seventh Streets, Murphy rose to prominence as a saloon owner before becoming a Tammany leader in 1892. Democratic Mayor Van Wyck appointed Murphy to the docks department in 1898, where he gained both power and wealth. From his new post, Murphy organized New York Contracting and Trucking with his brother and Alderman James Gaffney in 1901.[75] The announcement that New York Contracting had been awarded the PRR excavation contract made it appear as though the franchise hold-up was not a legitimate expression of public concern over the terms of an important grant to a powerful private company but an exhibition contrived entirely by Murphy and Gaffney, who had used their influence over a decentralized political body to extort bribes from a legitimate business.

The second franchise hold-up only strengthened that impression. Under Murphy, Tammany candidates did very well in the 1903 elections, winning the mayoralty and reducing the anti-Tammany contingent on the board of aldermen from thirty-nine to twenty-one. When the board refused to grant the PRR a franchise for the Connecting Railroad submitted in June 1904, it appeared that Murphy intended to extort another pay-off. Cassatt escalated the battle by vowing not to give a penny to the aldermen to secure the franchise—even as the New Haven awarded New York Contracting a $6 million contract for track improvements between the Bronx and New Rochelle in February 1905: a move correctly interpreted as an effort by the railroad to get under Murphy's umbrella to avoid its own hold-up.[76]

That in itself would probably not have inspired more than a familiar muckraking tirade against corrupt politicians and the corporate executives who cooperated with them had it not been for the mayoral race of 1905. But that year, politically ambitious newspaper mogul William Randolph Hearst pressured Murphy for Tammany's mayoral nomination. Murphy, of course, had no intention of supporting the willful Hearst, and his refusal prompted the tycoon to run as an "independent" candidate under the banner of the Municipal Ownership League. The league, led by the redoubtable Judge Samuel Seabury (later distinguished for his investigation of Tammany's Tin-Box Brigade), advocated public ownership of utilities and transportation companies—a cause that Hearst had championed since his days crusading against the Southern Pacific Railroad in San Francisco. Tammany, in general, and Murphy, in particular, were vulnerable to the criticism that they sold public franchises to line their own pockets, and the recent revelations by the Hughes Commission of fraud and corruption in the state's insurance industry only made Hearst's accusations more believable. Using his newspaper employees as a campaign staff, Hearst ran a caustic and effective campaign and indeed appears to have won the election, but Murphy's men waylaid ballot boxes from Hearst wards and dumped them in the East River. The sodden ballots were retrieved by Hearst's forces, but Murphy had others printed, marked for Tammany candidate George B. McClellan, and delivered to the board of elections. In the end, Hearst lost the official count by only 3,468 votes (out of 605,298 cast).[77]

In this heated political atmosphere, the debate over the franchise evolved into a struggle between Brooklyn and Manhattan within the context of the consolidated city. By March 1905, the hold-up became more and more a dispute between business interests in Brooklyn and Murphy's forces on the board of aldermen. The four-thousand-member Transportation Reform League of Brooklyn, headed by Republican J. Edward Swanstrom, Brooklyn borough president during the Low administration, portrayed the hold-up as a blow against Brooklyn's vital interests, with Murphy playing politics to help himself and showing little regard for other parts of the metropolis. Increasingly, Tammany Democrats appeared incapable of looking beyond their own parochial political concerns in Manhattan to address the broader needs of the greater city, especially the desires of business interests in the outer boroughs.[78]

As the Connecting Railroad hold-up evolved into a struggle between Manhattan and Brooklyn, it reopened the old debate over which government institutions could represent the interests of the consolidated city. The commission that helped draft the charter for Greater New York had urged that a variety of financial and policy functions be transferred from the board of aldermen to the board of estimate—that is, from a decentralized to a centralized political body. The board of estimate was composed of the mayor, the comptroller, the president of the board of aldermen (all elected by the city at large), and the five borough presidents. Voting power assured that at-large officials had a greater voice on the board of estimate, with each at-large official casting three votes, Manhattan and Brooklyn borough presidents with two each, and one vote for the presidents of Queens, Richmond, and the Bronx. This arrangement made it more likely that officials concerned with the city as a whole would have at least as much influence over public policy decisions as those concerned with particular boroughs or particular party organizations within the boroughs. The aldermen, who represented separate districts within each borough and who could not, or would not, take a broader view of the city's interests, were seen as too parochial to deal with many of the issues that faced the new metropolis.[79]

The PRR franchise hold-up thus represented a new twist on the long-standing debate over whether legislative bodies should control the granting franchises. The city had a blemished history of franchise grants, and the board of aldermen was but the latest legislative body to succumb to corruption. As one student of the city's franchise history put it, "The conclusion from experience is that the municipal council is incompetent, that the legislature is equally incompetent, and that legislative bodies in general are not the proper authorities to grant franchises." Typically, legislative bodies had given away important public franchises too readily, but here the aldermen failed to grant a franchise to a railroad that was addressing an important dimension of citywide transportation planning, rather than exploiting a public resource.[80]

The PRR encouraged and capitalized upon this line of argument by having a charter amendment drafted that would transfer authority over franchises from the board of aldermen to the board of estimate. The bill was authored by Edward Morse Shepard, the PRR's special counsel in New York City, with the assistance of his protégé, George McAneny, future borough president of Manhattan and a leader in the city-planning movement. Shepard, who was once described as "a courageous apostle of reform," provided the sort of intimate knowledge of New York politics that the PRR needed, just as it relied on engineers for their understanding of local construction matters. Shepard was born in New York and raised in Brooklyn. His father, Lorenzo, had been a grand sachem of Tammany Hall during the mayoralty of Fernando Wood in the 1850s, but had been an opponent of corruption within the Democratic Party. Edward Shepard became a member of the Young Men's Democratic Club of Brooklyn and developed a keen understanding of city politics through his participation in the Civil Service Reform Association of Brooklyn during the 1890s, where he worked with McAneny, the association's secretary. Shepard served as counsel to the rapid transit commission for nine years before accepting the Democratic nomination for mayor in 1901, a race he lost to Seth Low. Even though he failed in his bid for mayor, Shepard proved to be an invaluable ally to the PRR, mustering Brooklyn's political forces to support the measure shifting franchise authority to the board of estimate—and, as it turned out, moving the bill through the state legislature took an all-Brooklyn effort, including the support of Brooklyn political boss "Long Pat" McCarren (so-called because of his stature), who rallied assemblymen from his districts.[81]

Centralizing franchise-granting authority in the hands of the board of estimate turned out to be a mixed blessing for the PRR, however. The company reapplied for the Connecting Railroad franchise in November of 1905, but did not obtain it until December of 1906, thanks to Mayor George McClellan, a Democrat and son of the Civil War general, who instigated this third hold-up. McClellan rose through the ranks of Tammany Hall as the pawn of Commissioner Murphy, who backed him for mayor in 1903 and again in 1905 in the horrendous race against Hearst, but the relationship between McClellan and Murphy deteriorated. As mayor, McClellan grew increasingly independent, making patronage appointments that the commissioner did not approve, in part because Murphy promised to support Hearst for governor in 1906 as reparations for the 1905 mayoral race. Most important of all, Hearst's strong showing in the 1905 election made public franchises a hot public issue. As the rift between Murphy and his erstwhile puppet widened, McClellan made his own effort to court Hearst's supporters. To do that, McClellan held up the PRR's Connecting Railroad grant, emphasizing that he wanted to protect the city from the avaricious company before bestowing the coveted franchise.[82]

Although the PRR did ultimately get what it wanted from the city, the franchise battles highlighted the limitations of partisan politics in dealing with the new scale of public policy required for the greater city. Patronage democracy facilitated distributional policies throughout the nineteenth century, but as Democratic Brooklyn borough president Martin Littleton emphasized during the Connecting Railroad debate, "The developments in modern municipal life have made it necessary that in the granting of franchises legal, engineering, and financial problems of the most profound and intimate character should be solved."[83] The average alderman had little training and little inclination to deal with those sorts of issues. Party bosses like Murphy had no such interest; even while Murphy ran a centralized political organization, he continued to operate from the perspective of a ward politician, with little sense that the new metropolis required a different approach to getting things done.

As management foresight and engineering prowess transformed the individual projects of the PRR and H&M into integral parts of a citywide transportation system, party politicians had to ask what their roles would be in the creation of the emerging regional commuter network. Acculturated in the school of party politics, local elected officials approached the transportation problem very differently than did the itinerant generalist engineers who planned and built railroad stations, tunnels, and bridges throughout the city.

The franchise battle allowed partisan battles to spill into discussions within the PRR and the city's engineering community, highlighting the cultural differences—differences in approach and perspective—between the two groups. From the perspective of the experts who crafted the plans of the PRR and the H&M, this intrusion brought nothing substantive to the process of infrastructure creation in the city and appears to have been motivated entirely by parochial political considerations. Party politicians like Murphy seemed ill-equipped indeed to coordinate the varied undertakings of private industry, or even look beyond Manhattan to recognize the interests of other boroughs, and their efforts to hold up the franchises needlessly lengthened the already decades-long process of linking the city to the region. The real problem here was not that politics involved give-and-take among multiple parties (decision making at every level involved give-and-take); nor was it that local politicians were valiantly representing local interests against the demands of big business, for the franchise hold-ups had very little to do with genuine local interests. Instead, party politicians in the city did not look beyond the needs of the machine in its strongholds and therefore failed to expand their appeal beyond a relatively small area of the greater city.

This is not to say that elected officials were completely incapable of enlarging their partisan perspective to consider the city as a whole. Murphy's protégé Al Smith, a ward politician of the first order, learned how to consider the broader needs of the polity as he rose through the ranks of elected office. Fiorello La Guardia showed that an elected official could combine a good-government ethos and respect for expertise with a human touch in politics that outdid even the ward heelers. In Pittsburgh, Democratic Party boss David Lawrence transformed himself into a modern municipal manager by grafting regional planning onto ethnic politics to rebuild the city after World War II. And in Chicago, Richard J. Daley spent his entire life working inside the machine, only to declare, once he became mayor, certain areas of municipal management, such as city planning and fiscal matters, the domain of experts. Indeed, as Daley's reign showed, machine politics itself could be a highly centralized affair, with party bosses using all the power at their disposal to keep precinct captains in line with the demands of a citywide organization with a citywide perspective.[84]

Neither were corporations blameless when it came to perpetuating the system of local politicians using public franchises to extort political and financial subsidies. PRR managers, like many business executives, took advantage of cooperative politicians when they could find them. Railroad leaders like Alexander Cassatt and Charles Mellen knew how to play distributive politics and did not show much hesitation when it came to pulling strings to get their way. There can also be little question that many of the businessmen who urged the passage of the Connecting Railroad franchise were thinking primarily in terms of their own interests in Brooklyn and Queens, rather than creating new institutions to address the needs of the city as a whole. Nevertheless, in the pursuit of its own profits, the railroad succeeded in promoting a wider range of interests within the greater city than Tammany Hall, thanks to the experts who were planning the PRR improvements; at least in this case, the self-serving railroad was not merely self-serving, whereas the machine appeared to benefit only a handful of political insiders.

But the process of centralizing franchise-granting authority in the board of estimate involved more than self-interest. The PRR succeeded in that effort because the franchise debate raised fundamental issues of policy making for the new metropolis. The consolidation of Greater New York was predicated on the notion that there were citywide interests and that the city needed a new institutional approach to address those interests. Charter commissions had repeatedly questioned whether the aldermen or any group of legislators could adequately adopt the new citywide perspective, attuned as they were to particular districts within each borough and to the demands of party politics. The franchise battles illustrated that the PRR had adopted a citywide approach to the transportation problem of the new metropolis, acting as an invisible planning agency for the greater city. Of course, it was not enough to leave regional transportation planning to the railroads, for in spite of the political machinations behind the franchise hold-ups (Tammany's drive for votes, Murphy's ploy for contracts, Hearst's ambition for office, and McClellan's search for political identity) there were legitimate local issues that could have been raised about massive corporate projects like Penn Station and the Hudson River tunnels. Instead, the city would have to find ways to cooperate with private companies to address the multilayered transportation planning challenges facing the consolidated city.

The People versus the Interests: The Case of West Side Avenue Station

In the case of the PRR franchise battles, local politicians showed themselves to be shortsighted when it came to dealing with a large corporation with a citywide approach to transportation. But in other cases of conflict between the railroads and the city, the planning perspective adopted by engineers and railroad executives bordered on real neglect of legitimate local concerns. The West Side Avenue Station case illustrated how establishing the conditions of local development within citywide, regional, and national systems became a central problem for public officials. With the growth of the citywide perspective, in other words, finding a way to relate the whole to the parts emerged as a primary issue in policy debates.

The West Side Avenue Station case highlights this problematic relationship as it developed between the city center in Manhattan and communities on the suburban periphery in New Jersey. The case arose from the fact that, despite the importance of the PRR as a private planning agency for the city, the company's executives in charge of the New York improvements consistently maintained that they had no interest in developing a transportation system to serve strictly local needs. Although they provided a commuter link between New Jersey and Manhattan and controlled the LIRR with routes throughout Brooklyn and Queens, they wanted no part of the subway or trolley business in the region, which tended to focus on a smaller scale of transportation.[85] Their vision of the city involved different levels of planning by different institutions, with the corporation creating long-distance connections to local distribution networks provided by subway and trolley companies. This differentiation between transportation levels created the potential for conflict as local communities fought for access to the emerging H&M/PRR rail network directed primarily toward the city center.

Facilitating commuter traffic generated by regional growth had inspired the engineers who designed the New Jersey stretch of the PRR route into Manhattan. This express rail system began at Manhattan Transfer, a switching station in Harrison, New Jersey (figure 6). There, steam engines hauling passenger trains were exchanged for electric engines to make the journey through the Hudson River tunnels. The PRR tracks from Manhattan Transfer joined an extension of the H&M tracks in Jersey City, allowing PRR electric trains to take the H&M tunnels to lower Manhattan. H&M trains could also take passengers as far as Manhattan Transfer using the same tracks.[86] Three stations, all in Jersey City, lined the route between Manhattan Transfer and the H&M terminal in lower Manhattan: Exchange Place, near the Hudson River, Grove Street, and Summit Avenue. The combined H&M/PRR service across this busy stretch of track created scheduling difficulties at rush hour, which required coordination of suburban New Jersey commuter trains at Manhattan Transfer with trains using the Hudson tunnels.

When the PRR began operating this new express system, it ended its old commuter service through the area. That service had included a stop at Marion, a small area of Jersey City with a population of 11,200 that lived mostly in tenements and a few single-family homes. Marion had thirty-nine manufacturing establishments employing about five thousand people, along with a wealth of undeveloped real estate, and local boosters hoped that improved transit facilities would lead to growth. When they learned that the new PRR system did not include a stop in their neighborhood, residents immediately appealed to the New Jersey public utility commission to compel the company to build a station at West Side Avenue.[87] Marion residents ultimately made four appeals to the utility commission—in 1913, 1915, 1917, and 1921—to prevent the emerging regional transportation network from bypassing them. All four times the commissioners disappointed them.[88]

We see in these repeated appeals to the utility commission a locality struggling to retain a connection between its own development and the growth of the regional economy via the transportation network created by the PRR. The lead petitioner for Marion, the Manufacturers' and Property Owners' Interests Association, argued that the economic future of the area required a station at West Side Avenue, a contention poignantly illustrated by business after business.[89] Thomas Maloney, president of P. Lorillard & Company, testified that he had constructed a "million dollar plant" in Marion, employing fifteen hundred people, only to discover that inadequate transportation upset his plans for moving six hundred clerical workers closer to the factory. The company had built an office for their white-collar employees near the new Marion factory, Maloney recounted; "but after a few months of actual use it was found necessary to move the office force to New York City, abolish the office plant at Marion and spend the enormous sum of $40,000 a year rent because of the lack of transportation facilities in accommodating the persons to get from their homes to the office." Frank Schlicher, a manager at the Mangel Box factory, testified that in stormy weather he had to use a truck to transport employees to the Summit Avenue station. The vice president of the Davis Bournonville Company had to rent automobiles to convey visiting engineers from Summit Avenue to his plant because there was no direct route from the station to the factory and strangers found it to be a confusing and difficult journey. The relative remoteness of Marion made sales calls a problem as well. H. G. Ketten, a manufacturer of light machinery, testified that a customer had lost his way from the Summit Avenue station and took more than an hour and a half to reach the plant. One of the managers for Lorillard & Company noted

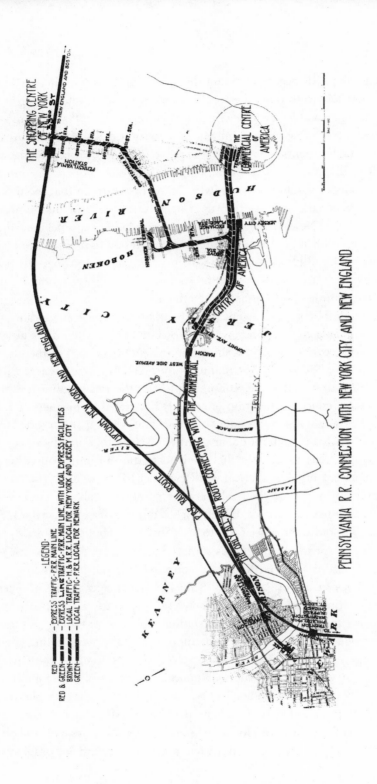

that he always got a better price for materials from salesmen who came to the factory: "When you go to see a man you buy at his price; when he comes to see you it is pretty surely going to be at your price." Without the West Side Avenue station, Marion remained off the beaten path in a region where access to transportation facilities meant crucial connections to labor, customers, and suppliers—and thus the difference between success and failure.[90]

In response to these compelling local demands, the PRR defended its role as a regional transportation planner. The company rejected Marion's request for a station at West Side Avenue primarily on the grounds that its system was conceived and designed "upon the principle of a High Speed Line" between Manhattan Transfer and lower Manhattan.[91] Engineer James Forgie characterized the PRR system as "the only all Rail Continental Line to the Commercial Capital of the United States," constructed for legions of commuters interested in rapid service, rather than for the little towns and businesses within the interstices of the vast region.[92] On Forgie's advice, counsel for the PRR emphasized to the utility commission that the question of adding a station to the new line was "purely scientific—a matter of operation" within a regional system. Simply put, another station stop increased the running time of the trains on the commuter line. To make up this time, the speed of the trains between stations would have to be increased to a level "higher than the limit of safety and comfort." With the added station, trains would have to reach fifty-seven miles per hour to maintain their scheduled running time, even though they were designed for a maximum speed of fifty miles per hour. Adding another station thus meant reducing the capacity of the line.[93]

With the technical limitations of the system firmly established, the question of adding another stop became a purely utilitarian matter. The PRR and its engineers presented the utility commission with the unenviable task of acceding to the demands of a handful of Marion residents at the cost of maintaining the integrity of a regional system and reducing service for millions of commuters.[94] As one of their engineers put it, "The greatest obligation that a transit company owes to the traveling public is best service to the greatest number of people. The development of an outlying section of the city, and the development of any community interest, is one of the most important things

Figure 6. Pennsylvania RR connection with New York City and New England, 1915. The little town of Marion fades into insignificance in this view of the new commuter rail system. *Source:* James Forgie Collection, box 5, Smithsonian Institution, Washington, D.C.

to be attained by the introduction of rapid transit, but such development should not be made by a sacrifice of the greatest number [of people using a rapid transit line]."[95] Local development was important, James Forgie acknowledged, but "it is not the function of a main line and a private investment to be called upon to compete within the five cent zone with trolley lines and with any future extension of rapid transit which may be constructed by City subsidy."[96] The PRR designed its system to address the regional transportation problem rather than strictly local needs like those of Lorillard & Company or the Mangel Box factory.

In the face of these arguments, Marion's boosters tried to frame the West Side Avenue Station case as a matter of the people versus the interests. A distant and aggressive company had, in effect, the power of life and death over a locality that depended on transportation services to link it with the larger regional economy. Like the battles among farmers, merchants, and railroads in the late nineteenth century that gave rise to state and national railroad regulation, the West Side Avenue Station case highlighted the apparent powerlessness of small communities in the face of corporate giants.

But the case in fact showed that the PRR did have a vision for the region, and that the company's refusal to provide Marion with a station grew from its efforts to create railroad infrastructure to serve commuter and long-haul passenger traffic into Manhattan—efforts that a majority of utility commissioners recognized and appreciated. This was more than an uncomplicated case of corporate indifference to public need, in other words. The battle between Marion and the PRR instead raised the problem of how to coordinate the various levels of planning in the region, for within the regional metropolis there were many defensible ways to formulate the public interest. Boosters in Marion had legitimate concerns, but so did the commuters who used the PRR/H&M system to get to work in Manhattan. For a metropolitan area of this scale and scope, there was no simple relationship between the whole and the parts. Negotiating the connections between Manhattan and New Jersey within a regional framework would constitute the central problem for railroads and public officials—and for planning advocates in general—for years to come.

Between Corporate and Civic Cultures

Assisted by experts like Samuel Rea, Charles Jacobs, and John Davies who understood regional needs and patterns of development, the PRR and H&M

provided unifying infrastructure for a city that lacked governmental organizations capable of thinking and acting in metropolitan terms. Indeed, the facilities created by the PRR and the H&M continue to play a vital integrative role in the New York region. The original Penn Station was torn down in the 1960s (so that the company could generate more revenue from its expensive midtown property—the very thing that Alexander Cassatt had resisted to build the grand civic structure), but the PRR tubes are still used by Amtrak trains, which travel under the Hudson, into midtown Manhattan, under the East River, through Queens, and over the Hell Gate Bridge, along the route created by Rea and the engineers at the Pennsylvania Railroad a hundred years ago. Although the Cortlandt Street Station was replaced by the World Trade Center (WTC) towers, the H&M tunnels were retained and are now known as the PATH (Port Authority Trans-Hudson) train. Even though the tragedy of September 11, 2001, temporarily closed the south tube between the PATH stations at the WTC and Exchange Place, passengers can still get to Manhattan using the north tube, which runs along Sixth Avenue to the Thirty-third Street Station; indeed, all stations between Newark and Thirty-third Street experienced a significant upsurge in passengers after September 11, prompting the Port Authority to add additional trains to meet the demand. The enduring utility of the H&M tunnel system as part of the Port Authority's regional commuter network proves just how prescient William McAdoo was.

The interaction of experts and businessmen lay at the root of this fortuitous combination of public benefit and private initiative. Walter Lippmann saw in this union of expertise and American industry the substitution of mastery for drift. Expertise leavened the pursuit of profit with rationality, thus producing a quasi-scientific voluntarism that allowed large-scale corporations to bridge the gap between self-interest and the public good.[97] This did not mean that rationality did away with social conflict. The engineers involved in the New York tunnel projects could have attested to the fact that disagreements plagued even the most expertly crafted plan, since cooperative arrangements between companies or between business and government generated disagreement even when mutually beneficial. They often proceeded nonetheless; but this depended on a process of mediation by experts built around a shared vision of urban development that came only with years of experience working in, around, and under the city.

In the early twentieth century, that shared vision of the city and the region as a whole created important openings for railroads to act in the public interest. By treating the city as a system, they could bring the values of engineers into civic life, infusing an often self-destructive political culture with a more enlightened approach to collective problem solving. Railroad engineers and executives could reasonably argue that they served the public better than the party officials who could not seem to look beyond their own political or pecuniary interests to acknowledge the value of the PRR system more expeditiously.

Thus, the success of the tunnel projects, especially in the context of the politicized fight over franchises, blurred the very real conflicts between corporate and civic interests. Not every railroad plan integrated smoothly with the larger public interest, even when based on the system ideal and crafted by farsighted engineers, as the people of Marion discovered. Arguably, the coincidence of public and private interests in the H&M and PRR projects represented a relative rarity; as will become clearer in the next chapter, even as they helped the city solve its passenger-transportation problem, the railroads frustrated efforts to solve the even more important freight problem. Engineers would not always live so happily in both worlds—civic and corporate—and some had to choose sides when the pursuit of profit came into conflict with the achievement of public objectives. But conflict did not always separate the two, and a great deal of the optimism of the Progressive Era, as evinced by thinkers like Lippmann, grew from concrete evidence that the rational pursuit of self-interest did not preclude the rational pursuit of the public interest. Given the alternative between local officials who could not seem to think beyond their electoral districts or railroad executives who got things done and solved the city's problems even as they fattened their own bottom lines, engineers could easily exaggerate the subtleties that separated the enlightened corporate leader from the shortsighted robber baron—it *did* seem like a new era with Cassatt, Rea, and McAdoo leading the way. The power of the system ideal—its potential to resolve the nagging conflicts of an industrial society that prevented the full realization of the great promise of abundance in the technological and organizational advancements of the age—made it appear as though expertise could devise the new terms of togetherness needed to render the vast metropolis truly livable by bridging the gap between corporate and civic.

In this sense, the fallout from the PRR franchise battles overstated the virtues of expertise when compared with local political culture. Enlightened self-interest could serve the city or the region in vital ways, but it also created new problems that did not yield so easily to technical know-how and system thinking. The city could be viewed as a whole, and very productively, indeed; that did not mean, however, that there was any "one best way" to relate the whole to the parts. The West Side Avenue Station case illustrated that the system builders, impressive though their accomplishments were, did not resolve the problem of aggregating the public interest in the new metropolis. By focusing on the region, the system builders opened the door to new types of conflicts; henceforth, they would have to defend their systems against claims by local interests and local governments seeking access to the larger networks they created in order to serve one important formulation of the public interest.

CHAPTER TWO

Beyond Voluntarism
The Interstate Commerce Commission, the Railroads, and Freight Planning for New York Harbor

For Calvin Tomkins, city docks commissioner from 1910 to 1913, reorganizing the freight system at the port of New York constituted the single most important planning problem of the era, just as it had been for so many of the merchants who had called for the creation of a consolidated city. New York's share of the nation's foreign commerce had fallen from 55 percent during its highpoint in the 1870s to 46 percent by the 1910s.[1] Other ports successfully chipped away at Gotham's commercial dominance because the freight "system" at the port was no system at all, Tomkins claimed; it remained what it had been in the nineteenth century—a conglomeration of uncoordinated facilities managed by competing companies. In place of this patchwork voluntary effort, Tomkins insisted that the "port must be developed as a unit." From that basic principle, a new era of publicly managed freight operations would emerge: "The port being once conceived as an organic whole—administered by the City for the benefit of all—there can be no thought of remaining in or returning to the chaos of jarring private rivalry and mutual obstruction from which we suffer; or of final dependence on the makeshift policy of separate subports constructed by great private corporations—no matter how perfect each may be in itself, or how welcome they may be as co-operators in a City system."[2]

With their plans to treat the port as a unit, both the docks department and the Port of New York Authority (the bistate agency created in 1921 in response to the transportation mess at the harbor during World War I) attempted to implement the idea of the city and region as a system, thus making the precepts of the civic culture of expertise the operating principles of public sector institutions. For the new breed of expert who championed the idea of public system building, the failure of enlightened voluntarism to address the freight planning needs of the region demanded government officials armed with new policies and ensconced in new institutions to look beyond the interests and achievements of individual firms to address the general interest of the port as a whole.[3] Neither Calvin Tomkins nor the Port Authority experts realized their plans for a publicly managed freight system at the port, however—a failure historians have generally attributed to the obstructionism of railroad executives too blinded by their own interests to act collectively.[4] Indeed, engineers, entrepreneurs, and public officials created so many ostensibly workable rail plans for the port of New York—from Gustav Lindenthal's Hudson River bridges to William Wilgus's subterranean freight subway and the Port Authority's regional beltway system—that it certainly *seems* like there should have been a way to fix the port system and sustain New York's commercial preeminence, had not powerful but shortsighted railroad companies resisted cooperation. Since technical solutions to the port's troubles abounded, the real issue in the freight debate must have been the battle between private greed and the public good.

While it is undoubtedly true that the railroads were, as usual, looking out for their own individual freight operations and bottom lines, that alone does not explain the course of institution building regarding freight transportation in the New York region. Instead, the evolution of freight planning at the port illustrates the difficulties of bringing the system ideal to the public sector—an arena that was affected by myriad institutions in different parts of the country with a multiplicity of agendas and constituencies, ranging from shippers and merchants to mayors, department heads, governors and legislators, U.S. senators, and interstate commerce commissioners. The effort to engage in citywide or regional freight planning was stymied by the fact that the port was suspended within institutional contexts in which policy debates were driven by considerations that had nothing to do with New York harbor or its transportation difficulties. The reason that this intergovernmental conflict was so frustrating to would-be planners, and so illustrative of the larger problem of planning the new metropolis, was that the ideal driving the process of centralization—the city as a system—seemed to point the way toward the reconciliation of the diverse interests within the region. Since experts could conceptualize the city and the region as a whole, since they could identify a public interest and specify the institutional forms necessary to act upon that interest, and since port planners, construction engineers, and railroad officials spoke the same language of system building, it looked as though conflicts besetting the harbor could be transcended in the interest of the greater good— in effect, repeating the H&M/PRR success story on a larger, public scale. In this sense, freight planning experts overestimated the power of the system ideal and underestimated the institutional problems involved in putting it into practice.

By following this debate through local, regional, and national contexts, this chapter shows how the process of institutional change generated not one approach to system building, but several—from the quasi-scientific voluntarism of the railroads to the municipal joint-venture planning of the docks department, the Veblenian planning of the Port Authority, and the regulatory balancing of the Interstate Commerce Commission (ICC). By the late 1920s, the state-building experience illustrated that while there was arguably a public interest, a citywide interest, a regional interest, and a national interest with regard to railroad policy, they were not necessarily compatible. Formulating the relationship between the whole and the parts—conceptualizing the terms of togetherness for the port—was thus far more conflict-ridden than the ideal of planning for the city and region as a whole would suggest.

Freight Geography: A Profile of Collective Institutional Failure

To gaze upon the frenetic activity in New York harbor on any morning in the early twentieth century offered a pointed lesson in the failure of private institutions to manage the city's commercial future.[5] The New York region served as both a great shipping center and a great railway terminus without any direct connections between trunk lines and steamships, the crucial element in the economical movement of freight. The majority of railroads—including the PRR, Erie, Central Railroad of New Jersey, DL&W, Lehigh Valley, and West Shore lines—terminated along the Jersey shore from Greenville to Weehawken.

Shipping lines were located primarily in New York City.[6] To move freight between New York piers and New Jersey rails, the New Jersey railroads employed an armada of ferries, lighters, and rail barges, converting the harbor into a giant floating railroad yard. On an ordinary day, between fifteen hundred and two thousand freight cars waited in the slips in lower Manhattan. Because of the lack of waterfront space, the railroads used the long (often 750 feet), narrow Manhattan piers as storage areas, and shippers themselves contributed to congestion on the docks by moving freight during the early-morning and late-afternoon rush hours.[7]

The separation of the city's many small industries from direct connections to transportation facilities further compounded the density of freight movement in Manhattan. The streets leading to the waterfront bulged with trucks and carts bringing goods from thousands of small manufacturers in Manhattan, Brooklyn, and Queens. Although lower rents for commercial property in other boroughs tempted manufacturers to locate outside of midtown, goods inevitably arrived at their final destinations more quickly when they were shipped from Hudson piers, even if that meant hauling them across the East River bridges and through the bustling streets below Central Park; as a result, the Manhattan waterfront remained the locus of freight movement even as other parts of the city developed industrially.[8]

Although the west side of Manhattan continued to serve as the nexus of freight movement for the entire city, it was as much obstacle course as entrepôt. The "West Side Freight Problem" resulted from the location of the tracks of the New York Central, which extended southward from Spuyten Duyvil (the waterway separating the Bronx from Manhattan) to Manhattanville (a railyard between 145th and 135th Streets) to the 60th Street yard, winding their way through city streets all the way down to Saint John's Park (figure 7). Despite the potential advantages of this route, the entire length of the Central's facilities was "wretchedly inadequate." Along the west-side line at all hours of the day and night, trains with from four to twelve cars blocked the streets that led from the waterfront to midtown, delaying the movement of freight and commuters between the piers and inland destinations, especially below 30th Street, where terminal facilities were so overmatched that the company loaded and unloaded cars as they stood on public thoroughfares. As engineer Charles Jacobs observed, "Notwithstanding the constant danger of these surface tracks, they are of the utmost benefit [to the company], and the

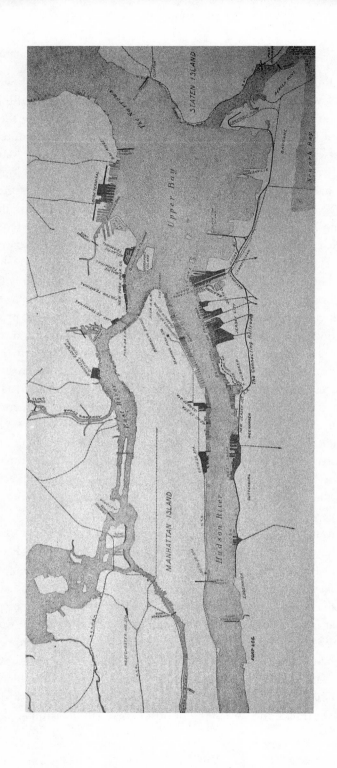

advantageous position of the New York Central in the handling of freight, due to its all-land position, gives a competitive zest to the operations of all the New Jersey roads with their amphibian methods."⁹

That competitive zest created an atmosphere in which every railroad and terminal company scratched and clawed for precious space along the waterfront. Rearranging the west side—whether by elevating the Central's tracks or building centralized collection and distribution facilities—might improve the overall efficiency of the port, but only at the expense of an individual company's competitive advantage. No one wanted to relinquish what they had for fear that other companies would gain another toehold along the Manhattan shore of the Hudson. From the perspective of public officials like Calvin Tomkins, it was only a matter of time before Baltimore, Boston, Philadelphia, New Orleans, San Francisco, Galveston, and, most important of all, the aggressive little cities on the far bank of the Hudson, such as Jersey City and Hoboken, created facilities that would attract foreign shippers who wanted to avoid the congestion and expense of the Manhattan waterfront.

Thus, by the early twentieth century it was undeniably clear that the invisible government of railroads and terminal companies at the port had produced an inefficient, often chaotic, freight-handling and distribution process that threatened New York City's long-term commercial prosperity. To be sure, businesses like the Bush Terminal Company in Brooklyn or the PRR in New Jersey made enormous investments to improve their own freight-handling capabilities. But their heroic individual efforts could not correct the inadequacies of the port as a whole. The port needed new facilities and a more comprehensive approach to planning to eliminate the bottlenecks on the west side and improve freight movement from Brooklyn and Queens. And while the port's problems were obvious to nearly everyone, transforming a collection of private companies into a public system to serve the port as a whole was no simple matter. The question was, what could the city do to replace the invisible government of private companies that managed (or mismanaged) the port?

Figure 7. Railroad terminals and terminal companies at the port of New York, 1917. Although New York had many impressive private freight facilities, they did not operate as a unit, thus eroding the overall efficiency of the port. Note the location of the Bush Terminal Co. and the New York Dock Co. along the South Brooklyn waterfront, where Docks Commissioner Calvin Tomkins envisioned the Brooklyn Marginal Railroad. *Source: New York Harbor* case, 47 ICC 643, foldout between 650 and 651.

The ICC and the Retained-earnings Controversy

During the early twentieth century when the railroads were the primary planning agencies at the port of New York, municipal efforts to move beyond the "jarring private rivalry and mutual obstruction" that the experts identified as the root of the harbor's shortcomings became entangled with larger national regulatory debates, even though those debates rarely considered the local consequences of federal policy. To understand why New York City's freight planning unfolded as it did, therefore, we need to shift the focus of inquiry to the federal level and examine regulatory discourse at the Interstate Commerce Commission, far afield though it may seem. From the perspective of officials at the federal level, the problem of the invisible government of railroads was not a lack of service or inadequate facilities, but the abuse of financial power by corporations with a national scope.

The key to understanding the conflict between federal policy and port planning lies in the ICC's pre–World War I decisions on the use of retained earnings: the revenue that railroads retained after paying operating expenses. As an attempt to strike a balance between shippers' interests and railroad profits, the ICC established the policy that investments in permanent improvements (stations, tunnels, bridges) could not be counted as operating expenses for purposes of rate regulation. This principle, developed in the *Advance Rate* case of 1903 and the *Central Yellow Pine Association* case of 1905, constituted the foundation of the ICC's subsequent arguments against rate increases in pivotal cases between 1910 and 1914. Those rate battles had a decisive influence on the railroads' willingness to cooperate with municipal plans to remake the port by gradually eroding their ability to invest in the permanent improvements called for by port planners. The rate cases thus illustrate just how differently federal and local governments conceptualized the problem of railroad regulation and, therefore, how intergovernmental conflicts shaped the course of state building where the port was concerned.

The ICC first enunciated its policy on operating expenses and permanent improvements in the *Advance Rate* case of 1903, when carriers asked for higher rates on products transported from the Midwest to the East Coast. The railroads had recently made extensive improvements—including eliminating grades, replacing wooden bridges with iron bridges, and building tunnels and stations—designating them as "operating" expenses. However, the ICC ruled that only insofar as these investments were dedicated to maintaining the existing capabilities of the railroads could they be counted as operating expenses. When such investments significantly upgraded or expanded a railroad's facilities, they increased its value and earning capacity and had to be counted as additions to the value of railroad property. In the eyes of the ICC, when a railroad made a permanent improvement by reinvesting retained earnings from freight rates, management shifted wealth from shippers to stockholders, effectively paying them twice. As the ICC put it, "Assuming that the stockholder is only entitled to exact from the public a certain amount for the performance of the service, he clearly has no right to both receive that amount in dividends, and add to the productive value of his property."[10]

The ICC thus approached the issue of railroad investment practices very differently than the would-be planners at the port of New York. From a local perspective, the need for railroad investment in new facilities to make the port more efficient was paramount, since new railroad bridges, tunnels, stations, and yards benefitted local businesses and, potentially, the port as a whole. For the ICC, however, railroad rates were the crux of the regulation issue, and rates had to be set to prevent railroads from transferring wealth from shippers to stockholders.[11] In the eyes of federal regulators, the invisible government of railroads abused its quasi-monopolistic position to transfer wealth by using freight rates to tax shippers, and the ICC saw itself in the role of protecting these shippers against the powerful investors who benefitted doubly from railroad investment practices.[12]

But the commission was not simply the instrument of shipping interests, nor was it trying to penalize the railroads because they were big businesses. Indeed, the commission acknowledged that railroads often made unremunerative investments to which the public should contribute. The commission referred specifically to the Pennsylvania Railroad's new Union Station in Washington, D.C. (which opened in 1907 and is still in use today), noting that although the new facility was an expensive permanent improvement, it did not add to the earning power of the railroad in proportion to its sizeable cost. Because the station was an enlargement of an outgrown facility (and because it was in the nation's capital), the public could be called upon to help defray its great cost.[13] The commission did not blindly restrict the railroads' earning power, in other words; instead, it tried to balance the need for investment with reasonable profits.

The commission continued that balancing act in the *Central Yellow Pine*

Association case (1905), where it reaffirmed the policy that permanent improvements could not be counted as operating expenses in rate determination. As in the *Advance Rate* case of 1903, the railroads cited increased "operating" expenses as one of the reasons for higher rates. Once again, however, the commission found that operating expenses had increased due to spending on tunnels, bridges, rails, ties, real estate, docks and wharves, and equipment—all permanent improvements. Though the ICC readily admitted that such spending was necessary, it emphasized that permanent improvements should be paid for over the life of the investment, rather than with retained earnings.[14]

Five years later, the ICC took its rulings on operating expenses and permanent improvements into perhaps the most celebrated rate battle of all time, the *Eastern Rate* case of 1910, where Louis D. Brandeis—the people's lawyer, economic adviser to President Wilson, and future Supreme Court justice—attacked the railroads for their inefficiency, using Frederick Winslow Taylor's principles of scientific management. While Brandeis's use of scientific management captured the public imagination, it was his concerns about the evils of big finance, rather than the inefficiencies of big business, that caught the commission's attention. The commission had a history of focusing on the role of investors in rate cases, and thus Brandeis, who would publish his ideas on finance in 1913 and 1914 under the title *Other People's Money*, struck a familiar chord when he criticized railroad investment practices.[15]

Although the *Eastern Rate* case was conducted under the authority of the recent Hepburn and Mann-Elkins Acts (which allowed the ICC to set maximum rates and shifted the burden of proof in rate cases to the railroads), it followed patterns established in earlier rate cases. Citing the *Central Yellow Pine* case specifically, the commission affirmed that it would not permit the railroads to treat permanent improvements as operating expenses, and it singled out the PRR for its violation of this basic principle of rate-making justice.[16]

The PRR's facilities-improvement strategies were widely known and admired: PRR managers considered it a matter of pride that they tried to invest as much as they paid in dividends. Under this so-called Pennsylvania policy, the company had plowed a staggering $262 million in retained earnings into new facilities between 1887 and 1910—precisely the sort of conservative financial policy that made the PRR the nation's greatest railroad. In the eyes of the ICC, however, that superb business strategy amounted to a massive redistribution of wealth from shippers to stockholders—and nowhere was this more evident than in the expensive projects railroads built in large cities. In an example first developed by Brandeis, the ICC took special exception to the PRR's investments in its New York City terminal and tunnels. Under Brandeis's cross-examination, the PRR president, James McCrea (who had recently succeeded Alexander Cassatt), proudly acknowledged that the company had invested more than $100 million on New York projects (mainly from the sale of stock) and did not expect the station to produce any revenue directly. Brandeis then demanded to know why the shippers along the Atlantic seaboard should pay higher freight rates to allow the PRR to pay a 6 percent dividend to stockholders who had invested in an extravagantly expensive station that generated no revenue. Creating a monumental civic vestibule to serve the nation's commercial capital thus came back to haunt the PRR, since, from a federal perspective, Penn Station was not evidence of corporate commitment to the life of the city but an illustration of just how profligate the company had been with the public's money.[17]

The commission even went so far as to admit that permanent investments in large cities would actually "damage" the railroads, thus exposing the anti-urban tendencies in agrarian-dominated federal regulatory policy. The ICC conceded the astronomical costs of terminals in densely populated urban areas, taking special note of the New York Central and PRR improvements in New York City. Although the commission had approved of the PRR investing in Union Station in Washington, D.C., the massive New York improvements of the PRR and the Central were too much to ask of the shippers who subsidized them. The commission therefore denied the railroads their rate increases by affirming its policy that the nation's shippers could not be held financially responsible for helping railroads defray the cost of these improvements. Here, then, was the crux of the conflict between national and local policy.[18]

By the eve of World War I, it had becoming increasingly clear that the ICC's consistent policy on retained earnings and permanent improvements had taken its toll on the quality of rail facilities, and the commission had to face the fact that the nation's railroad system as a whole required additional revenues. In the *Five Percent* case of 1914, 112 railroads claiming an "urgent" need for more income petitioned the ICC for an across-the-board increase in freight rates, forcing the commission to sacrifice its retained-earnings policy or run the risk of further discouraging needed railroad investment. The commission admitted that, taken together, the railroads did deserve additional revenues, since their operating ratio—the percentage of gross revenues absorbed by operating expenses—as a group had risen steadily from 64 percent in 1900 to nearly 72 percent in 1913. But the problem for the commission lay in the fact that some lines were financially strong while others suffered from a lack of revenue, a situation that would become known as the "weak railroad problem." Nevertheless, following its previous pattern of investigation, the commission noted that operating expenses had increased due in part to unproductive betterments made in dense urban areas; in this regard, the ICC again cited the PRR and New York Central improvements in New York City.[19]

Rather than granting an across-the-board rate increase, therefore, the commission decided to grant only the weakest roads (those operating primarily in the old Northwest) a rate hike. In making this decision, it is clear that the majority of commissioners had become increasingly fixated on big projects like Penn Station and Grand Central Station. Such improvements stood out as reminders that some railroads had operated with extravagant incomes in the previous decade. Furthermore, these projects were for the benefit of railroads passengers, and the commission emphasized that it did not want to burden freight shippers with the cost of facilities they did not use—that was too obvious a form of redistribution of wealth from shippers to stockholders (and from shippers to passengers).[20]

Shortly after the commission rendered its decision in the *Five Percent* case, World War I broke out in Europe. This raised the immediate prospect of financial trouble on world markets and threatened to weaken the financial condition of even more carriers. At the urging of railroad officials and investment bankers, the ICC reversed its earlier decision and granted a general 5 percent increase in freight rates, prompting some members of the commission to argue that the stronger railroads were using the war to gouge the public for more profits—an impression strengthened by Samuel Rea, who became PRR president in 1913.[21] In his testimony during the *Five Percent* case, Rea took a particularly harsh attitude toward the ICC's rulings on permanent improvements. He knew just how important investing retained earnings had been to the PRR, which, thanks to its conservative investment policy, had superb credit, high-quality facilities, and a long-term vision of its role in the national economy and the urban areas it served. The company had used its financial strength to build well in advance of traffic demands. If the commission denied them rate increases to provide sufficient retained earnings for investment purposes, Rea told the ICC, railroad managers would narrow their vision, limit their investment horizon, and cease planning in advance of traffic needs in order to protect their ability to pay stockholder dividends. "It is evident that without higher rates the Pennsylvania system cannot afford to continue the policy of improvement involving nonrevenue earning, or deferred revenue earning, facilities and accommodations," he declared.[22]

Rea argued, in other words, that the railroads' vision depended largely on a level of profitability that violated the ICC's philosophy of rate making. The PRR would invest only if it were allowed to earn consistently high dividends for its stockholders *and* fund improvements with retained earnings. Company executives would hesitate to invest in non-revenue-producing or deferred-revenue-producing projects unless they could simultaneously guarantee investors an immediate dividend payable from retained earnings, since investors would not wait for their investments to generate income. In short, railroad executives could afford to be broad-minded only when they were assured of making their stockholders rich.

The Local Consequences of the National Regulatory Battle

Unfortunately, at the very moment that the ICC took steps that prompted railroad executives to limit their perspective, the City of New York needed visionary corporate leaders. The conflict between ICC policy and railroad vision thus had a direct bearing on the planning efforts of city officials who were trying to encourage railroad investment in new freight facilities—a point illustrated by the failure of the Brooklyn Marginal Railroad (BMR).

Docks Commissioner Tomkins planned the BMR as a city-owned rail line to be jointly operated by the major railroads at the port. The line would have extended from the Brooklyn Bridge to the PRR terminals at Bay Ridge, a route that took it through the largest private terminal companies on the Brooklyn waterfront—the Bush Terminal Company and the New York Dock Company. The BMR would have improved freight service from Brooklyn by allowing the various terminal companies serving the borough to consolidate outbound shipments rather than sending them piecemeal through the already packed streets of Manhattan to west-side piers.[23]

The BMR was thus an integral part of Tomkins's effort to reconfigure the entire port—a vision he had developed through his own experiences working in and around the harbor. A native of Orange, New Jersey, and a descendent of one of the original settlers of Newark, Tomkins manufactured building supplies. As president of the New Jersey Waterways Association, he had watched with interest as the H&M and PRR completed their tunnel projects. Like the engineers who built those tunnels, Tomkins saw them, along with the bridges over the East River, as the essential transportation pathways for moving goods and people in an orderly fashion between the outlying areas of New Jersey and Long Island to the regional center in Manhattan. The only thing that could halt the inevitable development of this metropolitan system was conflict between the city and the private transportation companies that created the infrastructure. When Tomkins became the head of the Department of Docks and Ferries in 1910, he set out to create a comprehensive freight distribution plan that would permit the city to supplement and ultimately replace the invisible government of railroad and terminal companies with a new public planning bureaucracy.[24] Even though he acknowledged that some individual and corporate developers did a superb job of moving freight, he proclaimed that "the day of individual railroad and steamship terminal development has passed." In Tomkins's view, the time had come to treat the city as a system, and to make all the railroads and terminal companies parts of that system through aggressive municipal planning.[25]

Tomkins's plans for transforming the uncoordinated private freight facilities at the port into a system required unifying all terminal facilities along the Brooklyn and Manhattan waterfronts and the reconstruction of the South Brooklyn waterfront under municipal control constituted the crucial first step in this effort. He urged the city to purchase the Bush Terminal Company, the New York Dock Company, and all the other private freight distributors between Bay Ridge and the Brooklyn Bridge. The city, in conjunction with the major railroads at the port, would then construct the BMR. Tomkins also called for a publicly owned elevated railway for the New York Central and New Jersey railroads, which could reach the west-side line via a planned Hudson River freight tunnel, providing direct connections to piers and joint freight terminals. The resulting network of rails, warehouses, and docks would transform the facilities of competing private companies into "*a system*, and substitute order for disorder" at the port.[26]

Tomkins intended the BMR as the keystone of this portwide reconstruction effort, led by an invigorated docks department. Toward that lofty goal, he began negotiations with the steam railroads on the BMR project in 1910. Detailed plans were drawn up and a tentative agreement reached in 1913. The city then fought for and won changes to its charter that allowed it to enter into a partnership with the railroads. But in 1914, negotiations came to a halt—thanks to the federal battle over freight rates and retained earnings. As Ralph Peters, the Long Island Railroad president, explained, the trunk lines "were feeling very seriously the effects of the general depression in business, they had failed to receive promptly the permission to increase their rates [in the *Five Percent* case], and it was necessary for all of the railroads to guard against any new undertakings." And while the trunk lines intended to resume negotiations when conditions improved, by 1915 Peters wrote to the city to say that the railroads were not going to undertake the BMR project until the ICC completed its rate hearings in the *Anthracite Coal* cases. In 1916, the railroads decided not to undertake the BMR at all and suggested that the city scale back its plans for possible consideration by the railroads at some unspecified future date.[27]

The collapse of the BMR joint venture should have come as no surprise to anyone who watched the railroads "retrench" in response to ICC rate decisions after the *Eastern Rate* case. As early as 1910, the New York Central canceled orders for new freight cars and postponed construction of new tracks, shops, terminals, and stations. The Central lowered its dividend in 1911 and laid off twenty-five thousand workers in the winter of 1913–14, with the heaviest cuts coming from the construction department. "All new construction except such as is absolutely necessary to safeguard the public has been stopped," reported the company's vice president in March 1914; "The completion of the Grand Central Terminal marked the end of the extension plans for the time being." The New Haven likewise cut spending on improvements to "the lowest possible limit"—"nothing new will be started," the company announced in 1911. The PRR began retrenching that same year, canceling orders for new engines and sending a shiver of fear through the industry, for if the greatest railroad in the country had to cut back, what hope was there for lesser companies? Conditions so deteriorated by 1914 that the company laid off one-tenth of its workforce (fifteen thousand men), bringing total workforce reductions for eastern railroads to one hundred thousand.[28]

By then, the PRR was using "every measure to conserve its resources and seriously restrict its capital expenditures." "Retrenchment must begin on betterments and improvements not directly necessary to the movement of trains," Samuel Rea declared. So severely did the railroads curtail their purchases of supplies and equipment that steel manufacturers and other suppliers began to suffer, prompting the New York Chamber of Commerce—certainly not an organization inclined toward freight rate increases—to urge the ICC to give the railroads the income they needed to maintain their facilities, else "the commerce of the Nation will be crippled." Everyone involved in railroad finance knew that investment in new facilities had come from retained earnings, and as those disappeared thanks to ICC rate policy the prospect of new facilities of any sort disappeared. As the venerable Charles Francis Adams wrote to President Wilson, "In New York City alone within the last ten years, two railroad companies have spent in the neighborhood of $300,000,000. Had [this outlay of capital] not been made, it would not now, under existing conditions be possible; fortunately, it has been made, and is secure. *That* we have."[29] With even the strongest railroads postponing needed capital expenditures on their own facilities to the detriment of the communities they served, it certainly made no sense for them to contribute to a visionary scheme dreamed up by Calvin Tomkins that offered them no direct financial benefit.

The combination of the ICC's retained-earnings policy and the railroads' demand for high profits thus had the effect of limiting the perspective of the corporate managers operating at the port of New York. Denied the ability to make investments out of retained earnings, railroad managers focused on consolidating their financial positions and protecting their existing facilities and operations, rather than pursuing ambitious new projects that did not immediately enhance their own competitive positions.[30] Because of the rate battles, the railroads did not become full partners with the city. Instead, at a time when the new metropolis needed visionary railroad leaders, it got reluctant partners.

The War Crisis and Institutional Reconstitution

World War I promised to break this deadlock between local and national policy, at least initially. As is well known, the exigencies of war prompted a wave of institutional innovation in the United States, and experts were at the forefront of those changes. Chemists proved themselves indispensable in the creation of munitions; psychologists convinced the U.S. Army to undertake the world's first large-scale application of intelligence testing; engineers and statisticians began a national industrial inventory to rationalize wartime production; and mining engineer Herbert Hoover, who oversaw the U.S. Food Administration, attempted to halt rapidly rising food prices while increasing agricultural production. For the leading political philosophers of the era—Walter Lippmann, Herbert Croly, and Thorstein Veblen, for example—these experimental applications of administrative technique and the natural and social sciences held the promise of a fundamental restructuring of American culture, with new values based on a rational, scientific worldview superseding the older, competitive, laissez-faire approach to managing social change. But as historian David M. Kennedy has argued, "to a striking degree [President] Wilson and his principal mobilization administrators tried to base American participation in [the] war on the old principles. [T]hey doggedly diffused administrative responsibility; they relied wherever possible on the time-honored principle of contractual agreement; and they affirmed repeatedly the *temporary* character of those few naked instruments of authority they were reluctantly required to grasp." Even Hoover, whom Frederick Winslow Taylor's protégé Morris Cooke called "the engineering method personified," eschewed the centralization of authority at the federal level. Nevertheless, even though the war did not usher in a full-scale institutional reconstruction of the American polity, it did provide an important context for state-building initiatives, both nationally and locally, particularly where the demands of mobilization compelled government officials to respond to the limitations of private management.[31]

Nowhere were those limitations more obvious than in the railroad industry. By 1917, the war had dramatically illustrated the inadequacies of both the national rail network and the freight system at New York harbor. The major ports on the eastern seaboard worked overtime to move American goods to France and Great Britain, but financial strain, a scarcity of freight cars, growing traffic congestion, manpower shortages, mounting labor difficulties, and increasingly evident maintenance deficiencies pushed the rail system toward crisis. The railroads' inability to keep up with the heightened demand for service threatened to interfere with production schedules in basic industries. Bad weather frequently disrupted freight movement at the port of New York, and the winter of 1916–17 was unusually harsh, slowing the already inefficient process of moving goods between rail cars and transatlantic steamers. Then, on January 31, 1917, Germany resumed unrestricted submarine warfare against Allied shipping. By the following winter, the combination of U-boats, high volume, maintenance deficiencies, and systemic inefficiencies took a toll on the port: ships loaded with war provisions hesitated to leave the harbor; empty ships did not enter. Trains from the Midwest began filling the freight yards on the New Jersey shoreline, causing traffic managers to hold cars in Buffalo and Pittsburgh, then Detroit and Chicago. Conditions at the port gradually brought the nation's freight system to a standstill. Faced with a transportation network frozen in its tracks, President Woodrow Wilson federalized the railroads on December 26, 1917.[32]

To be sure, the brevity of federal control (which lasted only until February 1920) and the return of the railroads to private management confirmed just how limited wartime changes were; but the postwar reevaluation of railroad policy on the national level proved to be one of the more important institutional changes to emerge from the crisis of 1917. Although the dream of a nationalized railroad system did not last for long, the sheer magnitude of the transportation failure demanded a rethinking of the prewar approach to interstate commerce policy. By 1919, the debate over the reconstruction of railroad policy was guided by an almost universal desire to create a cooperative approach to regulation. Rather than continue the acrimonious and ultimately self-defeating battle over rates, reformers hoped to develop a more rational and constructive approach to transportation conflicts that would create "an efficient industry that would meet future national economic requirements."[33]

The war crisis also inspired renewed interest in regional freight planning policy. The dreadful conditions at the port in 1917 confirmed what Calvin Tomkins and others had been saying for years, and it now seemed clearer than ever that the absence of comprehensive organization and direction of port facilities had exacerbated the wartime emergency. Just as federal officials had taken dramatic steps to salvage the national railroad system in 1917, local officials had to take the initiative and create an integrated regional freight transportation system for the decades ahead. The time had come, in other words, to treat the port as a unit and establish a cooperative approach to securing the future economic needs of the region.

On the surface, at least, the local and federal approaches to the postwar reconstruction of railroad policy seemed complementary. Both emerged in an atmosphere that emphasized cooperation, rather than competition, interdependence over individualism, and change over the status quo. Both produced significant institutional developments—in the form of the Transportation Act of 1920 at the federal level and the Port of New York Authority at the regional level. In practice, however, federal policy proved to be incompatible with the ambitious objectives of planners in the New York region. Just as before the war, conflicts between federal and local policy would thwart regional freight planning efforts.

The crux of this disagreement between the Port Authority and the ICC emerged from their distinctive state-building milieus. While both local and national policy seemed to be guided by the same set of ideals, the Port Authority and the ICC produced dissimilar responses to the perceived failure of previous regulatory efforts to encourage private enterprise to meet the shipping needs of the public at large. Although they appeared to proceed under the same conceptual banner, regional and national attempts to encourage cooperation and integration turned out to mean very different things in practice. When, in the late 1920s, the Port Authority turned to the ICC to carry out its goal of integrating railroad facilities at the port, the limits of the ICC's new powers as written by Congress had already insured that the two regulatory bodies would approach the problem of railroads and the public interest in incompatible ways. While the Port Authority pursued an approach to port planning in which the railroads' interests took a distant second place to the interests of the region as a whole, the ICC developed a balancing approach to harmonizing the interests of carriers and shippers. Ultimately, this conflict over the meaning of cooperation and integration illustrated just how indeterminate the system ideal was as a state-building strategy.

The Veblenian Reconstruction of Local Freight Policy: The Port of New York Authority

Nineteen seventeen marked the beginning of what might be termed the Veblenian reconstruction of freight policy in the New York region. That year, a battle over railroad rates between New Jersey and New York led to the creation of the Port of New York Authority, the agency that took the lead in regional freight planning in the postwar period. Port Authority planners intended to unify both terminals and rail lines, converting all the railroad facilities serving New York harbor into single system in which the traffic of every railroad company would have equal access to all other parts of the system, including the facilities of their competitors. Under the Port Authority's comprehensive plan, every railroad at the port could use the PRR's expensive new improvements in Queens and Brooklyn, as well as the New York Central's strategic west-side tracks. The Port Authority, rather than the railroads acting individually, would take responsibility for the smooth functioning of the consolidated regional freight system.

In its pursuit of these ambitious plans, the Port Authority illustrated just how radical the system ideal could be when moved to the public sector. Taken to its logical conclusion, this central precept of the civic culture of expertise—the city and region as a system—took on a distinctly Veblenian tone. Just as Thorstein Veblen had described the emerging national "industrial system" and its "indispensable General Staff" of production experts in *The Engineers and the Price System* (1921), Port Authority planners called for an end to the sabotage of the port's efficiency by the captains of industry who mismanaged the freight facilities around New York harbor. Port Authority planners could direct the use of freight facilities without supervision from railroad executives or stockholders; once freed from their subservience to those pecuniary interests, the experts would integrate rail facilities regardless of ownership, thus creating a port system that functioned efficiently for the benefit of the community as a whole. Although Veblen's vision of an industrial system managed by experts alone seems far-fetched (historian Samuel Haber called it a "technological hallucination"), the idea of the port as a system was a completely plausible state-building goal in the New York region, embodying a widely held belief about the trend of institutional development in the greater city.[34]

At least initially, however, defending the idea of the port as a unit was less a strategy than a counterstrategy. In fact, the Port Authority emerged from another unsuccessful attempt to separate the New York and New Jersey sides of the port—this time, by those aggressive little cities on the far bank of the Hudson. Since 1877, when the major trunk railroads signed an agreement ending price competition on freight between Chicago and eastern port cities by dividing the East Coast into rate groups, the port of New York had been considered a single destination, with rates to New York and New Jersey fixed at two cents higher than rates to Philadelphia and three cents higher than rates to Baltimore. New York merchants had challenged this differential system three times (in an effort to end the rate advantage that Philadelphia and Baltimore had), but the ICC rejected their plea in 1898, 1904, and 1912.[35] In 1917, the threat to the rate differential came not from New York merchants but from New Jersey business and civic groups who initiated the *New York Harbor* case with the ICC in a bid to divide the port into separate regions with different freight rates.[36] In other words, they wanted the differential system partially dismantled, just enough to encourage more shippers to do business in New Jersey at the expense of New York. Faced with this threat to their livelihoods, the New York State Chamber of Commerce and Merchants' Association responded by mounting a defense of the differential system and, along with it, the idea of the port as a unit.[37]

Although it originated as a counterstrategy, the idea of the port as a system quickly became a centerpiece of a regional state-building effort, due in large measure to the work of Julius Henry Cohen, counsel for the New York merchants. Cohen had been active in industrial disputes in New York City, working with Louis D. Brandeis, Felix Adler, and Morris Hillquit to create new mechanisms for arbitrating conflicts between garment manufacturers and workers. Because of his familiarity with the plight of the city's small manufacturers, Cohen was chosen by Irving Bush (head of the Bush Terminal Company) and Eugenius Outerbridge (then chairman of the chamber's committee on harbor and shipping) to represent the city before the ICC.[38]

Rather than dismantling the rate differential system at the expense of New York City, Cohen told the ICC, the federal government should recognize the long list of engineering, legal, and institutional factors that portended even further unification of the port, such as the PRR/H&M tunnels. Most significant in this regard, during World War I not only had federal control of railroads shown the need for better coordination of private facilities, but the director general's representative at the port, Frederick Williamson (later president of the New York Central), had allowed the New York Central to use the PRR's new Hell Gate Bridge to transport goods to and from Long Island with great success. The railroads were too competitive, Cohen argued, but their own actions illustrated the trend toward cooperative use of facilities. So convinced was Cohen of the trend toward unification that he collaborated with Bush and Outerbridge to initiate the creation of a Port Authority for New York harbor to act as a regional planning agency for both sides of the Hudson. With this move, Cohen transformed a defensive legal strategy into an affirmative state-building strategy, allowing him to present the ICC with evidence that legislation was being considered in the New York and New Jersey legislatures to develop the whole port along cooperative lines. Rather than simply defending the rate-differential agreement, therefore, Cohen showed the ICC that the states themselves were acting to give new institutional form to the historic unity of the port region.[39]

The ICC rejected New Jersey's petition to end the differential system in part by affirming Cohen's argument that the port was a unit. In what would become a cherished and portentous declaration for Port Authority planners, the ICC noted that "historically, geographically, and commercially New York and the industrial district in the northern part of the state of New Jersey constitute a single community." The commission also agreed that competition between the railroads caused the dismal conditions at the port, leading them to invest vast sums in their individual facilities while preventing them from using those facilities jointly to solve the urgent transportation problems of the port, especially in time of national emergency. In words that could have been written by Tomkins or Cohen, the commission emphasized that "it is necessary that the great terminals at the port of New York be made practically one and that the separate interests of the individual carriers, so long an insuperable obstacle to any constructive plan of terminal development, be subordinated to the public interest."[40]

With this *apparent* blessing from the ICC, Cohen's merchant sponsors created the New York, New Jersey Port and Harbor Development Commission in 1917 to recommend a cooperative, bistate solution for the port. The commission included business groups, engineers, and public officials from throughout the region. After four years of exhaustive inquiry (which included interviews with almost everyone involved in freight transportation at the port, including railroad executives) the commission concluded, not surprisingly, that the port problem was essentially a railroad problem and called for the creation of a single Port Authority—illustrating just how widely held the system ideal was.[41]

Although the solution to the port's problems was clear enough to the harbor commission, institutionalizing the system ideal was another matter. The railroads resisted efforts to create the Port Authority, pulling strings in the New Jersey legislature to block the passage of Cohen's enabling acts.[42] While they agreed with the principles behind the comprehensive plan, railroad executives did not want to turn their properties over to Cohen, and they continued to block attempts to enlarge the Port Authority's regulatory reach. "The Port Authority is now trying to secure broad powers so as to practically condemn the terminals of all the steam railroads at New York—certainly on the New Jersey side—and thus revolutionize the situation," Samuel Rea wrote to engineer William Wilgus in 1924.[43] Since they remained locked in their struggle for scarce space along the Manhattan waterfront, railroad managers showed little interest in sharing the facilities they used to win those battles with their competitors, even in the name of regional efficiency, especially when those facilities had cost them millions of dollars and years of planning.[44]

The centralization of freight planning authority also meant disrupting the existing distribution of public power at the port. "*Authority* must mean something," Cohen insisted, "hence it must mean a *supergovernment!*" And a supergovernment necessarily usurped authority exercised by the plethora of local governments that already had a hand in port planning. "At bottom," recalled Cohen,

> the opposition of the Port Authority was based on local and party politics. We became a sort of Halloween spectre to the municipal officials of all the cities within the [port] District. Like the witch in "Hansel and Gretel," we were ready

to capture all the little children, push them into our oven and make gingerbread out of them. New York's valuable piers and waterfronts and railroads, even its subways, would surely be gobbled up!

Although Cohen eventually won the support of the two state governors, New York's Charles Whitman and New Jersey's Walter Edge, the Port Authority was vigorously opposed by New York Mayor John Hylan and Comptroller Charles Craig, as well as a host of local officials in New Jersey. Like the railroads, Hylan continued to oppose any extension of Port Authority powers as a "grab" and characterized Cohen as a "cunning counsel and silent scrivener" bent on "the creation of a tyrannical power" that would reduce "elected local officials to puppets." As a consequence, when the state legislatures created the Port of New York Authority in 1921, they did not endow it with all the powers envisioned for it by Cohen and his allies. As political scientist Jameson W. Doig has argued, the Port Authority emerged from the state legislatures a "toothless giant," with none of the regulatory authority over the railroads that Cohen had intended. The Port Authority thus had exceedingly ambitious goals with little power to achieve them.[45]

Nevertheless, the Port Authority's statement of fundamental principles declared that "terminal operations within the Port District should be unified," which Port Authority planners recognized would discourage the railroads from investing in their own new facilities. By using all the existing facilities intensively, however, fewer new ones would be needed. Most significant of all, once the Port Authority made maximum use of all existing facilities, it could use its own power to float bonds (one of the few powers the legislatures granted it) to make the needed investments in new ones. The Port Authority, in other words, was finally taking command of freight policy in order to transform the disconnected private facilities at the port into a unified, efficient transportation machine.[46]

Turning these goals into enforceable policies dominated Cohen's attention through the late 1920s, and it was this unenviable task that prompted him to turn to the joint-use provisions of the Transportation Act of 1920. Although Port Authority officials had received from company executives what they thought were verbal assurances that the railroads would conduct their operations in accordance with the comprehensive plan outlined in 1921 (which certainly made sense, given that the system ideal was largely the product of engineering culture nurtured by the railroads), increasingly it appeared that they would not cooperate with Cohen and his colleagues. Denied plenary authority to enforce the plan by the state legislatures and rebuffed by uncooperative railroad executives, the Port Authority then turned to the only institutional ally it had left, the ICC. Because of the commission's favorable remarks in the *New York Harbor* case (which only added to the overwhelming evidence of the fundamental unity of the port of New York), Cohen interpreted the reconstruction of federal railroad policy just as he understood the course of institutional change at the port: cooperation, unification, and efficiency were the new guiding principles of railroad policy. As Cohen saw it, the obvious inefficiencies and high costs of doing business at the port could be overcome "if there were cooperation among the railroads and if there were unification and joint use of terminals, consolidation of shipments, and coordination of the use of carloads—all of this was accepted national policy by the terms of the Transportation Act of 1920."[47]

Postwar Federal Policy: The Transportation Act of 1920

There can be little question that the Transportation Act of 1920 inaugurated a new era of federal railroad regulation. Rather than focusing predominantly on protecting shippers from powerful financial interests, the ICC turned its attention explicitly to creating and maintaining an adequate national transportation *system* in order to avoid a devastating collapse of service like the one that occurred in 1917. In this sense, the reconstruction of railroad policy at the federal level *seemed* to be compatible with the goals of the Port Authority. Indeed, with its new emphasis on cooperation, system building, and national service, the revitalized ICC appeared to be a natural partner for Cohen and his team of regional planners.

The Transportation Act spelled out the ICC's new mission in unequivocal terms, obliging the commission to consider "the transportation needs of the country and the necessity of enlarging such facilities in order to provide the people of the United States with adequate transportation" as it made its rulings—a positive duty that had appeared in neither the original Interstate Commerce Act nor its subsequent amendments. To carry out this expanded mission, the ICC would henceforth work alongside the railroads, rather than as their adversary.[48] As U.S. Supreme Court Chief Justice Taft noted in 1921, the primary mission of the ICC prior to the act was to "prevent interstate railroad carriers from charging unreasonable rates and from unjustly discriminating between persons and localities," but the new act "seeks affirmatively to build up a *system* of railways prepared to handle promptly all the interstate traffic of the country."[49]

The Transportation Act included several key provisions to allow the ICC to fulfill this revised and expanded mission. First, the commission would set rates to guarantee the carriers 6 percent profit as long as they operated "under honest, efficient, and economical management."[50] Generous though it appeared, this guarantee came at the price of increased federal intrusion into railroad management. As Chief Justice Taft observed, the act "puts the railroad systems of the country more completely than ever under the fostering guardianship and control of the Interstate Commerce Commission," which had expanded authority to monitor the issuing of railroad securities and supervise the construction and abandonment of rail lines.[51]

Far more intrusive, however, was the second key provision of the new act, which gave the ICC extraordinary authority to "recapture" retained railroad earnings. If a railroad earned more than the specified 6 percent, the ICC could take the additional revenue, placing one-half of these "recaptured" earnings in a reserve fund, which the railroad could draw upon to pay dividends, interest, and rent if it failed to earn its 6 percent the next year. The ICC placed the other half of the "recaptured" earnings in a contingent fund, used to make loans to *other* carriers for capital expenditures—a policy that railroad executives like Samuel Rea objected to as the first step toward outright government ownership.[52]

The recapture provisions of the Transportation Act were designed to remedy the "weak railroad problem" that confronted the ICC in the *Five Percent* case back in 1914. In that case, the ICC had hesitated to raise freight rates for all carriers because it determined that financially strong railroads would benefit unduly from the increases. However, by 1917, the need for rail service took precedence over the threat of excessive corporate profits. William Gibbs McAdoo, the entrepreneur behind the H&M tunnels whom President Wilson had appointed director-general of the railroads, promptly raised freight rates by a whopping 25 percent in an effort to jump-start the rail system in March 1918—a move that only strengthened the argument that the ICC's prewar regulatory approach had not allowed a proper balancing of fairness and profitability.[53] The recapture provisions remedied that problem by allowing the ICC to grant general rate increases to bolster weak railroads while capturing the excess earnings of stronger roads.[54]

For Cohen, however, the joint-use-of-terminals provisions of the Transportation Act constituted the most important new element of federal regulation. Section 3, paragraph 4 of the act gave the ICC the power to require terminal sharing agreements among the railroads "if the Commission finds it to be in the public interest and to be practicable, without substantially impairing the ability of the carrier owning or entitled to the enjoyment of terminal facilities to handle its own business."[55] Armed with this new authority, the ICC could promote cooperation among former competitors in precisely those areas that Port Authority planners needed. In this crucial respect, the ICC and the Port Authority apparently drew the same lessons from the brief period of federal control. During the war, federal managers had used railroad facilities intensively, with little concern for who owned the needed terminals or bridges; the sharing of facilities irrespective of the prerogatives of private ownership, in other words, was necessary to make the national railroad system more responsive to the urgent demand for improved rail service. Now that the war had ended and the railroads had been returned to private control, the federal government seemed prepared to continue that policy of cooperation through the joint-use provisions of the Transportation Act. Given the language of the *New York Harbor* case, given the operating experiences of the railroads during the war, given the terminology of the Transportation Act, there seemed to be no doubt that the Port Authority and the ICC were working from the same set of values toward the same set of objectives. Those shared values provided the basis for mastering the problem of freight transportation nationally and regionally since terminal sharing seemed to be the first step along the road to terminal unification. Even though local and state governments in the New York region could not bring themselves to institutionalize this movement toward cooperation, the Port Authority could still rely on the ICC to push the railroads toward a more rational and enlightened approach to port planning—or so it seemed.

The Evolution of the Joint-use Provisions: Terminal Sharing versus Terminal Unification

Initially, it appeared as though the joint-use provisions of the Transportation Act could be used to achieve the Port Authority's goals—again because of Cohen's skillful efforts before the ICC. In 1922, the commission heard its first joint-use case, brought by the little town of Hastings, Minnesota (population 4,571 in 1920). The Hastings Commercial Club wanted the ICC to compel the

Chicago, Burlington & Quincy Railroad to use the tracks and terminals of the Chicago, Milwaukee & St. Paul Railway to better serve local grain merchants—an arrangement neither railroad wanted. In this opening test of the joint-use provisions, Hastings was represented by none other than Julius Henry Cohen, who recognized that the Port Authority needed a favorable decision here if it wanted any chance of acting as a "supergovernment" for regional freight policy in New York. And Cohen's efforts seemed to pay off since the commission delivered an eight-to-three decision in favor of joint use and ordered the two railroads to work out terms for terminal sharing. The *Hastings* decision thus seemed to provide yet another undeniable piece of evidence that cooperative port management was now on the verge of becoming a reality for New York harbor—thanks to the ICC. Only four years later, however, the *Hastings* case was reheard by the ICC and its earlier decision reversed (by a six-to-four vote), thus robbing Cohen of his crucial precedent, and this parting of ways exposed the differences between regional and federal state building.[56]

That the Transportation Act embodied a different approach to cooperative regulation can be seen especially in the modest goals the ICC had for terminal sharing, which were themselves a response to the failures of prewar regulation.[57] Prior to 1920, the ICC had developed a distinction between "open" and "closed" terminals: if a company kept its terminal facilities closed to other carriers, the ICC could not compel the opening of those terminals to any other carrier; but if a company had opened its terminals to another carrier, the ICC could oblige it to offer similar access to all other carriers.[58] Because this distinction proved cumbersome and confusing in practice, the commission recommended that Congress give it authority to promote "a more liberal use of terminal facilities" that "would undoubtedly bring about agreements between competing carriers."[59] Commissioner Edgar Clark told Congress that since most previous cases involved discrimination—where a railroad asked for use of facilities already open to other railroads—requests for terminal sharing would likely come from the carriers themselves. A revised terminal sharing policy would therefore help the ICC avoid the uncertainty of the open/closed distinction and facilitate cooperation already desired by the carriers.[60]

However, just as the Port Authority emerged from the state legislatures as a "toothless giant," the joint-use provisions of the Transportation Act emerged from the congressional bargaining process as a clumsily written, poorly conceived tool for encouraging cooperative use of terminals, particularly where the issue of compensation was concerned.[61] In the final text of section 3, paragraph 4, Congress stated that the ICC could compel joint use "on such terms and for such compensation as the carriers affected may agree upon, or, *in the event of a failure to agree,* as the Commission may fix as just and reasonable for the use required, to be ascertained *on the principle controlling in condemnation proceedings.*"[62] In its early decisions, like first *Hastings,* the ICC tried to preserve the utility of the joint-use provisions by leaving the question of compensation to the railroads, in part because of the commission's conviction that clarifying terminal sharing policy would free the railroads to pursue cooperative relationships that they desired but had been afraid to undertake. But in the absence of a voluntary agreement, an unwilling railroad could simply demand too high a price for the use of its terminals, transforming terminal sharing into an act closer to eminent domain.[63]

Faced with these built-in limitations, the ICC was compelled to develop an approach to enforcing the joint-use provisions in light of the larger purposes of the Transportation Act of 1920. The act had endowed the ICC with the elaborate new mechanisms for monitoring railroad earnings in order to strike a balance between publicly acceptable profits and needed investment in the national transportation system. Given the institutional history of the act (its emergence from the strong/weak railroad debate, the dickering over the language of the joint-use provisions), federal policy in the postwar period necessarily placed great emphasis on encouraging investment in order to strengthen the national transportation system. That national system was, in effect, the sum of all the private railroad systems in the country. During the prewar period, the ICC's failure to grant rate increases led to underinvestment in individual systems, and that ultimately meant a national railroad system unable to handle the demands of the war. To strengthen the national system and insure its capacity to carry goods during peak periods, the commission would henceforth help those individual systems. That meant encouraging the railroads to invest in needed facilities (especially terminals in urban areas) while restricting their ability to gouge the public through excessive earnings. Formulating the public interest, therefore, now required the ICC to consider the financial needs of the railroads, including their legitimate desire to protect their investments in terminals from their competitors.

Within that institutional context, the commissioners developed a balancing approach to applying the joint-use provisions, weighing the competing interests of shippers, railroads, and the public at large to determine whether to exercise its powers to compel terminal sharing. "In determining what is 'in the public interest,'" the commission noted, "we must take into consideration the interests of the particular shippers at or near the terminal considered, but also the interests of the carriers and of the general public."⁶⁴ The commission would not compel terminal sharing in all cases, but only in those instances where gains derived from joint use by other carriers, shippers, and the public at large outmatched the losses to the railroads.⁶⁵

This balancing approach, as developed through a series of cases decided after first *Hastings*, showed that the commission was willing to compel terminal sharing where multiple parties derived clear benefits. In the *York Manufacturers Association* (1926) and *Port Arthur Chamber of Commerce* (1928) cases, for example, the commission reversed earlier decisions and ruled in favor of the joint use of terminals. But rather than approving of joint use universally, it weighed the costs and benefits of joint use for shippers, carriers, and the traveling public for each specific case of terminal sharing. In that process, it preferred to see that a locality was ill-served by the existing use of terminal facilities and it looked for evidence that substantial improvements in service would result from the cooperative arrangement—substantial enough to justify placing burdens on one or both of the railroads involved as directed by the compensation requirements of the joint-use provisions. The commission also found it easier to conclude joint use to be in the public interest when carriers, not merely shippers, wanted access to closed terminals. Taken together, the ICC's joint-use rules did not unequivocally serve any group. Instead, they represented a genuine effort to wring some utility out of a poorly written statute within the broader terms of the Transportation Act.⁶⁶

The ICC reversed its *Hastings* decision precisely because the case met none of its criteria for finding joint use of terminals in the public interest. Upon rehearing, the commission learned that terminal sharing in Hastings would lead to little improvement in service for the city since it was intended to help just one local business, and existing service was already fairly good. This slight improvement in service came at a significant price. Neither railroad wanted the arrangement, which meant that the awkward compensation requirements of joint-use provisions would be triggered. And because the Burlington did not want to use the Milwaukee's facilities in Hastings, the arrangement would have required extremely complex and expensive terms for compensating *both* railroads: the Milwaukee for allowing its terminals to be used and the Burlington for being forced to use those terminals. In this sense, the entire Hastings case was a very bad precedent for the joint-use-of-terminals policy at the national level. For Cohen, however, the case provided a perfect example of the problems he faced in New York since he wanted a precedent that said terminal sharing should go forward regardless of the desires of the railroads and irrespective of the immediate benefits resulting from any particular instance of joint use.[67]

Even though the general trend of joint-use cases at the national level supported the idea of cooperative use of facilities, second *Hastings* strongly suggested that federal policy did not endorse the sort of terminal unification sought by the Port Authority. To be sure, both were interested in cooperation and efficiency; both employed the language of system building and the public interest. However, each effort at centralization was a response to a particular course of institutional change, one national and the other local. That they partook of the same discourse, the same language of system building, was, in the end, deceptive. For the Port Authority, institutionalizing the system ideal meant replacing private railroad managers and turning individual facilities into a unified public transportation network. In the case of the joint-use provisions, institutionalizing the system ideal meant linking the desire for cooperation with the larger goals of the Transportation Act—primarily the need to encourage private investment in new facilities to sustain a viable national rail system. In that context, terminal sharing was not the Veblenian goal that it had become for the Port Authority.

National versus Regional Policy: The Hell Gate Bridge Case

The conflict between the Port Authority's approach to regional planning and the ICC's policy of maintaining a national railroad system emerged decisively in the *Hell Gate Bridge* case, argued before the ICC in 1927. Here, the Port Authority attempted to use the joint-use-of-terminals provisions to force the PRR to open its Hell Gate–New York Connecting Railroad–Long Island Railroad route to all other railroads operating at the port—effecting a continuation of the joint-use policy initiated during the war by the future president of the New York Central.[68] This policy would have applied particularly to New York Central traffic destined for points on Long Island, which the company usually sent via car floats from its Sixtieth Street yard, around the Battery, and up the East River to Long Island City in Queens. According to a Port Authority study, the new route would allow the Central to avoid peak hours delays at Long Island City that averaged thirty-five hours. Even more important, opening the Hell Gate route would have represented a crucial step toward transforming the various private railroad facilities at the port into a communal rail highway. Cohen, frustrated with his long battle against the railroads, saw the case as a matter of corporate selfishness versus the welfare of the city: "Those are the two alternatives, and there can be no compromise between them. Either the public interest must govern or railroad policy must govern. There can be no sharper conflict between the two."[69]

During the *Hell Gate* case, that conflict was highlighted in a remarkable exchange between Cohen and PRR vice president Albert J. County—an exchange that included the "John Wanamaker Store" argument and the *Charles River Bridge* case, revealing just how differently the railroads, the Port Authority, and the ICC conceptualized the relationship between private corporations and the public interest and underscoring the conflicts among quasi-scientific voluntarism, Veblenian planning, and regulatory balancing.

The "John Wanamaker Store" argument captured the essence of the railroads' objections to compulsory terminal sharing. Under Cohen's cross-examination, County explained that the Port Authority's plan to open the Hell Gate route would discourage the PRR (and other railroads) from investing in new facilities: "it [would be] like taking John Wanamaker's store in Philadelphia and saying that it is in the center of things, and every other dry goods dealer going to his counter and saying 'I want to sell my goods there.'" One commissioner quickly responded that Wanamaker's was not a public service corporation like the railroads. County acknowledged this, but he insisted that the effect of the decision would be the same for either a department store or a railroad, since both businesses built facilities to gain a competitive advantage over their rivals. County also acknowledged (and the ICC agreed) that the PRR would actually make money by permitting other railroads to use the Hell Gate route—which was, in 1926, not intensively used by PRR traffic. But PRR managers were not merely interested in a quick profit. In the same way that they intended Penn Station to meet the company's passenger needs for fifty years, they intended the Hell Gate route to meet the company's freight needs for decades to come. Why build for the future, County argued, if public authorities could make those facilities available to other railroads?[70]

Cohen responded to County by adapting an argument made in the *Charles River Bridge* case—the landmark U.S. Supreme Court decision of 1837, which pitted Supreme Court Justice Joseph Story, representative of republican legal tradition, against Roger B. Taney, spokesman for Jacksonian economic dynamism. In 1785, the Massachusetts legislature granted a charter to the Charles River Bridge Company to build a toll bridge between Boston and Charlestown. In 1828, the legislature granted a charter to the Warren Bridge Company allowing it to build a toll-free bridge very close to the old bridge. Although the specific legal issue was whether the original charter included an implied monopoly that prevented the legislature from permitting other bridges across the Charles, the case also raised a larger public policy issue: how governments could encourage private investment for the public good. As legal historian R. Kent Newmyer has observed, Judge Story believed that "if the Court permitted state legislatures to go back on their promises to investors in corporations [like the Charles River Bridge Company] then no one would invest; progress would come to a standstill; entrepreneurial genius would be stifled and with it economic progress and national greatness." Roger Taney, on the other hand, argued that established property rights should not be allowed to stand in the way of investors interested in building new bridges. Taney spoke for a new generation of Jacksonian entrepreneurs who were ready to invest and build new bridges; the old guard of investors, like those who built the Charles River Bridge, thwarted their advance. Taney was thus "striking a blow for democratic capitalism" by arguing that the old bridge charter should not deter new investment.[71]

Cohen saw a parallel between Story's desire to protect the willingness of capital to invest in projects like the Charles River Bridge and County's insistence that the PRR would continue to invest in new facilities only if it had exclusive use of them. Cohen admitted that opening the Hell Gate route would discourage the PRR from making additional investments, but the whole purpose of the Port Authority was to prevent the railroads from gaining a competitive advantage, since destructive competition for space led to lack of coordination in the interest of the port as a whole. Instead, the Port Authority would open up all facilities to all railroads, thereby promoting the intensive use of existing facilities and postponing the need for new investments. As Cohen observed, "Every bit of sound advice that the Port Authority has received has been to use existing facilities and use them intensively before going in for large capital expenditures" on new projects—that was the policy written into the Port Authority's charter. The ICC did not need to worry about discouraging the railroads from investing, Cohen insisted, because intensive use of existing facilities would postpone the need for investment and because the Port Authority stood ready to makes new investments once they were needed.[72]

From a technical standpoint, the only thing the *Charles River Bridge* case had in common with the *Hell Gate* case was that they both involved bridges; beyond that, there was really no reason to bring up the ninety-year-old decision. But Cohen was more interested in the institutional change the *Charles River Bridge* case represented, and here he revealed the Veblenian goals of the Port Authority's terminal-sharing plans. Cohen wanted the ICC to use the *Hell Gate* case to make the transition from transportation management by the railroads to transportation management by government authorities in the same way that Roger Taney had used the *Charles River Bridge* case to make the transition from republican capitalism to Jacksonian capitalism. Just as Taney argued that protecting the existing bridge charter would deter economic development by a new breed of entrepreneurs, so Cohen argued that protecting the PRR's exclusive rights to the Hell Gate bridge would block a new era of publicly managed rail development. For Taney, Jacksonian entrepreneurs were ready to replace patrician investors. For Cohen, Port Authority bureaucrats were ready to take on the role of investor heretofore played by railroad managers.

With this transformation in mind, Cohen maintained that the principles behind the Port Authority's Hell Gate Bridge policy were precisely those enunciated in the *New York Harbor* case. He emphasized that the Port Authority was "the direct child of the policy of this Commission as announced in the *New York Harbor* case," and he forcefully reminded the commission what it had concluded in 1917:

> that the competitive situation under which the terminal facilities had been created in this Port District resulted in waste and excessive overhead, that the effort of the carriers each to secure strategic locations had resulted in the increasing cost of terminal service, that the political lines between the two states had contributed to the failure to coordinate these facilities, and that it was necessary to use intensively existing facilities and to bring about a unification of such facilities.

In response, the Port Authority had treated the harbor as a single administrative unit; it developed a plan to end the rivalry among the railroads; and it intended to eliminate the wasteful spending of competing companies by allowing all roads to use all existing facilities intensively before investing in new tunnels, bridges, and terminals. The Port Authority also instituted what Cohen saw as the principle feature of the Transportation Act of 1920: creating an adequate national rail system by encouraging cooperative use of terminal facilities (broadly interpreted), achieved in this case by the Port Authority's comprehensive plan for the port region. The whole trend of institutional change represented in ICC decisions and Port Authority actions portended the withering away of private port mismanagement and the ascendence of public planning by newly empowered regional officials.[73]

The ICC rejected Cohen's sweeping arguments, not by discussing public policy history and institutional change but by referring to its own legal powers as established by congressional acts. It dismissed Cohen's *Charles River Bridge* case analogy as inapplicable to the ICC's regulatory authority. And while acknowledging certain "criticisms and suggestions" made in the *New York Harbor* case, the commission did not even remotely suggest that they constituted a statement of policy. As it dispensed with these crucial elements of Cohen's approach, the commission finally dispelled the illusion that its own regulatory balancing approach and the Port Authority's Veblenian planning were compatible.[74]

The policy that did guide the ICC was, of course, its own joint-use rulings. Citing the balancing test it established in the *Hastings, York,* and *Port Arthur* cases, the commission focused on whether the railroads in the port region would substantially benefit from using the Hell Gate route. The notable but occasional delays experienced at Long Island City were simply not enough for the commission to conclude that the Hell Gate route would save the railroads either significant time or money over their existing methods of reaching Queens and Brooklyn since car-float operations were not the primary problem. Commissioner Clyde Bruce Aitchison observed instead that "the delays after the traffic reaches the Long Island Railroad are by far more serious, from the standpoint of shippers, than are the comparatively short delays directly attributable to the car-float operations"—a fact admitted by the Port Authority. The commission was also impressed by the efforts of the PRR to create facilities for the benefit of Long Island residents, noting that the PRR had constructed the Hell Gate route (figure 8) not merely with an eye to immediate profit but to have sufficient facilities to handle traffic to and through Queens and Brooklyn well into the future. The PRR was also pouring vast sums of money into the LIRR in order to upgrade its facilities. Because the company was going to great lengths to improve its own system and thereby provide better service to the residents of Long Island, the ICC had little choice but to support its efforts under the terms of the Transportation Act.[75]

Figure 8. Hell Gate Bridge, c. 1915. The PRR's Hell Gate Bridge—with a length of one thousand feet the longest steel-arch bridge in the world—still connects Queens to the Bronx via Randall's and Ward's Islands. It stands two hundred feet high and weighs a stupendous eighty thousand tons. The Port Authority tried to force the company to allow other railroads to use it as part of a comprehensive program of publicly controlled freight management at the harbor, precipitating the *Hell Gate Bridge* case before the ICC in 1927. *Source:* American Art Publishing Co., New York City; postcard in author's possession.

There was, of course, a bigger picture that the Port Authority tried to get the commission to see. That picture suggested that the real benefits of the Hell Gate route would have become apparent only after the Port Authority had control over terminals and traffic in the entire harbor district. The Port Authority could then have used its power to float bonds to enlarge inadequate yards, eliminate difficult grades, renovate interchange facilities, and add service tracks. The Port Authority could also have begun routing traffic throughout the region, coordinating truck and rail service and eliminating bottlenecks, thus yielding substantial gains in service for Long Island shippers via the Hell Gate route. But from the perspective of the ICC, the considerable operational difficulties of the particular joint-use arrangement for the Hell Gate bridge made the Port Authority's plans seem quite fanciful (Cohen would probably have preferred use of the word *visionary*).[76] Indeed, no one on the commission found the Port Authority's argument convincing: the decision was unanimous. Cohen undoubtedly sensed this, and he responded by using the *Charles River Bridge* case in an attempt to persuade the ICC to see the much larger change in transportation policy envisioned by the Port Authority. That change, however, was well beyond anything contemplated in the joint-use provisions of the Transportation Act.

The *Hell Gate* decision thus revealed the basic conflict between federal policy and the Port Authority's comprehensive plan. The Transportation Act was designed to encourage railroads to invest; the comprehensive plan, had it been successfully implemented, would have discouraged the railroads from investing. That the Port Authority would have made up the investment deficit in the region did not resolve the problem on a national level, and the ICC had to think in national terms. As Albert County asked, what incentive was there for railroads to build new facilities (in New York or elsewhere) if public authorities could open those facilities to competing companies? Just as before the war, the goals of national railroad policy were incompatible with the objectives of planners in the New York region.[77]

Port Planning, Civic Culture, and the Public Interest in the New Metropolis

The *Hell Gate* decision showed definitively that freight planning for New York harbor would not be accomplished by a dramatic reconfiguration of the relationship between the port and the railroads. Ambitious bureaucrats like Calvin Tomkins, Julius Henry Cohen, and the Port Authority experts felt that the only way to plan for the port's economic future was to make regional government the railroads' master, but the states were not willing to donate their regulatory powers to the Port Authority, the ICC was not ready to make cities the primary beneficiaries of its regulatory policies, and the railroads were not prepared to allow local planners to direct the use of their facilities. In spite of the best efforts of the most determined local planners, the era of "separate sub-ports constructed by great private corporations" was not over.

The Port Authority did not have sufficient regulatory power to integrate the railroad facilities at the port, but it did have the power to float bonds to build bridges and freight terminals—and it could use these powers to free regional policy from both corporate and federal institutional entanglements, at least in some measure. After its defeat in the *Hell Gate Bridge* case, the Port Authority focused its attention more fully on policies that involved building rather than regulating: constructing a union freight terminal in Manhattan in 1932 and four automobile bridges connecting New York City with New Jersey by 1931—Goethals Bridge (1928), Outerbridge Crossing (1928), Bayonne Bridge (1931), and George Washington Bridge (1931). The Port Authority did not entirely abandon its railroad vision after the *Hell Gate* decision, but motor transportation became an increasingly important part of its freight plans—a development that had a major impact on other regional planners during the 1920s.[78] Although this signaled the defeat of the comprehensive plan, it also partially liberated local planners from their subservience to the national regulatory battle that had thwarted their efforts for nearly thirty years. By planning for automobiles rather than railroads, regional planners could do *something* to provide for the port's future. In this sense, automobile planning represented a step toward increased local autonomy, giving planners in the region a measure of influence over the provision of facilities (and thus the movement of goods) that their efforts at rail planning never permitted.

As a result of this conflict between local and national systems, the centralization of authority prompted by Progressive Era state building succeeded in creating only a supplementary locus of governmental power (the Port Authority), adding to the complex of public and private institutions that already had a hand in managing the harbor. Port Authority planners joined the mayors, docks commissioners, chambers of commerce, railroad executives, governors, state legislatures, state utility commissioners, local, state, and federal judges, representatives, senators, and interstate commerce commissioners who dabbled in port management from time to time. Although the Port Authority would, over the years, become a leader in that group, its policies were dictated by its institutional weakness and its success depended on the virtuosity of its top managers—first Cohen and, later, Austin Tobin—in overcoming the diffusion of freight planning power.

This process of institutional development called into question the very values that made it possible. In 1917, real change for the freight system at New York harbor had seemed possible because state builders appeared united on the basic principles of railroad management and regulation. The system ideal highlighted the deficiencies of regulated private management and sustained the promise of a more rational approach to public management. Indeed, the idea of port planning owed far more to railroad engineering culture than the railroads' resistance to the Port Authority's plans would suggest; Samuel Rea (whom Cohen characterized as "a really great railroad leader," even after their battles over Hell Gate) and the Port Authority planners spoke essentially the same language of system building.[79] But that was not enough.

Bringing the system ideal into the public arena illustrated just how problematic it was as an approach to defining the general interest for a regional metropolis. Most obviously, system building could become draconian planning in the blink of an eye. Like Thorstein Veblen, Cohen could imagine an industrial system that worked best when freed from the influence of private owners; trained experts would then manage the port system exclusively for the benefit of the public, as Cohen conceived it. The notion of private property (at least where railroad terminals were concerned) could be discarded as a vestigial nineteenth-century concept that only stood in the way of the new public interest. Under Port Authority management, private companies would gradually wither away, leaving public organizations free to make freight policy in the public interest. Indeed, the Port Authority could formulate the public interest without reference to the interests of the public's component parts—something the ICC, given its prewar experiences, could not afford to do. To have that kind of influence over private companies, however, the Port Authority needed the power of a supergovernment, and here it became a radical threat not only to railroad executives but to local elected officials and department heads, who had their own conceptions of their interests and the public's interests. Even while politicians came to admire the goals of the Port Authority, they never gave it the regulatory power that Cohen demanded.

The emergence of the Veblenian strain of system building thus exposed the nascent contradictions within the civic culture of expertise. The coincidence of private and public interest formed a mainstay of that culture. Specialized knowledge and technical skill, a common vision emerging from shared problem solving—these had held the promise of reconciling private and public in the new metropolis. But even the detailed and widely accepted understanding of freight movements at the port could not overcome the conflicts of interest and institutional perspective that beset the debate over harbor policy. As that deadlock persisted, enlightened voluntarism lost its appeal; private interests and alternative formulations of the public interest became mere obstructions. Just as the dream of reason produced monsters, the dream of expertise produced Veblen. The railroads thus nurtured an ideal that threatened to make them obsolete. When the system ideal emerged in the public sector in the form of the Port Authority, the railroads were put in the same position that the little town of Marion played within the emerging PRR system. The new terms of togetherness—the new articulation between the whole and the parts—did not include them. Unlike Marion, however, the railroads had the resources to fight the public system builders and win.

Although the railroads' resistance to the system ideal in public sector hands suggests a certain hypocrisy on their part, they were not the only ones hesitant to follow systematization to its logical conclusions. Cohen and Outerbridge were looking out for the interests of New York businesses when they used the notion of a regional freight rail system to battle the railroads; but taking the idea of regionalization seriously leads to the obvious conclusion that the New Jersey shore of the Hudson is the best place for shipping terminals. A real commitment to regional system building would compel those New York merchants to concede the necessity of moving the port across the river—where the Port Authority began moving it in the 1950s and where it remains today. Indeed, applied to land-use planning, the system ideal might compel us to abandon once-important areas of the city when they do not fit with some emerging conceptualization of the interests of the city as a whole. Taken a step further at the national level, we might even abandon whole cities in the name of some new configuration of the national interest. Under the system ideal, historical claims to importance (whether of individual companies, industries, neighborhoods, cities, regions, or states) are less important than the new concerns of the whole.

Even more disturbing was the fact that the system ideal seemed less definitive in practice than it appeared in theory. There were many ways to put the system ideal into practice and not all of them were compatible. The federal government focused on the creation of a national rail system, but that did not mean that the ICC and the Port Authority worked toward the same goals—a point painfully illustrated by the history of the joint-use provisions of the Transportation Act. The system ideal had been so appealing because it promised to reconcile the divergent interests of the greater city in ways that the partisan political system could not (as the PRR franchise battles portended). But there were so many different interests in the new metropolis and so many different configurations of interests—local, regional, state, and national—that the system ideal could not in practice produce a definitive collective interest on which public institutions could act. By the late 1920s, it seemed as though the system ideal produced as much interinstitutional conflict as its absence had permitted partisan politics.

Herein lay the fundamental difficulty of creating policies and institutions to deal with the new scale and scope of urban life represented by the modern metropolis. Greater New York City was so vast and so complicated, with so many conflicting interests at work in the region, that there was no clear or easy way to determine the relationship between the whole and the parts. The system ideal was so alluring because it suggested that the whole and the parts could be reconciled through the application of expertise, making it easier to conceptualize the whole (like the regional freight system) and organizing the activities of the parts (such as the individual railroad companies) so that they contributed to the good of the whole. With the system ideal, everyone would be concerned with the smooth, efficient functioning of the whole system, nationally, regionally, and locally.

Relating the whole to the parts was more problematic in practice, however. For the railroads, relating the parts to the whole was essentially an aggregative process; enlightened voluntarism would guide individual investment efforts so that private companies advanced both their own and the public's interest—that was the message of the PRR and H&M tunnel projects. From the perspective of the Port Authority, relating the whole to the parts was a process of disaggregation. Expert port planners visualized the whole and then worked from the whole to parts. Where individual rail companies helped realize the plan, they would be welcome partners; but the interests of the parts could not stand in the way of the whole. This problem constitutes one of the basic dilemmas of bureaucracies in a democratic society, as political scientist Pendleton Herring showed in *Public Administration and the Public Interest* (1936). "Groups must be willing to recognize that the state has a purpose which transcends their own immediate interests"—a point that Cohen believed the railroads never accepted. At the same time, Herring noted, "the bureaucracy cannot assert a state purpose against the united hostility of groups that basically comprise the source whence its authority springs"—a point that Cohen, in his enthusiasm to serve the public interest, seemed too willing to overlook when it came to the interests of the railroads.[80] The Port Authority, given its approach to port management, was never quite able to overcome the gap between those two extremes.

The ICC took yet another approach to relating the whole to the parts. Relying neither on the quasi-scientific voluntarism of the railroads nor the Veblenian planning of the Port Authority, the ICC attempted to balance the interests of the whole (the national rail system) and the parts (the individual rail companies) by expanding its regulatory control over rates and facilities. The disastrous consequences of prewar regulation forced the ICC to modify its posture in formulating the public interest under the Transportation Act of 1920. It had to consider the railroads' interests as an element of the public interest since their individual systems provided the building blocks of a viable national rail network. It did not have the luxury of ignoring the railroads' concerns, as the Port Authority did, and it had to consider the cumulative results of individual railroad efforts in ways that corporate executives did not. This federal balancing act did not help Cohen, however, since from the ICC's perspective, the New York region was just a small piece of a much larger system, and when it came to freight planning, the national system took precedence over the local system.

By the close of the 1920s, therefore, the promise of the civic culture of expertise had waned. In the end, the experts could not deliver. The quasi-scientific voluntarism that had so inspired Walter Lippmann had revealed its limitations in the arena of freight planning, but the alternative to voluntarism—bringing the system ideal to the public sector—had created a new set of problems; the drive to institutionalize the experts' perspective produced additional pockets of limited governmental power, working at cross-purposes, rather than a clear centralization of authority capable of reconciling whole and parts at multiple levels of aggregation. The result was neither drift nor mastery, but a shift in the locus of conflict to new settings, with the great city further entangled in multiple institutional webs, unable to control its own fate.

Part 2 / Public Infrastructure, Local Autonomy, and Private Wealth

CHAPTER THREE

Buccaneer Bureaucrats, Physical Interdependence, and Free Riders

Building the Underground City

Seven-year-old Emil Miller had just made it to first base—a manhole cover at the corner of Tenth Avenue and Fifty-first Street on the west side of Manhattan—when the ground exploded underneath him. Thrust upward by a jet of gas and fire, the manhole cover lifted Emil ten feet into the air. The boy fell headfirst onto the pavement, landing so close to the manhole that the flames emanating from the ground enveloped him, fatally burning his face and body.

For ten terrifying minutes on the afternoon of October 7, 1909, nearly eighty manhole covers launched as high as forty feet into the air along Tenth Avenue and its side streets between Forty-second and Fifty-third. The heavy metal discs flew in all directions, smashing through windows, crashing into lampposts, landing in doorways. Pedestrians fleeing from the "tongues of flame shooting from the earth" collided with the frightened occupants of stores, factories, and tenements who, convinced the shaking signaled an earthquake, were running into the streets for safety. Tremors from the explosions, which rang out "like broadsides from a fleet of warships," could be felt as far away as the Times Building, on Forty-second Street between Seventh and Eighth Avenues. When the peal of eruptions stopped, a thick layer of smoke hung over Tenth Avenue, crowded with dazed pedestrians, frantic parents, four squads of anxious policemen, and several teams of puzzled firefighters.

Benjamin Curry, a foreman with the city's sewer department, realized immediately what the explosions meant. He heard the blasts while working at Forty-second Street and Eighth Avenue and ran toward Ninth Avenue just in time to see another manhole cover sail over the terrified crowd. Curry knew that a volatile combination of gasoline and grease had pooled in the sewer lines under Tenth Avenue. The sewer department had repeatedly reprimanded garage employees in the area for degreasing their machines with gasoline and washing the mixture into the sewers. The combustible garage runoff usually worked its way through an ad hoc network of pipes leading down to the waterfront, spilling from drains underneath the piers into the Hudson River. But high tides covered the drains and stopped the flow, forcing a slurry of garage waste, street filth, household sewage, and urban jetsam to thicken in the pipes. As it happened, the tide was high that Thursday afternoon, creating the opportunity for a carelessly discarded match or a spark from a trolley car to ignite the pools of greasy gasoline that had slowly accumulated in the Tenth Avenue sewer line, with lethal results.[1]

On that fateful October afternoon in 1909, the city beneath the city suddenly and violently announced its presence to the pedestrians along Tenth Avenue—as though the underground city were reaching up to remind that corner of Gotham what had been forgotten in the midst of the bustle of private activity: that there were public spaces in the city, the nooks and crannies of community infrastructure, that had been taken for granted. The experts who built the underground city did understand just how important it was to the new metropolis. The "many-storied towering hives of commerce and industry" that made New York's skyline unique depended upon equally dramatic but invisible subterranean achievements uniting neighborhoods, boroughs, and cities within the region. "To consolidate the different [boroughs of the city] by legislative enactment into a political unit was a comparatively simple matter," observed Nelson Lewis, chief engineer for the board of estimate, "but to make them a physical entity is a very different proposition, and one vastly more difficult." In 1898, Brooklyn Bridge served as the "one great ligament" binding Manhattan to Brooklyn.[2] The consolidated city added three more bridges across the East River by 1909—the Williamsburg in 1903, the Queensboro and Manhattan six years later—but that was only the most visible set of public linkages created during this crucial period in New York's history. Between 1898 and 1914, the city issued debt totaling $1.182 billion, much of it for underground infrastructure that did even more to integrate the boroughs into a single metropolis.[3]

For municipal engineers, just as for their counterparts in the railroad industry, this vision of urban interdependence had technical roots. During the nineteenth century, developing new water and sewer technologies for individual houses had encouraged engineers to think in terms of larger urban systems, and interconnectedness became the central principle of their understanding of urban life.[4] The city-building process in New York reinforced and expanded that recognition of physical connection; as private developers pushed Gotham upward (with skyscrapers) and outward (into suburbs), municipal engineers were compelled to extend the city not only downward but also throughout the region. Facilitating this multidirectional expansion thus provided a strong impetus for regional planning, which, for many experts, became a necessary component of civic life for the new metropolis.[5]

But the creation of the underground city in New York is not a story of technological or economic determinism or the triumph of functional problem solving in the face of growth. Since the experts had been thinking in terms of interconnected, citywide systems for more than a generation, the question was how to realize the potential in these technologies. Understanding the underground city is therefore really a matter of exploring how New Yorkers confronted the political implications of physical interdependence, as experts attempted to bring the system ideal, long a component of their professional culture, into the public realm. Here, they encountered two key problems—centralization and financing—that determined the extent to which underground systems could be treated as public goods.

Defining the scope of underground systems raised basic issues of inclusion and exclusion, and establishing centralized institutions to create and manage those systems raised basic issues of power and authority. Like the Port Authority, the organizations necessary to build subway, water, and sewer networks crossed well-maintained political boundaries, predicated as they were on the experts' notion that citywide or regional planning, rather than local autonomy, private initiative, or low-cost government, should be the primary focus of public policy. As such, they required acceptance of new configurations of power in which citywide institutions usurped the functions of borough politicians or regional organizations dictated to municipal officials. This concentration of power, even when it promised to achieve widely desired public goals, disrupted vested interests at every turn.

Paying for the underground city was equally problematic. The construction of citywide systems had been thwarted during the nineteenth century by the difficulties of finding politically acceptable ways to distribute the costs of public goods. Differences of class, ethnicity, and geography discouraged taxpayers from footing the bill for public amenities that did not directly benefit them. This did not mean that large-scale public works were not built; instead, the ideology of privatism resulted in localized construction of public works, as real estate developers and property owners who were willing and able to pay for sidewalks, street pavement, or sewers built them, while leaving poorer neighborhoods to do without. This privatized system worked very effectively to "segment" the distribution of public services geographically through the use of special assessments and user fees: no one paid for public works they did not want; and only rarely did someone get to use public works they did not pay for themselves. In this way, the segmented system effectively minimized the problem of "free riders."[6] Overcoming the limits of privatism meant raising taxes of every sort to pay for multiple, expensive, citywide infrastructure systems that, because they were directed at large-scale problems, often did not benefit particular neighborhoods, businesses, or boroughs in direct proportion to what they paid. Enlarged to a regional scale, this problem became extremely thorny since addressing regional sewage and water problems meant imposing huge costs on other cities for the resolution of problems that did not touch them directly.

To effect the change from an ideology of privatism to an ideology of public goods, therefore, would constitute one of the more dramatic transformations of urban politics between the nineteenth and twentieth centuries. But making the experts' vision a commonplace notion in metropolitan policy making was neither a straightforward expression of humanitarianism nor an act of class domination (by the elite in the interest of moral order, for example).[7] Instead, to apply the system ideal to the public sector and to realize the potential in underground infrastructure technologies, experts became deal makers, finding ways to link the short-sighted demands of socioeconomic groups with the long-term interests of the city as a whole. This allowed the powerful interests who benefitted directly from subway and water projects to spread the cost of large public works to the broader city; at the same time, these projects made improved public services available to a much larger (and often poorer) group of people. In other words, rather than focusing on municipal efficiency or new privatization strategies to realize their vision of an interdependent metropolis, the experts turned the free-rider problem to their advantage. Greater New York became, in effect, a whole city of free riders—and in this sense, the experts were accomplishing what historian Charles Beard would later admonish them to do: creating a partnership with "the economic groups that may be enlisted in virtue of their practical interests on the side of a comprehensive community scheme."[8] The experts thus grafted the elements of a new civic culture onto distributive politics, sustained by the hope that physical interdependence would bring about a broader public reconceptualization of civic life in the new metropolis.

Subways: Machines, Money, Monopoly, and the Metropolitan Planning Perspective

It is something of a paradox that New York, as the quintessential *private* city (founded not for religious freedom, as were Boston and Philadelphia, but for the unabashedly self-interested purpose of making money), should have the longest *public* transit system in the world—722 miles of it. Indeed, modern New York would be inconceivable without the subway. Although Manhattan had emerged as a commuter city before the opening of the first subway in 1904, even the best combination of street cars, elevated railways, ferries, and bridges still made getting to the central business district a chore. The construction of six hundred miles of subway track between 1904 and 1925 substantially reduced commuting times to every borough except Richmond, opening the way for suburban development in Queens, Brooklyn, and the Bronx. Thanks in large part to the subways, the central business district in lower Manhattan hosted a huge transient population. By 1924, 2,181,000 people entered Manhattan on a typical workday, with 1,226,100 coming by subway from Queens, Brooklyn, and the Bronx. The skyscraper district at the very tip of Manhattan (only one-third of a square mile bounded by Fulton Street and the Hudson and East Rivers) was served by twenty subway stations—more than any other section of the city.[9] By providing linkages between the central business district and rapidly growing suburban districts in the Bronx and on Long Island, the subways helped decongest the most crowded areas of Manhattan—a goal long sought by urban reformers and planning advocates of almost every stripe. As a result,

Manhattan (which had grown steadily since 1790) actually declined in population from a peak of 2.33 million in 1910 to 2.28 million in 1920, to 1.87 million in 1930.[10]

These changes are all the more remarkable given the obstacles to expanding the subway beyond its first leg—the Interborough Rapid Transit (IRT) Company's line, begun in 1900 and opened in 1904, from Brooklyn, through downtown Manhattan to Times Square, and along the upper west side to the Bronx. Four issues converged in the policy debates that surrounded the extension of the subway beyond its beginnings with the IRT: distrust of machine politics, the city's limited financial strength, distrust of corporations, and citywide transit planning. That New York built such an extensive subway system during the early twentieth century was due to the fact that, for a brief moment, the need for citywide transit planning won out over political resistance to the institutional changes necessary to make it a reality. To create a citywide system meant centralization of some sort—either in private hands, in public hands, or by means of a public/private partnership, none of which were acceptable within the prevailing political and fiscal limitations of the day. Key public officials overcame this fear of centralization by capitalizing on the "free rider" problem. Although this approach did not fully transform the city's political culture by making the subway a true "public good" (publicly owned and operated, subsidized by general municipal funds to make up for operating deficits), it did provide New York with the world's longest subway system—which was no small feat.[11]

The idea of underground transportation to solve Manhattan's crowding problems and boost real estate development emerged during the 1880s and 1890s when opposition to Tammany Hall served as a key element for successful reform political coalitions. The thought of millions of public dollars being made available to the machine discouraged the idea that the subway could be an exclusively public system—funded, built, and operated by local government. After a decade of delays, the rapid transit commission finally agreed to complete the first section of the subway only because a private businessman, financier August Belmont, would take charge of construction and operation. While the city contributed almost $37 million to the project (Belmont had to raise the funds for equipment and operation in the bond market), that money was presumably safer in the hands of a profit-seeking capitalist than it would have been in the hands of profit-seeking politicians. The need for a private/public partnership of this sort had become increasingly clear during the 1890s, since neither private investors nor city government had the financial means to build a citywide system by themselves. The merchants who backed the subway had hoped that private companies would seize the opportunity on their own, but even a 999-year lease could not attract a bidder to the first franchise offered in 1892. Subways were that risky.[12]

While fear of Tammany Hall certainly played an important role in the decision to partly privatize the subways, it is also clear that the city did not have the financial ability to go it alone. Although it has been said that New York was the only city rich enough to rebuild itself every ten years, it was, in fact, not rich enough to build itself once. The city needed to partner with private capital because the state constitution restricted its borrowing capacity to 10 percent of its assessed property value. Consolidation did boost collective borrowing power, but the city was going into debt for so many "public goods" (like schools, which could not generate revenues through user fees) that there was little margin upon which to draw—certainly not enough to fund the construction, operation, and maintenance of a truly citywide system—and whether a margin existed at all was the subject of debate.[13] Therefore, the city needed to invest in joint ventures with private investors who could bring capital to the table. To be sure, the city could exempt subway debt (at least partly) from the constitutional borrowing limit by claiming that it would eventually generate user fees. However, by the time the IRT had built its first stretch of subway, the idea of additional public investment tended to scare away private capital. When, by special amendment to the constitution, the city exempted self-sustaining subway debt from its borrowing limit in 1909, it actually made it harder to lure additional private companies into the subway business since it was widely believed that the city would use that increased borrowing power to create new subways to compete with privately run lines like the IRT—thus hurting their ability to pay off bondholders and reducing the likelihood of a successful partnership with the city.[14]

And there was good reason to believe that the city's subway policy would be driven by the desire to limit the profitability of private businesses: distrust of traction companies was even more potent than distrust of machine politics in the state-building process. In 1906, the state legislature passed the Elsberg Bill, which reduced the length of subway leases to twenty-five years in an effort to restrict transit companies' ability to profit from control of franchises. Said one observer: "The Elsberg Bill in effect has been an obstructive legislation as though it had been designed for the purpose of preventing any further construction of Rapid Transit Lines."[15] At the same time, William Randolph Hearst, who narrowly lost the mayor's race to George McClellan in 1905, during the 1909 campaign continued his crusade for public ownership of utilities by promising to build a gigantic municipal subway to compete with the IRT. Though he lost the election to William Gaynor, Hearst's continuing role in city politics through the 1920s assured that the evils of privately owned traction companies would remain highly visible.[16]

Belmont and the IRT only contributed to that visibility. As the sole subway line through the most congested part of the city, the IRT was so crowded that conditions bordered on the inhumane. Profits were extravagant: just over 17 percent in 1910. When Belmont purchased the IRT's closest competitor, the Manhattan Elevated, in 1906, it was clear—and Hearst's *American* and Joseph Pulitzer's *World* made it even clearer—that the financier was bent on monopoly.[17] As in Los Angeles, where poor service and high profits contributed to that city's turn toward the automobile, and as in San Francisco, where Hearst's papers hounded streetcar companies and the Southern Pacific Railroad as enemies of the public, the very real threat of a powerful private company exploiting a municipal franchise put enormous pressure on local political leaders to frame the debate over subways as a fight against monopoly rather than as an effort at citywide planning.[18] The rapid transit commission (RTC) and the public service commission (PSC) that replaced it in 1906 were thus caught up in the same dynamic that shaped the ICC's policies on rate regulation and delayed the PRR from getting its franchises from the board of aldermen. In each case, the problem of providing large-scale transportation systems was lost in a political battle against concentrations of private economic power. As one observer noted, "The truth of the matter is that every man in public life in New York seems to be afraid to favor anything whatever that would appear to be to the advantage of the Interborough System—notwithstanding it might be the best thing for the City—by reason of the unpopularity of the Interborough."[19]

Between 1904 and 1913, the battle between policymakers (at the RTC, the PSC, and the board of estimate) and Belmont was thus driven by the public's desire "to punish the Interborough and limit its profits as a primary purpose."[20] Belmont and the IRT board wanted to add new lines to their subway, but only into areas where profits would be high (like built-up sections of upper Manhattan) since, as Edwin Hawley, an IRT director, noted in 1910, the company "could not profitably carry out the plan for a Brooklyn connection and branches in the Bronx without endangering [our] entire investment."²¹ Meanwhile, IRT directors felt as though the PSC was trying to blackmail them into building unprofitable lines into unpopulated areas to please "real estate boomers in the outlying districts."²²

Of course, the same goal of a metropolitan transit system could be achieved if the city guaranteed the company a profit by pledging public credit to cover operating deficits on new routes, but that alternative was even less politically palatable than giving the IRT the routes it wanted.²³ Therefore, the PSC tried to coax competitors into the subway business by offering new routes that paralleled the IRT through densely packed lower Manhattan and extended into the outer boroughs. But no one wanted to compete with the IRT or risk their money with the threat of a municipally owned and operated line. And even when potential competitors came forward, like the Brooklyn Rapid Transit Company (BRT) in 1910, the city's offer of a large role in the emerging subway scheme was considered a "bluff" to frighten the IRT into being more cooperative since the BRT's ability to raise capital was in question (not just anyone had the technical and financial wherewithal to carry out a subway contract) and their profitability on routes into the outer boroughs could be assured only with guarantees of public money—the very thing the city refused the IRT.²⁴ It also made little sense from a planning perspective to spend a lot of money building new subways merely to punish the IRT when such routes would clearly serve the city better if they were extended into Brooklyn, Queens, and the Bronx. Belmont was thus in an ideal position to wait for the city to accept one of his offers, and public officials could do little more than complain that he should try to emulate H&M impresario William Gibbs McAdoo, "who tried to please the public," rather than exploit it.²⁵

Pressure to put the citywide transit planning perspective at the forefront of the policy debate grew during this extended battle with Belmont, thanks in part to the PRR. PRR managers like James McCrea, Samuel Rea, and A. J. County desperately wanted *someone* to begin work on the planned Seventh Avenue subway route from downtown, into Penn Station, and along the west side of Manhattan, which, County observed in 1910, remained "the only element that is lacking to make our New York Station perfect as a transportation machine."²⁶ That route, they repeatedly reminded city officials, had been promised by the RTC when the PRR obtained the franchises for its tunnels, and the company had gone to no little expense to accommodate linkages between its system and the city's emerging transportation network—building

Penn Station forty feet deeper than necessary to provide clearances for subways.[27] The company's multi-million-dollar civic vestibule had begun dumping twenty million passengers a year into "the heart of New York" without any mass transit connections nearby, while the IRT already served the PRR's rival, the New York Central.[28] In spite of the "moral obligation" of the city to the PRR, however, the Seventh Avenue subway, along with routes into other boroughs, was held up by the deadlock between public officials and the IRT.[29] Indeed, it is a remarkable testament to the diffusion of decision-making power in the city that the PRR, with its considerable political and financial resources and despite all the money it poured into its station and tunnels, could exert no effective pressure on the PSC or the board of estimate to proceed with the Seventh Avenue subway.[30]

The problem here was not the validity of the PRR's claims or the appropriateness of the Seventh Avenue route, but the relative importance of the political battle against the IRT as compared with the strength of the PRR or the need for a citywide subway system, both of which public officials readily conceded. "We have had quite enough of friendship and suggestions," County grumbled in 1911, "now we want votes to carry it [the Seventh Avenue route] through."[31] Where subway policy was concerned, in other words, public officials held real power, and the overriding public desire to wage war on the IRT had the most influence over their views. In this sense, Hearst had more pull in the subway debate than the mighty Pennsylvania Railroad.

Unable simply to pull political strings to get their way, PRR officials tried to mobilize a constituency for a citywide transit planning perspective. Rea rallied civic and real estate groups along the Seventh Avenue route, encouraging them to make their demands for subway service known to the PSC. At the same time, the company worked through Ralph Peters, then head of the Long Island Railroad, to drum up support for subways in Queens and Brooklyn. And in words that could have been written by Walter Lippmann, PRR President James McCrea told the chamber of commerce that the company's station and tunnels have "made our system to even a greater extent than heretofore a corporation of the state," since they were "so designed as not to interfere with, but to feed, your rapid transit lines, existing and prospective"—if the city would only hold up its end of the bargain.[32] McCrea and Rea were convinced that the PRR's interests coincided with those of "the City at large"—both in Manhattan and the outer boroughs—and they hoped that encouraging groups with a potential interest in new subway routes would bring city officials to the realization that these larger issues in the transit debate should take precedence over the fight against the IRT—as had happened in its tunnel franchise fight.[33]

Because Rea conceived of the city as a system, he was prepared to accept the institutional changes necessary to realize that vision. Although he had no love for the IRT (and in fact criticized the company for antagonizing the public), Rea believed that a partnership with a profit-oriented private corporation in a quasi-monopolistic position should not be an obstacle to the completion of additional subway routes, and he urged the city to abandon the idea of competition among subway providers as a means of checking the power of the IRT. The "competitive idea has not a peg in its favor to the benefit of the City or the people," Rea counseled; "co-operation among railroads, working agreements, pools and whatever else is for the general good, with proper Governmental regulation, as against unreasonable and destructive competition" offered a better way to meet the needs of the city as a whole. From Rea's perspective, the city had so many needs that were far more compelling than avoiding a partnership with unscrupulous traction companies that the proper institutional set-up could control the evils of monopoly and encourage private capital to invest while fulfilling a metropolitan transportation vision: "The need for subways to be promptly built, the protection of the City's credit, the extension of its subway ownership on the best terms, the welfare of the wage earner for the greatest travel for a five cent fare, and a substantial profit to the City at the earliest date, all make me feel that the guarantee proposition [a guaranteed return on investment] ought to be revived for both Companies [the IRT and BRT]." Rea and his colleagues thus recognized that the reluctance of the financial community to lend money to private companies (due to the instability of subway politics, particularly the threat of "an active new subway program by future city governments"), when combined with the city's limited borrowing capacity, made it essential that the public sector offer some sort of assurance of future income for subway operators, if New York was ever to get a transit system that served the city as a whole.[34]

But it took a politician closer to the real seats of power than Samuel Rea to realize the political implications of physical interdependence. George McAneny, who was elected president of the Borough of Manhattan in 1909, finally broke the transit policy deadlock in 1913 by proposing the so-called dual-contract system (involving contracts with both the IRT and BRT).[35] In many ways, McAneny was the ideal man for the job of overcoming the rift between the city and the IRT. In the first place, he had unimpeachable good-government credentials. After seven years as a journalist for various New York papers, he had become deeply involved in civil service reform in Brooklyn in the early 1890s. He rose to secretary of the National Civil Service Reform League in 1894, served as an executive officer on the New York Civil Service Commission in 1902, and was elected president and chairman of the City Club ("the meeting-ground of experts in government administration") in 1907. Secondly, McAneny knew the managers at the PRR. He had studied law with Edward Shepard (his mentor in civil service reform) from 1903 to 1906, during the time that Shepard had helped the PRR secure its tunnel and Connecting Railroad franchises. Because he knew how to deal with the railroad, McAneny could use the company very effectively as a go-between in negotiations with the IRT. Most important of all, however, McAneny had a passion for city planning (he demurely called it his "hobby"). "I got many of my ideas on city planning from Shepard," McAneny recalled, and those ideas would guide much of his subsequent career. As borough president, McAneny would lead the zoning movement (along with his colleague Edward Bassett, who was a member of the PSC and very active in subway planning), which produced the nation's first comprehensive zoning ordinance in 1916, and later helped found the Regional Plan of New York. McAneny considered planning to be the "main concern" of government officials in New York by the second decade of the century since it involved all the elements—transportation, land use, fiscal, and aesthetic—necessary to resolve the city's paramount problem of overcrowding. He also knew as much about subway financing as anyone in the city and used that knowledge to promote transit as a planning tool.[36]

As a member of the board of estimate's special committee on subways, McAneny was appointed to find a way beyond the impasse in the subway debate. In that process, creating "a complete [transit] system for the purposes of city-building" remained foremost in his mind. To be sure, McAneny had no illusions about the IRT. "Its engineering department is efficient," McAneny told his colleagues on the board of estimate, but "its traffic department is stupid." While he acknowledged the company's exploitative tendencies, McAneny predicated his efforts to resolve the conflict between the city and the IRT on the idea that "the old subway system [the original IRT lines] should be taken as the nucleus of the comprehensive system of the future that would serve as a basis for a logical and well-ordered future city plan."[37] City planning would be the primary consideration in subway policy. This meant, as a first step, guaranteeing both the BRT and the IRT a profit in order to entice them to build new subway routes to sparsely inhabited areas, since neither company wanted to spend money on lines into the middle of nowhere just to help the cause of city planning.[38] By itself, this guarantee (which was originally offered only to the BRT, since the IRT was already making a profit on its existing lines) would have aroused suspicion of complicity with monopoly. But McAneny capitalized on the free-rider problem to overcome that. Taking Rea's vision a step further, McAneny proposed a subway system that was so large and so comprehensive that it affected real estate interests all over the city, while offering the working class access to a genuine mass transit system for a mere five-cent fare. No longer was it a matter of deciding between more routes to the Bronx or a new one to Brooklyn: there would be new routes everywhere, literally hundreds of miles of them (bringing the city's total transit mileage from 181 to 614), in order to facilitate the redistribution of population in a more rational pattern throughout the greater city. True enough, the city had to join forces with the hated IRT and guarantee it further profits; but so many New Yorkers benefitted from the deal—there would be so many "free riders"—that the danger of a cozy relationship with monopoly capital was worth it. McAneny thus used the basic technique of distributive politics to overcome the strong political pressure to punish the IRT—but it was a distributive politics in the service of a citywide system. In this way, he gave the general interest (as conceived by planning advocates like himself) a broad appeal by connecting it with parochial interests all over the city.[39]

Although it seems foolish to argue with success, it is nonetheless true that using distributive politics to push metropolitan planning ahead of antimonopoly as the basis for state building did have unfortunate long-term implications for subway policy in New York City. In particular, it provided an unstable foundation for a new approach to transit planning, since distrust of business proved too durable a political issue to be overcome by better mass transit service. Only four years after the dual contracts were signed, urban populist John Francis Hylan won the mayoral race on a municipal ownership platform. Although Hylan ran as a Tammany Democrat, his views on the subway were pure Hearst, thanks in part to the fact that, after a decade of self-defeating battles, Hearst and Charlie Murphy had agreed to pool political resources and Tammany became the party of municipal ownership. Once in office, Hylan pushed for the creation of an independent transit system—the IND—a municipally owned and operated subway line designed to compete with the IRT and BRT: exactly what the financial community had long feared. After bitter public battles with McAneny and the subway companies, who vehemently opposed Hylan's plans as a threat to the careful deal embodied in the dual contracts, the IND was constructed into already developed areas of the city with existing transit lines. By failing to appeal to real estate interests in unsettled areas of the city (as the dual-contract system had), the IND created no constituency for further subway expansion. At the same time, the free-rider principle became associated with Hylan's anticorporate cause. When postwar inflation eroded the value of the five-cent fare, raising the price of a subway ride effectively returned the debate over transit to its pre-dual-contract deadlock: increasing the fare meant subsidizing traction companies; but punishing private subway operators by denying them new revenues meant no new subway routes (and only encouraged them to put greater emphasis on cost cutting to stretch the value of their nickels).[40] This return to deadlock had serious financial implications for the city: by 1927, New York had spent $351 million on the subways, and municipal authorities anticipated an additional $674 million would be needed to complete the system; during the 1920s, subways surpassed water supply, schools, streets, and sewers as the most expensive municipal infrastructure project.[41]

Water: The Regional Implications of Tolerated Inefficiency

Providing water for three and a half million people presented one of the greatest challenges facing the consolidated city. Although the need for more water was obvious, how to organize collectively—institutionally and financially—to secure additional supplies was no easy matter. Here, too, functional problem solving does not explain the course of public policy change. Instead, as with the subways, creating a water system adequate for a vast and growing metropolitan population forced New Yorkers to confront the limits of their togetherness. Although the experts saw the water problem as a clear case of metropolitan interdependence, only gradually and incompletely did their approach become the guiding force in the development of New York's water supply. Even then, efficiency—which they saw as a necessary component of interdependence—had to be sacrificed to gain the support of a city of "free riders."

Although Gotham had a diversity of water resources in 1898 (private wells in Staten Island and Queens; Long Island ponds and water-bearing sands for Brooklyn; the Bronx and Byram watersheds and the Croton aqueduct system for Manhattan and the Bronx) no borough had a supply adequate to meet the prodigious requirements of the modern city.[42] Skyscrapers were the most visible evidence of this new level of demand: the Empire and World Buildings consumed almost twenty times more than the structures they replaced; the water bill for the American Tract Society Building was nearly ten times greater than its low-rise predecessor; the Waldorf-Astoria paid $30,000 for just six month's supply.[43] These buildings housed legions of nonresidents who commuted to the central business district each day, and this "large floating population" disproportionately taxed the city's resources: a 1903 study found that a tenement district with nearly 130,000 residents used 6.8 million gallons of water a day, whereas an office district with an estimated population of 114,000 used 9.5 million gallons a day.[44]

The most apparent concern for Manhattan at the turn of the century, however, was water pressure—a problem typical of late-nineteenth-century cities. By 1895, water pressures at the northern end of the island topped out at 130 pounds per square inch, whereas pressures in downtown fire hydrants dipped as low as six pounds. In buildings throughout the most congested parts of the city, water did not rise beyond the first floor, and in some cases not beyond the basement, prompting engineers to close water-main valves in the downtown area to boost pressure in midtown buildings (a game also played in Brooklyn). Those who could afford it (and it could be very expensive) installed pumps to move water to storage tanks on the roofs of buildings—one of the more notable features of the Manhattan skyline in the 1890s: in other words, what could not be achieved beneath the city had to be moved aboveground.[45] This lack of pressure was all the more noticeable because of the general increase in building heights in commercial districts, and here Manhattan's status as an entrepôt became the driving force for the expansion of the underground city. "Lower New York has become a store-house of costly merchandise and a lofty workshop of from eight to twenty stories," a group of real estate agents and owners emphasized in 1895, and the lack of water pressure left the "immense destructible wealth" stored in those warehouses and factories at great risk.[46] These medium-sized commercial structures especially worried insurance executives, who knew that they were too tall to be reached by water from hydrants without the aid of pumps.[47]

The water resources of the outer boroughs were also under extreme strain. During the 1890s, Brooklyn residents consumed virtually all of the 93 million gallons available to them daily, leaving no margin for population increases or dry weather.⁴⁸ In Queens, the completion of the East River bridges threatened to overwhelm the patchwork of public and private suppliers. Officials limited pumping from municipal wells because of saltwater infiltration, forcing the city to contract with private companies, some of whom charged "monopoly rates." In newly developed sections of the "high" Bronx—so-called because of its elevation—pressure was so low that property owners in University Heights, Morris Heights, and Woodlawn Heights had no water at all, leading to complaints that the city was denying them a basic right.⁴⁹

Growth therefore posed a real danger to city dwellers in the New York region. There were simply too many people and too many businesses demanding more and more water. The engineers who knew the limits of existing supplies lived in dread of below average rainfall, since a dry season meant water famine. That almost happened in November 1891, when Croton reservoirs came within forty-eight hours of being completely drained; fortunately, the city was saved by a downpour. Should it happen again and the city not be so lucky, one expert warned, "there would then be nothing left to do but defer washing, shut down all steam plants, factories and transportation, treble the fire patrol, move out into the country, and wait for rain."⁵⁰

If obvious need and political clout were all it took to get things done in the new metropolis, then the threat of water famine and the demands of more than a thousand merchants, insurers, and building owners from Manhattan and Brooklyn would have prompted dramatic steps toward securing a new water supply much earlier than 1905, when the consolidated city finally began work on the Catskill system. But to address the water problem required more than a clear mandate and a powerful constituency since public authorities faced a number of institutional constraints that delayed their response to the crisis.

First, New Yorkers were effectively barred from simply seeking water from adjacent counties. When Brooklyn authorities recommended drawing additional supplies from Suffolk County on Long Island, estate owners and farmers, acting out of the well-founded fear that the city's needs would deplete trout streams and ponds, secured legislation preventing such expansion in 1896. Seeing the writing on the wall, Dutchess County, to the north of the Bronx, likewise secured legislation in 1903–4 to prevent New York from tapping its water. Manhattan and Brooklyn were thirsty monsters and everyone knew it, which meant that the nation's two largest cities were increasingly hemmed in by rural communities that had no interest in becoming municipal watersheds.

To resolve Gotham's water crisis, engineers would have to extend their underground systems far beyond the city limits to reach new supplies, pushing the cost of a solution toward $100 million.[51]

Second, as with the subways, limited borrowing capacity shaped every decision officials made regarding water. Even though water debt was exempt from the constitutional limit, the city could borrow only so much before it had to offer unreasonable interest rates on its bonds (assuming anyone would loan money to a municipality drowning in debt, as it were). Before consolidation, neither Brooklyn nor Manhattan had the financial resources to build elaborate new water systems, and Greater New York inherited their problem. As the commissioner of water supply observed, "this great Municipality is confronted with a financial condition which makes such bond issues, for water supply purposes alone, impossible, unless it be done with the exclusion of all other necessary improvements, which are payable from bonds, such as parks, schoolhouses, bridges"—and, he could have added, subways.[52]

Third, neither Brooklyn, Manhattan, nor the consolidated city had engineering staffs capable of the institutional innovation necessary to resolve the water crisis. The Croton Aqueduct Commission was "an old and more or less moribund outfit," its staff constrained by ossified civil service and purchasing restrictions that prevented them from hiring the most capable workers and forced them to accept low bidders on contracts even when it meant slow, shoddy work.[53] The problem was not so much corruption—although Tammany periodically dabbled in staffing issues—as a restricted view of the civic role of expertise. Staff engineers were men "with many figures and few results."[54] By and large, they were caretakers of existing systems, rather than visionaries. The Brooklyn Water Department posed an even bigger problem. In the words of one disgusted consultant, it was "a gallery of professional malpractice where deceit, magic, and mystery have long been substituted for professional honor, competence, and truth."[55] Without "new and unprejudiced talent," existing staff stood little chance of solving the water problem.[56]

It was therefore with great relief that public officials embraced an offer from the Ramapo Company to bring water to Gotham—under the false hope that privatization could resolve the difficulties of managing the new metropolis.[57] Unlike the subway problem, where no one came forward to build the system, here a private company provided a solution that immediately satisfied a vocal constituency by transcending the institutional and political constraints faced by decision makers: there would be no addition to the public debt, no chance for a Tammany boondoggle, and no more water problem, with one simple decision—or so it appeared. Not surprisingly, when the company proffered a contract in 1899, the commissioner of water supply, supported by a majority of the board of public improvements, gleefully recommended that the city accept.

But Ramapo was not quite the private sector in shining armor that merchants and insurers had hoped. Instead, their deal turned out to be a Steffansesque collaboration between business and political machine. Formed in 1887 by a group of upstate politicians, the company had power to acquire water rights far exceeding those of any municipality, thanks to two leading Republicans—S. Fred Nixon, the state assembly speaker, and Benjamin F. Tracy, right-hand man to Thomas C. Platt, the Republican Party boss, in the consolidation process. Nixon and Tracy persuaded the legislature to limit the consolidated city's power to acquire upstate watersheds, virtually guaranteeing that New Yorkers would have to turn to Ramapo for their water—at a price per gallon exceeding that of the Croton system.[58]

Once the terms of the Ramapo deal were exposed, the water issue became a political battle against a franchise-exploiting company in league with corrupt politicians, rather than a debate over the best way to deal with the problems of citywide growth. The city comptroller, Bird Coler, an independent-minded Brooklyn Democrat whose political ambitions were rivaled only by Hearst's, seized upon the Ramapo scheme as a cause célèbre, and his investigation of the contract scam provided him with the perfect issue to launch a campaign—perhaps for mayor, probably for governor, even for president. In 1900, Coler placed himself prominently in the running as gubernatorial candidate on an anti-Ramapo platform, and in 1902 he actually won the Democratic nomination (losing to Republican Benjamin O'Dell in the general election). But while Coler's success in stopping the Ramapo deal was important—since the company showed remarkable tenacity in its attempt to hold on to the watersheds it acquired, engaging the city in a legal battle that was not resolved until a U.S. Supreme Court decision in New York's favor in 1915—his ongoing fight against franchise graft did nothing to increase the city's supply of water. As the Ramapo controversy was subsumed by other campaign issues, newspapers no longer paid much attention to the water problem, leaving merchants and property owners, in the words of one insurance executive, waiting "to see if we cannot get the matter stirred up again."[59]

As institutional failure contributed to the growing resource crisis, some members of the insurance industry decided to push for a partial solution to the water problem—in effect, shifting to a more clientelistic approach to getting action from the city. They began lobbying for a saltwater pumping system that would draw water for fire protection from the rivers encircling Manhattan. No one could drink the water, of course, but it would make all that flammable, insured wealth in midtown a little safer. At the same time, however, the groundwork for a long-term approach to the water problem was being laid by experts working outside of the formal channels of government. When Bird Coler began his investigation of the Ramapo contract in 1899, he asked Continental Insurance Company executive F. C. Moore for advice on evaluating the water supply problem: Moore recommended that he hire consulting engineer John Ripley Freeman. Together with allies at the merchants' association, Freeman established a shadow water-policy agency to undertake the sort of comprehensive study of the water problem that seemed to be beyond the competence of either elected officials or municipal staff.

Freeman certainly had the credentials for such a job. After graduating from MIT in 1876, he had gone to work for the Essex Company, a waterpower operation that employed some of the leading lights of hydraulic engineering in New England. In 1896, he became president and treasurer of the Massachusetts Mutual Fire Insurance Company (a position he held until his death in 1932), and there he emerged as one of the nation's foremost authorities on fire prevention. Over the course of his long career, he consulted on the Owens River Aqueduct in Los Angeles, planned the Hetch-Hetchy water system for San Francisco, advised the Chinese government from 1917 to 1920 after the devastating Yellow and Hwai River floods, and did tours as director, vice president, and president of the American Society of Civil Engineers.[60]

It was his work on the New York City water supply for which Freeman was best known, however. His 1900 report to Coler provided the background studies on water sources that led the city to construct the vast Catskill reservoir-and-aqueduct system. Cautioning his fellow insurance executives that the saltwater fire system was "a delusion and a snare," Freeman advocated a public water system that would begin in the mountains north of the city, extend to every borough, and provide no less than 500 million gallons per day at very high pressures. Freeman also called for improvement of the efficiency of the existing distribution system with water meters and expansion of the operational capacity of the city's lackluster engineering staff. The goal, he insisted, was not to solve the problem on the cheap (the primary appeal of the Ramapo contract) but to think of the long-term needs of the entire city—at least fifty years ahead.[61]

Although another five years passed before a new board of water supply put these principles into action, Freeman and his allies took immediate steps to remake the city's engineering staff, beginning with the Croton Aqueduct Commission. Realizing that it would take a decade or more to build the Catskill system, Freeman recommended that the city expand the Croton system with new reservoirs to avert water famine. To do that, the commission hired a new chief engineer, J. Waldo Smith, who emerged from the same hydraulic engineering milieu that produced Freeman. In the mid-1880s Smith worked at the Essex Company, and it was there that he met Freemen, who inspired Smith to follow in his footsteps and attend MIT. After graduating in 1887, Smith took a position at the East Jersey Water Company, where he developed a reputation as an expert in masonry construction and gained familiarity with regional water problems. Because of that experience, Smith was the ideal candidate to lead an emergency reservoir construction program. Smith reconstructed his staff with engineers like W. B. Fuller, "the most expert man in the country" on concrete reservoirs; Alfred. D. Flinn, who had worked at the Associated Factory Mutual Fire Insurance Company during Freeman's tenure; and Horace Ropes, who assisted Freeman on researching locations for tunnels and aqueducts for the Coler report. Freeman and Smith thus grafted their informal network of engineering colleagues onto the formal bureaucracy of city government, expanding its capabilities and changing its organizational culture.[62]

When, in 1905, the city obtained permission from the state legislature to create a new board of water supply, its engineering culture and policy approach had already been established. Smith became the board's chief engineer, with Flinn overseeing dam and aqueduct construction and Freeman serving as consulting engineer, a position he held until his death. Although the board was not given total power over water issues (its plans for watersheds had to be approved by the state water commission), it did have authority to plan and build new dams and aqueducts in the Catskills. The board was also separate from New York's water department—an arrangement that, even though it meant the board did not have control over water policy within the city itself, did keep it from becoming "a political machine." Drawing on the studies done by Freeman and others, the new, politically insulated engineering staff recommended that the city create watersheds along the Esopus, Rondout, Schoharie, and Catskill creeks more than a hundred miles north of the city (figure 9); within three months, the experts had plans for the system and were ready to let contracts.[63]

The Catskill water system they devised was engineering on a grand scale—both in terms of expertise and financial commitment—and represented the institutionalization of part of Freeman's vision of the underground city. By 1911, the board of water supply had an engineering corps of 1,348—a total since unmatched. At its peak, the staff directed the work of an army of twenty-five thousand contractors and their employees. By 1917, the city had spent $175 million on the system. Thanks to this massive commitment of expertise and public money, the city had a water system that included a drainage area of nearly 800 square miles and a delivery aqueduct almost 160 miles long extending from the Schoharie Reservoir in the Catskills to the Silver Lake Reservoir on Staten Island.[64]

These upstate works were actually less impressive than the engineering beneath Gotham. Rather than laying a new system of water mains in the streets (already too crowded with pipes, or designated for subways), engineers determined that it would be cheaper for the city and less disruptive to local businesses to approach from below. They built a pressure tunnel that ran 250 feet underground for nearly six miles from the Hill View Reservoir in Yonkers (just north of the city line) to the Harlem River; there, it sloped downward, "bobbing and ducking" to avoid layers of faulted and decayed rock that could not support a high-pressure water tunnel, descending to 752 feet below the street level at the tip of lower Manhattan. A series of shafts connected the tunnel to existing distribution mains near the surface. Shaft 21 (at Clinton and South Streets) extended two feet deeper under the city than the Woolworth Building rose above it. So great was the pressure that water rose sixteen floors above ground. With additional pumping, firefighters could send a steady stream of water to the top of any skyscraper.[65]

Although the upstate and underground systems created by the board of water supply represented a clear triumph of expertise, they did nothing to solve the biggest problem with the city's water system: New York leaked like a sieve. In 1900, Freeman estimated that underground leaks wasted between a half to two-thirds of the water delivered to Manhattan, which meant that each week literally hundreds of millions of gallons of water trickled (or, rather, gushed) into the ground. Fifteen hundred miles of cast-iron main pipes and two hundred and fifty thousand service pipes in Manhattan and Brooklyn

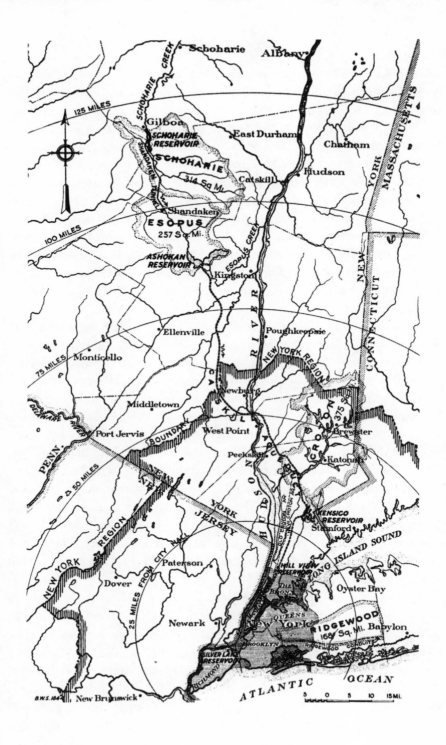

alone would have to be replaced to use new supplies effectively.[66] But that was only part of the problem, since much of the leakage occurred once the water reached the buildings where it was consumed. This meant that in the same way that engineers on the board of water supply were expanding the public side of the water system, building owners would have to upgrade the private side—a process that would be achieved by the universal installation of water meters.

Nothing illustrated the implications of physical interdependence more so than the debate over water meters. At the turn of the century, the vast majority of New Yorkers paid for their water according to a water-rents formula based on the frontage of the lot and the number of stories in a building, not on actual usage: a building could be leaking thousands of gallons a day, but that was never reflected in the owner's water bill.[67] From the engineers' point of view, meters were the ideal approach to integrating consumers more fully into the water system; by apportioning the cost of water according to actual usage, meters would encourage owners of buildings with leaky pipes to fix their plumbing to avoid paying for wasted water. For that very reason, however, meters represented a distributive nightmare. With meters, there would be no free riders—everyone would have to pay for what they used, one way or the other.[68]

Metering thus represented the internal cost of physical interdependence just as the new Catskill system represented the earthworks of metropolitan imperialism. A city so large necessarily imposed on its hinterlands, and as the city grew upward, outward, and downward, it necessarily claimed resources within the region. Freeman and his colleagues recognized that, and they were fully committed to assisting the city's expansion. But there were obligations that the city incurred when it reached into the region. It was both bad politics and bad engineering to let the waste continue, according to the experts. New York City, Freeman observed, was accustomed to bullying its way around and condemning property without much thought for the people it trod under foot; city officials' general approach had been "to rob a man of his land and to pay what

Figure 9. Catskill Water System, begun 1905. The vast Catskill system saved the city from water famine, though a great deal of the water it provided was wasted. Hydraulic engineer John Ripley Freeman persuaded officials to build a giant public water system to supply the city for at least fifty years, but could not get them to address the problem of leakage in the distribution network. *Source:* Harold M. Lewis, *Physical Conditions and Public Services* (New York: Regional Plan of New York and Its Environs, 1929), 37.

they ought to for it a long time afterward."⁶⁹ This understandably incurred the enmity of rural communities and upstate voters, inspiring those legislative efforts to prevent New York from siphoning water from Dutchess and Suffolk Counties. From Freeman's perspective, the city would have to expand into the region, but New Yorkers would have to bear a big part of the cost for that expansion—by reducing waste, by paying for what they consumed, and by using resources wisely for the future. From his point of view, the underground city connected office workers in skyscrapers, tenement dwellers, and suburban homeowners with landowners in the Catskills who had lost their farms to help Gotham survive; metering would insure that city dwellers paid their fair share for being part of that system. But from the perspective of distributive politics, this commitment to efficiency translated into new fees on legions of citizens. Because it forced New Yorkers to use water efficiently or to pay the price for their inefficiency, metering was the apotheosis of the experts' vision of the underground city and the antithesis of the distributive approach to public policy.⁷⁰

Not surprisingly, the experts' metering plans—which included a staff of engineers, inspectors, assistants, and clerks to conduct a "rigid house-to-house investigation" to determine the condition of plumbing fixtures—met stiff resistance.⁷¹ Political opposition arose, in part, from the potential conflict between efficiency and public health, since any system of metering that increased the cost of water to consumers might also have the effect of decreasing water use—a major concern for Mayor William Gaynor (1909–13), who initially endorsed metering. As he told one metering advocate: "If dwellings were metered, the head of the house would realize every time his wife or children were taking a bath that he had to pay for it, and there would be a grumbling and bickering over the matter which would lead to a depreciation in the use of water and to uncleanliness."⁷² Gaynor thought the best public policy would create an abundant supply of water and allow everyone to use it; "Every one in the family should feel free to take a bath at any time," he affirmed.⁷³ For the experts, however, such resistance to metering to help the poor was simple ignorance. "In the houses of the rich and well-to-do much more water per capita is used than in the houses of the poor," Freeman insisted. "The meter fixes the charge accordingly. The cry of the uninformed and of the demagogue, that the poor man's water supply would be impaired by metering, has no real foundation, for he can be guarded against rapacity of landlords of tenement houses by individual meters and minimum rate."⁷⁴

But it was not just demagogues and politicians who opposed metering; real estate groups likewise had no interest in incurring additional costs for an expanded water supply. Owners of metered tenements paid water bills that were on average twice as large as the water bills paid by owners of identical but unmetered tenements, which meant that large apartment buildings would undoubtedly have paid more when metered.[75] Predictably, the Real Estate Board of New York announced that it opposed universal metering because it would increase water bills for property owners. Metering was seen, therefore, as yet another form of property tax, rather than as a means to municipal efficiency.[76] Despite the best efforts of civic groups like the merchants' association and water supply experts, therefore, water metering did not get past the experimental stage.[77] Although city officials periodically revived the issue, real estate owners and tenants fought it every time. So great was resistance that it was not until the 1980s, under Mayor Ed Koch, that the city finally began installing meters in all residential buildings.[78]

Due in part to the failure of metering, engineers soon saw the need for additional extensions of the water system. The city approved plans for another deep aqueduct to carry water to Queens and Brooklyn in 1927—estimated to cost nearly $68 million. And in 1921, the board of water supply called for the creation of a tristate commission to explore the use of water from the Delaware River, which ran through New York, New Jersey, and Pennsylvania, thus extending the city's underground system even further into the region and across state lines. New Jersey and Pennsylvania rejected the plan, citing, among other things, the city's "extravagant use of existing supplies." Not until 1931 did the U.S. Supreme Court rule that New York could take water from the Delaware. Even so, the inefficiencies that existed before the construction of the Catskill system allowed water to leak from the distribution network before reaching consumers. The Catskill system, intended to meet water needs for fifty years, had sufficed for less than a decade for a city that continued to waste unconscionable quantities of water.[79]

This failure to resolve the leakage problem brought to the fore the distinctions between the civic culture of expertise and the system of distributive politics the experts faced. Metering represented both efficiency and interdependence and was as much a part of the experts' vision as the upstate dams and aqueducts they planned and the politically insulated institutional structures that finally succeeded in getting them built. But water meters pushed the cost of the water supply problem onto individual building owners who would no doubt pass it on to tenants. Wasteful usage and leaky pipes would lead to enormous water bills, which could be lowered only by investing in new plumbing. Efficiency and interdependence therefore did not translate into free riders but into higher costs for key constituencies in the political process. Focusing on an increased supply, on the other hand, meant the whole city got a free ride—on the backs of upstate landowners whose property was taken for the benefit of the giant leaking metropolis. Those supplies would have to be tapped eventually, of course, but leaving the waste problem unaddressed insured that more would be needed sooner and would not last as long as they should. In short, residents of the Catskills (and, in turn, other groups who relied on the Delaware River) were subsidizing New York City's inefficiency and resistance to physical interdependence.

Sewage: Resisting Regionalism

"In the important field of sanitary engineering," admitted engineer Alfred Flinn in 1913, "New York must confess to being laggard, at least in so far as the disposal of its sewage is concerned"—a statement that applies to the entire period from consolidation through the 1980s, when the city finally got around to completing a comprehensive sewage treatment system.[80] Even more so than addressing the transit problem, dealing with sewage required new, centralized administrative structures that were politically unpopular; in fact, sewage treatment involved institutional breakthroughs equaling, and in many ways surpassing, the establishment of the Port Authority. To an even greater extent than resolving the water problem, sewage treatment also imposed significant new costs on citizens for which they would receive little direct benefit; if water metering threatened to minimize the number of free riders, then sewage treatment would do away with them altogether. The battle over sewage policy thus epitomizes municipal resistance to the institutional implications of physical interdependence—within the city and the region.

New York's abject failure with sewage treatment emerged from its enviable success in the provision of sewers. The city's sewage system—not that its ill-planned network of sewer lines deserves the name—was both extremely simple and marvelously effective, at least in narrow terms: each borough laid hundreds of miles of pipes to move untreated effluent into the nearest body of water, making the rivers that posed such an obstacle to passenger and freight transportation a boon for waste disposal. By 1913, Brooklyn alone had more than 925 miles of sewers.[81] By 1922, Manhattan had 180 outlets (many located under piers) discharging 750 millions gallons of raw sewage into the Hudson, East, and Harlem Rivers every day. By 1930, that total has risen to an astounding 1.35 billion gallons. To move this huge volume of waste safely away from the city, everyone counted on the cleansing power of the tides in New York harbor.[82] In this sense, New York harbor supported a whole region of free riders.[83] The waters of the bay served as a free sewage-treatment plant for millions of residents in New York and New Jersey—a natural waste-disposal service for which they did not pay (and in fact worked assiduously to avoid paying for) but upon which they relied every day to keep their cities free from the aesthetic and health consequences of untreated human waste.[84]

This is not to say that New Yorkers were unconcerned about sanitation. For nearly a century, experts had encouraged public spending on water and sewer infrastructure to reduce epidemic disease, especially waterborne illnesses like typhoid, cholera, and diarrhea.[85] But the last major cholera and typhoid epidemics in New York occurred in the 1860s. Although diarrheal diseases were the city's third leading cause of death in 1900 (behind tuberculosis and pneumonia), they had been cut in half between 1870 and 1900 and would decline by a factor of nearly eight between 1900 and 1930, when no progress was made on sewage treatment and the harbor became even more polluted.[86] Abundant water and miles of sewers undoubtedly accounted for the drop in waterborne diseases, even though they exacerbated the condition of the bay, but there appeared to be no connection between sewage in the harbor and public health. Thus, in spite of the growing pollution problem lapping at the edges of the city, New York had a low death rate (fourteen deaths per thousand in 1912), making it "one of the healthiest of the great cities of the world."[87]

As with freight planning, only after other cities in the region began to use the harbor to dispose of their own waste did New York see sewage as a matter worth addressing. In the 1880s, growing industrial centers in New Jersey began emptying more effluent into the Passaic River, which flowed into Newark Bay and thence into upper New York Bay.[88] But Newark Bay was a very poor drainage basin—the tide ebbing and flowing without ever allowing the Passaic River to empty completely, resulting in inordinately high rates of death from typhoid and diarrhea in Newark and Jersey City. Industrial wastes became so concentrated in 1894, in fact, that acidic fumes from the river blistered the paint on houses along its banks. Finally, in 1902, after much bickering over how to distribute the costs of the cleanup, the state created the Passaic Valley

Sewerage Commission, which recommended a joint outfall sewer running under Newark Bay and discharging forty feet underwater in the middle of the Upper Bay (figure 10). The theory behind this system was deceptively simple: New York harbor had a powerful tidal flow that moved water into the Atlantic Ocean through the Narrows; if the Passaic River communities could get their sewage there, Mother Nature would do the rest.[89]

Faced with the threat from the Passaic Valley project, New Yorkers began the slow process of coming to terms with the institutional implications of their regional sewage problem. They took their first tentative step in 1903 when the legislature created the New York Bay Pollution Commission, which quickly concluded that the only concern of the communities using the bay "has been to conduct the sewage by the cheapest possible route and means to tidal waters, practically regardless of the effect upon them." Like the competing railroad companies who could not bring themselves to cooperate to improve freight movement around the harbor, cities in the region sought only to resolve their internal sewage problems by moving them downstream. No one saw the bigger picture, or at least no one cared to act as though they did. To remedy that problem, the pollution commission recommended that the legislature create a new commission to study the bay from a regional perspective—and here the first suggestion of the demands of the experts' vision of the underground city emerged. For even while they emphasized the need for further analysis, the commissioners suggested that the logical institutional outgrowth of that investigation would put all of the sections of New York and New Jersey that contributed to harbor pollution "under the direction and control of a permanent interstate commission, with plenary power to control the discharge of all sewers hereafter constructed, as well as the evolving of a comprehensive plan for ultimately rendering the present chaotic and systemless method of sewage disposal, sanitary and suitable for all future." Like the Port Authority later envisioned by Julius Henry Cohen, this "central authority" would be empowered "not only to direct, but also initiate such great public works" as would be required to clean up the harbor. In other words, solving the sewage problem required regional government.[90]

Before New York City could cooperate in a regional sewage program, however, it had to develop its own metropolitan sewage policy. Inspired to reach that goal by the New York Bay Pollution Commission, in 1906 the legislature directed the city to create the Metropolitan Sewerage Commission (MSC)—a group of engineering and sanitation experts whose job it was to conduct a

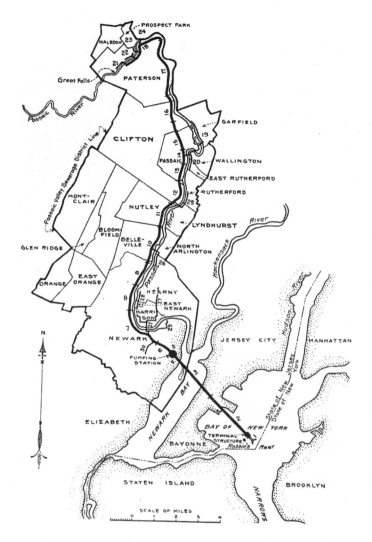

Figure 10. Passaic Valley Sewer Project, completed 1924. The communities of eastern New Jersey dumped their untreated sewage into the Passaic River and thence into Newark Bay, but the absence of a strong tidal flow caused the filth and industrial waste to stagnate, leading to very high rates of waterborne disease. In response, the Passaic Valley Sewerage Commission proposed this outfall sewer into Upper New York Bay. Only then did New Yorkers become concerned about harbor pollution. The metropolitan sewerage commission attempted to block the Passaic Valley project in court, but the U.S. Supreme Court sided with New Jersey. *Source:* Harold M. Lewis, *Physical Conditions and Public Services* (New York: Regional Plan of New York and Its Environs, 1929), 70.

thorough, scientific study of bay pollution and make recommendations for public action. The MSC included some of the most notable water and health specialists in the country, beginning with the young sanitary engineer George A. Soper, who started his civil engineering career at the Boston Water Works, graduated from Rensselaer Polytechnic in 1895, and earned a doctorate from Columbia in 1899. From there, Soper took his expertise back into the field, helping Galveston, Texas, with sanitation works after a catastrophic storm in 1900, before accepting an engineering position with the New York City Health Department. In 1904, Soper joined the New York State Health Department, where he was in charge of suppressing a typhoid epidemic in Ithaca. Prior to becoming president (later chairman) and director of scientific work for the MSC, Soper studied air quality on subways for the rapid transit commission.[91]

By the time they had completed their work in 1914, Soper's team of experts had dramatically advanced the state of knowledge about the bay. In their most important experiment, they set floating markers adrift in different sections of the harbor and followed them by boat for several days to see whether tides in fact washed water out of the bay. They found that currents instead caused sewage "to oscillate back and forth near its points of origin," which meant that the bay did not act as a dilution basin—thus undermining the theory behind every sewage program in the region.[92]

Armed with this new information, the MSC decided to set a standard of cleanliness for bay waters (based on the quantity of oxygen dissolved in the water) that all of the municipalities in the region would have to cooperate to achieve. In the most polluted sections of the bay—in the lower East River around the Brooklyn bridge, especially—the lowest oxygen levels varied from 13 to 25 percent, hovering around 50 percent in the Hudson, and fluctuating between 60 and 70 percent in the middle of the bay and the Narrows. The experts thought that all bay waters should maintain at least 58 percent dissolved oxygen, except around docks and piers where sewers discharged. With this regional benchmark in mind, the MSC then determined how much sewage each community could dump into the bay without lowering saturated oxygen levels. Because the bay was already polluted, they concluded, New York City would have to start treating its effluent before dumping it into the Hudson and East Rivers.[93] The MSC thus established that the regional dog should wag the municipal tail where sewage matters were concerned. By putting the regional goal first—bay waters of a specific measurable level of cleanliness—the experts required cities to invest in treatment systems, rather than spending minimal (but nonetheless large) amounts of money to move untreated sewage to the bay. Regional sewage policy would thus determine local budgetary decisions.

As New York pushed into the region, therefore, the region was pushing back onto the city. The skyscrapers and suburbs of the metropolis relied on underground systems like subways to sustain their growth. As the city moved upward and outward, new residents demanded water, compelling the city to extend its reach into the Catskills. As residents used more water, they generated more sewage, requiring more sewer lines to transport increasing volumes of effluent into bay waters. With so many cities engaged in that process, the region strained to assimilate the underground systems that made collective living feasible at this new scale. According to the MSC, citizens now had to come to terms with the implications of their collective impact on the bay. A single regional sewage policy carried out by multiple localities had to replace the haphazard collection of separate sewage policies. Henceforth, municipal officials had to think regionally and act locally, instead of thinking locally and acting in their own short-term interest—thus effecting a transition from an aggregative to a disaggregative approach to sewage policy.

The experts on the MSC were not, it should be added, unmindful of the budgetary constraints faced by municipalities in the region. Their work occurred at precisely the same time that the water and subway debates were raging in New York, and they knew the role that the city's restricted borrowing capacity played in those decisions. They therefore recommended a *low-cost* plan (totaling more than $50 million), which included a series of screening and filtration plants to remove solid wastes from sewage, interceptor sewers to collect effluent from the lower Bronx and upper Manhattan for more thorough treatment at a plant on Ward's Island, and interceptor sewers in Brooklyn and southeast Manhattan to channel sewage away from the East River to a fourteen-mile tunnel for discharge at a twenty-acre ocean outlet between Coney Island and Sandy Hook (figure 11). Most important from a financial point of view, rather than attempting to build every component at once, construction would be staggered over time, with a truly citywide system emerging within a few years.[94]

To propose that the city spend so much money, even a few million at a time, in order to implement a regional policy was almost recklessly ambitious in spite of its apparent restraint, especially so since the MSC had no constituency to speak of. Bay pollution did not create an immediate public health problem.

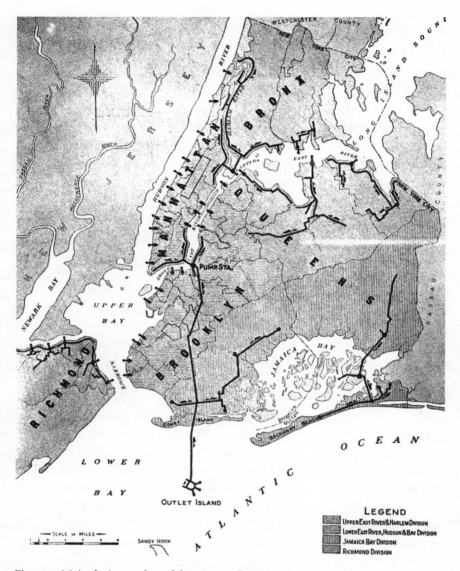

Figure 11. Main drainage plan of the metropolitan sewerage commission, 1914. New Yorkers dumped more than a billion gallons of raw sewage into the Hudson and East Rivers by the late 1920s. The commission designed this low-cost (about $50 million) network of screening and treatment plants to reduce the enormous impact this had on the bay. It included the world's largest sewage-treatment plant, on Ward's Island—a project that was completed in 1936. *Source:* Harold M. Lewis, *Physical Conditions and Public Services* (New York: Regional Plan of New York and Its Environs, 1929), 63.

Unlike the water shortage, a deficient sewage system did not lead to a crisis for businesses or residents. Real estate developers were clamoring for new sewers, of course, but not for new sewage treatment plants. And because the city struggled to fund the underground systems it really wanted, little borrowing power existed for sewage treatment plants that almost no one wanted.[95]

The centralization necessary to carry out such a plan also came into direct conflict with the city's fragmented approach to sewage policy. According to the charter, borough officials, rather than a centralized citywide authority, controlled all matters of sanitation. Borough engineers tended to see sewage problems in strictly local terms, and borough presidents wanted to maintain control over the distribution of resources devoted to such problems—which inevitably meant conveying their untreated sewage to the bay, irrespective of its consequences for other boroughs or the region.[96] The MSC experts saw things very differently. "The sewage problem is essentially one problem and not an aggregation of more or less loosely related parts," they responded to recalcitrant borough officials. "The pollution is not only local, but general, and the system which is to correct the condition should be general also. Such divisions of the work as are necessary should [not depend] upon political boundaries." The whole purpose of a metropolitan sewage policy, they insisted, was to compel "each city and each State [and each borough] to do its share toward the reasonable protection of the harbor," whether officials wanted to or not.[97]

This is not to say that no borough made an effort to treat sewage. By the early 1930s, the city had eleven screening plants at locations recommended by the MSC. But these facilities were not elements of a coherent plan. George Soper even called them "inconsequential."[98] Indeed, by the 1920s it was quite clear that there was an enormous difference between a handful of screening plants constructed at the whim of borough officials and a systematic program of facilities development carried out in stages by a citywide authority. The deteriorating condition of the bay proved it: in September 1921, the saturated oxygen level in the lower East River reached zero for the first time, and by 1926 levels averaged a mere 23 percent in the Hudson River and only 26 percent in the Upper Bay. The city had more sewage treatment, but also a bigger sewage problem. Before New York could even begin to participate in a cooperative regional sewage program, therefore, it needed to develop a unified municipal policy, rather than five different uncoordinated, sewage policies.[99]

By 1929, a confluence of environmental and political factors did allow the city to overcome part of this legacy of decentralization. Although the concentration of sewage in the bay cannot by itself explain the course of institutional change in the city, by the mid-1920s the pollution problem was becoming more apparent to more people than it had been for the previous three decades.[100] In 1926, a state legislative committee declared the Harlem and East Rivers to be "practically open sewers," so soaked with sewage that it was oozing into Long Island Sound and threatening beaches at City Island and Pelham Bay.[101] During the Walker administration (1925–32), the city health commissioner warned of a rise in waterborne illnesses among swimmers at contaminated beaches at Gravesend, Flushing, and Jamaica Bays and declared the waters around City Island "decidedly unsafe" due to the threat of typhoid.[102] The city comptroller, Charles Berry, who had worked as an inspector for the health department and as an epidemiologist for the state and who taught public health courses at New York University and Long Island Medical College, likewise made sewage a major issue by calling for the centralization of sanitation functions and for more funding for sewage projects, particularly the Ward's Island treatment plant.[103]

But perhaps the most important impetus toward centralization came from the sewer scandal in Queens in 1928. It had long been recognized that sewerage in Queens was "distressingly inadequate," and the completion of the East River bridges and the subways only made that problem worse by boosting the borough's population from 152,999 in 1900 to 284,041 in 1910, to 469,042 in 1920, and a stunning 1,079,129 in 1930. It was also well-known that Queens was the "breeding ground of a surviving Tweedism," especially in Long Island City, "that ancient seat of graft," where memories of bosses "Battle Axe" Gleason and "Curly Joe" Cassidy (who ended up in Sing Sing) were still quite fresh. When Maurice Connolly, the long-time borough president, launched an ambitious postwar sewer construction program, it turned out to be a marvelous opportunity for graft. In cahoots with Frederick Seely, an assistant engineer in charge of sewer bids, Connolly rigged construction specifications to guarantee that contracts would be awarded to John Phillips, a disreputable character whose latest venture was manufacturing precast sewer pipe. When sewer assessments skyrocketed in the mid-1920s, the "sewer ring" was uncovered.[104]

The exposure of the Queens sewer scandal gave Mayor Jimmy Walker the perfect opening to do away with the system of "five little mayors" that characterized the city's sanitation policy. In January 1929, Walker proposed a legislative measure, drafted by Comptroller Berry, that transferred control of sanitation matters from the boroughs to city hall. The obdurate borough presidents lobbied against the bill, particularly Republican George U. Harvey, who had first publicized the sewer assessment problem and subsequently replaced Connolly as president of Queens. "I have gone to the trouble of cleaning up the borough," declared Harvey, "and now I want to run it." Harvey favored continued borough "autonomy" where sanitation was concerned, and he had some very good reasons for doing so. The street-cleaning department in Queens had about two thousand employees, half of the personnel under the borough president's direction. Centralization would have meant a dramatic cut in patronage positions for Harvey, and he did not hesitate to suggest that the centralization plan was really a patronage grab by Walker and Tammany Hall. That argument, coming as it did from one of the few Republican officeholders in a Democratic city, had immediate appeal in the Republican-dominated state legislature. Even though the bill was endorsed by the Regional Plan Association and the indefatigable Soper, it failed to win support in Albany, along with similar Walker administration measures calling for unified operation of the subways, the creation of a city plan commission, and the establishment of a bridge and tunnel authority.[105] Undeterred, the politically savvy Walker immediately launched a campaign for a citywide referendum on his centralized sanitation program under the slogan "partisan patronage versus the city's health." With a few modifications to appease the borough presidents, and backing from the merchants' association and shore-front developers on Long Island (not surprisingly), the sanitation measure won overwhelming public approval—by a margin of 644,320 to 114,652—in November 1929. (This occurred just one week after the stock market crashed: it is unclear whether Walker's campaign slogan or the last gasp of 1920s optimism deserves more credit for the victory since voters passed two other referenda by similarly large margins.)[106]

This victory for centralized sewage policy came too late to act on the citywide planning perspective, however: a problem illustrated by the Ward's Island treatment plant.[107] Designed as the world's largest sewage-treatment plant, this $30 million facility was intended to process 180 million gallons of sewage from eastern Manhattan and the western Bronx. First conceived by the MSC but not authorized by the board of estimate until 1928, the plant was not begun until it emerged, with considerable fanfare, as the first piece in a $300 million program by the newly centralized sanitary commission in 1931.[108] However, only about $7 million worth of work was completed before the city's mounting financial difficulties stopped construction in 1933. Most distressing of all, the construction that had been completed did not include the giant interceptor sewers along the Harlem River, under the East River, and to Ward's Island: even if the city found the money to complete the plant itself, there was no way to convey sewage to it ("I suppose they expected to get it over there by wireless," quipped Fiorello La Guardia). To correct this embarrassing situation, city officials petitioned the federal government for money, which Mayor La Guardia finally secured in 1935 after prolonged negotiations with the Public Works Administration.[109]

But the massive Ward's Island plant was only one piece of a much bigger puzzle. The world's largest sewage-treatment plant processed effluent for 1,350,000 New Yorkers, or just about one-fifth of the city's population, and thus represented only a large drop in the bucket. Even Soper, who had first conceived of the plant nearly twenty-five years before, admitted that it would do little to clean up the East River, since its real value would only emerge after other treatment facilities were built and a regional sewage authority was established. And because the thickening blanket of sludge that covered the bay waters had begun to slip out of the harbor and around to the few remaining bathing beaches in the region, the meager resources available for sewage treatment were directed away from the bay itself. The situation had so deteriorated by 1934 that Robert Moses began an emergency program of swimming-pool construction in response to the polluted condition of local beaches, saying "it is one of the tragedies of New York life, and a monument to past indifference, waste, selfishness and stupid planning, that the magnificent natural boundary waters of the city have been in large measure destroyed for recreational purposes by pollution through sewage." Coney Island, the Rockaways, and a handful of other beaches on Long Island remained the only ones clean enough for public recreation, and even those were threatened by raw sewage discharged into nearby streams. Therefore, when monies became available (either from the city or the federal government), plants were begun at Coney Island (completed in 1936, a year before Ward's Island opened) to protect the beaches, and at Tallman Island, near Flushing Bay in northern Queens, to clean the waters near the World's Fair of 1939 and prevent the rest of the world from seeing just how backward the great city was when it came to sewage disposal.[110]

The twin problems of centralization and financing made for similarly slow progress at the regional level. Regional cooperation on sewage matters meant devoting local resources to sewage-treatment plants, and local officials could always point to the lack of progress by neighboring cities as reason for dragging their feet. This did not prevent them from building new sewers, of course. By the mid-1920s, dozens of communities had completed outfall sewage projects that flowed into the harbor: the Bronx Valley sewer, for all the little communities from White Plains to Mount Vernon; a union outlet for Montclair, Orange, Bloomfield, and Glen Ridge; a similar outlet for Vailsburg, Summit, Elizabeth, and the smaller cities near Newark; another for Cranford and Rahway; another for Newark and East Orange; one for Jersey City and Bergen; in addition to the Passaic Valley outfall. The MSC tried to forestall this trend by bringing suit against the Passaic Valley project in 1908, but the U.S. Supreme Court ruled against them in 1921, opening the way for the project's completion in 1924.[111] By the late 1920s, nearly three decades of debate had only confirmed that neither the combined efforts of state and city health departments nor attempts at intercity agreements nor defensive litigation could solve the problem, since there were too many local governments (around two hundred municipalities in all) contributing to the mess.[112]

Against this background of institutional failure, George Soper, working through the New York State Chamber of Commerce, helped establish a Tri-State Anti-Pollution Commission to revive the study of the regional sewage problem. With the help of then-Congressman La Guardia and then-Governor Franklin Roosevelt (both of whom would be instrumental in securing federal financial aid for sewage treatment), the commission commenced a "war" on pollution in 1931. Under the leadership of New York real estate developer Joseph Day, who argued (not inaccurately) that there was not "a scintilla of oxygen" in the East River, the commission proposed the creation of an Interstate Sanitation Commission (ISC) to monitor and enforce a regional sewage-treatment program.[113] Modeled after the Port Authority, the ISC was endorsed by the New York legislature in 1931 but rejected by the New Jersey legislature because of well-founded concerns from Hudson River communities that a "foreign body" would have too much power to impose the costs of sewage treatment on local governments. New Jersey officials also hesitated to join a regional organization that would charge them with half of the costs of harbor cleanup when New York City was to blame for three-quarters of the effluent pouring into the bay, in spite of claims by New York officials that reducing pollution would benefit both states equally. Not until 1936 did both legislatures pass the ISC legislation, and it took until 1941 for Connecticut to join.[114]

After nearly four decades, the creation of the ISC inaugurated a new era in regional sewage policy. Even though, like the Port Authority, the ISC had limited powers, it was not a toothless giant. The ISC could not construct new sewage works on its own, but it could bring suit against any of the 103 municipalities within its jurisdiction if they failed to undertake pollution-abatement efforts. Of course, the ISC could not force municipalities to spend money they did not have, especially when wartime material and labor shortages made capital projects very expensive.[115] Consequently, the ISC initially focused on "persuasion, cajolery, and exhortation," which often worked against smaller cities when they had the resources. When suasion failed, the ISC turned to threats, issuing eighty cleanup orders between 1937 and 1950; when that failed, the ISC exercised its power to enforce its orders by court action. And that power could indeed be burdensome to local officials. In 1947, the ISC won a suit against Long Beach (on Long Island) that forced the city to spend $250,000 on a sewage-treatment plant—a sum that nearly pushed the city over its constitutional debt limit. In 1948, the ISC won a suit against Weekhawken, Union City, and West New York (in New Jersey) that enjoined them from dumping untreated sewage into the bay; in 1938, the commission had ordered the three cities to construct a joint outfall sewer and sewage-treatment plant, but after eight years they had bothered only to complete the outfall, not the plant.[116] The ISC also issued four cleanup orders to New York between 1940 and 1948, but it did not press the city in court because of wartime shortages. The commission lost patience in 1948, however, when city officials continued to delay action on treatment plants while laying new sewage lines. The city did not have the borrowing power to build the plants (as usual), but that would no longer suffice as an obstacle, and the ISC forced city officials to adopt a "sewer rent" financing program, whereby sewer-fee revenues were set aside specifically for treatment facilities.[117]

This combination of timing (depression, war) and local resistance to regional goals meant that progress toward cleaning the bay occurred very, very slowly. New York City did not complete its comprehensive sanitation plan until 1941 and did not get around to fully implementing it until 1987. Real progress in water quality was delayed until the 1970s, when the federal Clean Water Act allowed cities to build long-overdue treatment facilities with 75 percent matching funds. Not until the completion of secondary treatment plants throughout the city did dissolved oxygen levels in the lower Hudson start to rise permanently. Even so, raw sewage still flowed into the river as late as 1986.[118]

That it would take so long for New York to address this aspect of the underground city was evidence of neither the difficulty of the problem, nor a lack of recognition of the problem, nor a dearth of scientific understanding of the problem. Try as they might, neither George Soper, Joseph Day, nor even Fiorello La Guardia could ever make sewage more of a public concern for local officials than it was. Soper attempted to connect pollution to property values and to point out the potential uses of a cleaner waterfront (along the lines of European cities) to no avail. Day became so exasperated that he half-hoped for "an old-fashioned epidemic of some kind to shake us out of our complacent attitude." Even closing beaches only belatedly provoked action on pollution; the thought of swimming in water that was the equivalent of taking a bath with a quart and a half of sewage in it did not seem to arouse much in the way of public indignation or interest. "No local election has ever turned on sewage disposal," the *New York Times* noted in 1934, and therefore local governments could always postpone spending money on sewage treatment when there were so many other things for which the public was clamoring. As ISC officials learned after more than a decade of observing bay communities, "elected officials often proved loath to bury large sums of tax dollars in sewers or treatment works" since they had already done a pretty good job of moving their sewage downstream, making it somebody else's problem.[119] A centralized regional authority would have to spend someone's money to resolve the pollution problem, and because money spent on sewage treatment advanced no one's agenda as much as doing nothing about it, there were always reasons to postpone action. That is undoubtedly why Mayor La Guardia, who had done his utmost to make pollution a priority, called for a $600 million federal postwar construction program to aid cities in the construction of sewage-treatment plants.[120]

The Political Limits of the System Ideal

It is now commonly believed (rising almost to the level of dogma, in fact) that the experts of the Progressive Era fell short when it came to politics. Unlike the party politicians whom they fought, the story goes, they never quite connected with the human side of policy issues and therefore never reached the average person in the street: hence, the conflict between expertise and democracy. If we look at their record with the underground city, however, they did pretty well when it came to politics. Given the limitations of what they had to sell and judging by the amount of money they persuaded the city to spend, we might even conclude that they were better at politics than the politicians who were more skilled at pandering to the voters. That New York had, by the mid-1930s, the most elaborate subway and water systems in the world shows that the experts' vision of the city, albeit in restricted form, could garner widespread support.

The real issue here was not the experts' willingness to deal with politics or to publicize their plans (Soper was unflagging in his efforts to put the sewage issue prominently before the public, and even a politician as popular as La Guardia had a very hard time making progress on sewage policy), but the obligations that came with those plans. Large-scale underground systems meant either close relationships with unsavory private companies or centralization of authority at the expense of local officials, and in every case it mean tapping private wealth for public purposes that did not necessarily advance private agendas. The public wanted the fruits of expertise—cheap transit and cheap water, publicly subsidized real estate development in the outer boroughs—but not the institutional implications of their systems. The experts required too much togetherness: their systems were necessary and beneficial, but embracing interdependence (its institutional and financial implications) was far less acceptable. Rather than transforming civic culture, therefore, the experts had to settle for working inside a system of distributive politics where everybody got paid off and judged the value of public policies in those terms. And when they could not link the interests of the city as a whole (as they saw them) to myriad particular interests (as in the case of sewage treatment) no one got a free ride and their vision garnered insufficient support.

Pollution abatement illustrated this problem especially clearly. The system of distributive politics that relied so much on the free-rider principle could not accommodate sewage treatment as it could subways or water. Unlike George McAneny (who traded centralization for growth) and John Ripley Freeman (who traded centralization for inefficiency), Soper could not cut a deal for sewage treatment that gave someone a free ride. Sewage treatment meant ending the free ride represented by the polluted harbor, especially since New Yorkers did not suffer from sewage in the streets or epidemics of waterborne disease or more Emil Millers—consequences of neglect that might have forced officials to act. Any investment in sewage treatment absorbed money that could have been spent to advance a more particularistic interest: a new school for some neighborhood; streets for developers in Queens. The fact that officials from Weehawken, Union City, and West New York would comply with an ISC order to build an outfall sewer while shamelessly ignoring the requirement that they build a treatment plant as well speaks volumes about the political obstacles that advocates of sewage treatment faced. In this sense, individual commitment to collective living rested precariously on the notion that the chief purpose of the city as a whole was to provide resources for the advancement of particular interests.

Although the experts perceived no necessary conflict between addressing the needs of the city as a whole and assisting developers or neighborhoods or taxpayers associations, they did not want an emphasis on parochial agendas to prevent government from acting on a citywide or regional perspective. From George Soper's point of view, cleaning up the harbor was a collective responsibility, a necessary condition of aggregate living, and a civic requirement for the citizens of the great metropolis. Like water meters to insure the efficient use of regional resources, or new subway routes to insure the proper distribution of population within the city, sewage treatment was part and parcel of collective life and an obvious response to the obvious fact of physical interdependence. For experts like Soper, the harbor had become so polluted because so many people chose to live in the metropolitan region; they could do that only because of the physical systems that made this new scale of life manageable; they were all part of the same regional system of water use; and, therefore, they bore the same responsibility for dealing with the consequences of their togetherness. But because sewage treatment represented collective responsibility for the consequences of collective life without advancing any particular interest, it could not easily be carried out by the system of distributive politics that continued to control important aspects of the underground city. In the case of sewage treatment, everybody's interest in general (as conceptualized by the experts) was nobody's interest in particular.

In this sense, the slow pace of development of the underground city in the New York region represented a "conspicuous" institutional failure, at least in the eyes of McAneny, Freeman, Soper, and their allies. With their specialized knowledge and citywide perspective, these experts chaffed against the restrictions imposed by locally oriented political officials. Of course, those officials reacted only to the concerns and interests of the groups who supported them. Citizens who wanted extensive subways but not subway companies, abundant water but not water meters, and miles of sewers but no sewage treatment got pretty much what they wanted from local government, even if that meant overcrowded subways, dwindling water pressure, and unsanitary bathing beaches. If democratic governance in the new metropolis were simply a matter of being responsive to the public, then the experts clearly pushed things that the people did not want. But collective living on this scale required something more than thinking locally and advancing one's own interest through responsive government. Communities of interest, whether ethnic, geographical, economic, or political, were not enough to sustain the conditions of coexistence in a city of three, five, or seven million people. Metropolitan living required other connections that such communities either did not see or would not organize to act upon. The experts who could see the linkages among the upward, outward, downward, and regional expansion of the city pushed for a recognition of that new sense of the city and region as a whole, and acting on that recognition required institutional change.

This cultural conflict resulted in a diffuse centralization of authority in New York by the 1930s. The subways had been extended, but subway policy remained divided among private companies, public transit authorities, and politicians still responsive to concerns that subway operators had too much power. The water supply had been expanded, but water policy remained divided among state and city water boards and local officials who resisted water metering. And the city's sewage-treatment facilities had been enlarged, but sewage policy remained divided between a regional authority and elected officials who preferred not to spend money on pollution abatement. Viewed within the longer context of the twentieth century, of course, New York City eventually did almost everything the experts recommended between 1900 and 1910: the city now has a centralized public subway authority, water meters, and a harbor clean enough to support a growing variety of fish life and water birds.[121] But it would take a new set of financial, institutional, and political circumstances to overcome the limits of Progressive Era state building and fulfill the experts' conception of the implications of the underground city.

CHAPTER FOUR

Taxing, Spending, and Borrowing

Expanding Public Claims on Private Wealth

In 1895, Nathan Matthews Jr., four-time mayor of Boston, declared that the "chief danger of popular government is not Corruption but Expenditure"—that is, "the demand of individuals, interests, classes, sections, and sometimes of the whole community for a systematic distribution of wealth by taxes." It is important to recognize, as Matthews did, that this was not merely a question of the downward redistribution of wealth through machine politics. The problem was, in fact, much bigger than that. Everyone had their hand in the public purse—suburban villagers demanded firehouses and police stations; speculative landowners called for streets and sewers; and constituents of all sorts wanted everything from pure water and better drainage to public transit, new schools, and museums. "A gradual change has come over the spirit of the people," Matthews concluded, "and a large part of a population once the most independent and self-reliant in the world is now clamoring for support, as individuals or in classes, from the governments of this country—federal, state, and city." Or as political scientist Pendleton Herring put it so picturesquely almost forty years later: "The 'voice of the people' sometimes suggests the squeal of pigs at the trough."[1]

In New York City, there were squeals aplenty. By 1913, the city had more than one billion dollars in long-term debt, and this predilection for borrowing rendered Gotham susceptible to periodic fiscal crises.[2] This chapter places the fiscal crises of the late nineteenth and early twentieth centuries—in 1871, 1907, 1914, and 1933—in the context of the city's changing philosophy of public finance. In the years between consolidation and World War I, an era of fiscal conservatism initiated by the downfall of the Tweed Ring gave way to an era of high taxes, increasing debt, and rising expenditures. This new pattern was a reaction against the post-Tweed low-tax, low-debt consensus, which, in the eyes of many experts, prevented municipal authorities from solving the problems of the modern city. New York veered toward crisis so often after 1898 because experts argued successfully that good public policy required greater levels of taxing, spending, and borrowing and then created new institutional mechanisms to make that the operating philosophy of modern fiscal management.

Institutionalizing this new philosophy created two interrelated dilemmas, one associated with increasing access to private wealth and the other associated with the close relationships with private financial institutions necessary to sustain that access. Because of the physical requirements of making a city this size livable and because of the expectation that consolidation would produce collective spending power capable of realizing many long-delayed projects, public officials faced mounting pressure to spend liberally for a wide array of what experts considered legitimate purposes. In response, a group of municipal bureaucrats, university economists, taxation specialists, and planning advocates fought to move the debate over municipal finances beyond the concerns that had dominated the late nineteenth century: taxpayers' insistence on low taxes and reformers' emphasis on government corruption.[3] This was not a simple case of good-government advocates restraining unruly urban democracy, in other words. Quite the contrary, experts in public finance played a central role insuring that the old bugaboos of corruption and inefficiency would no longer stand in the way of necessary expenditure. Efficiency and ability-to-pay became the keynotes of their reform efforts since both facilitated greater public claims on private wealth. Indeed, the experts argued that properly managed fiscal policy would permit the city to spend huge amounts of money, provide a very broad range of citizens with modern services and infrastructure, and increase the value of the tax base—with spending thus paying for itself. Having partially succeeded in overcoming the old restraints, however, they unleashed an enthusiasm for expenditure they found difficult to control. Along with their "legitimate" claims on the public dollar came a host of questionable demands. How to separate the legitimate from the spurious, parochial, and extravagant became a problem for which there was no simple solution.

This challenge was made all the more difficult because public officials were not completely free to make such decisions on their own. Taking a fresh approach to municipal financial management required a new level of cooperation between public and private institutions that experts built around the idea that the metropolis was a financial system. Where the engineers and sanitarians who constructed the underground city saw physical interdependence as the touchstone of metropolitan policy, the experts who found the money to pay for Gotham's infrastructure likewise saw financial interdependence—the interconnections among real estate values, public improvements, and borrowing capacity—as the necessary starting point for modern fiscal management. But the relationships that allowed the city to spend so much also provided a lever by which powerful financial institutions could, under the right circumstances, apply the brakes to municipal spending, reasserting a simplistic fiscal conservatism fundamentally at odds with the civic culture of expertise and making financial policy outside of the public sphere. How to maintain these relationships without being controlled by the private sector posed the second great financial dilemma faced by the city.

New York's periodic fiscal crises can therefore be understood as key moments in the city's ongoing attempt to develop a more sophisticated, deliberative approach to putting private wealth to public use. The experts who are the protagonists of this chapter wanted to make that process more reflective of genuine collective needs (as they saw them) in order to avoid the irrationality of unrestrained municipal spending (associated with extravagant machine politics and undisciplined local boosterism) and of arbitrary limitations on the public purse (embodied in late-nineteenth-century fiscal conservatism). Finding an alternative to either extreme, and to the equally undesirable prospect of cycling between the two, meant creating a decision-making process that prioritized the multitude of demands associated with large cities and providing access to resources beyond the reach of preconsolidation New York.

Taxes, Debt, and Politics in the Late Nineteenth Century: The Roots of Fiscal Conservatism

In 1871, New Yorkers embraced fiscal conservatism, thanks to Tammany boss William Marcy Tweed. In response to machine extravagance, the city dramatically curtailed its outlays; but this retrenchment simultaneously prevented local government from playing its customary role as economic booster. As developers, party politicians, and fiscal conservatives wrestled with this ambiguous legacy, they set the stage for the emergence of a new philosophy of public finance.

The real danger of Tweed was not merely the level of corruption he allowed but the degree of consent his profligacy represented. Although the boss fed the machine through sinecures and contract graft, his spending spree benefitted almost every interest group in the city—from the real estate speculators who applauded public funding of street paving to the merchants who cheered municipal investment in docks and wharves. To provide so many with so much, Tweed relied on borrowing, allowing the city to spend aggressively while keeping tax rates low (in fact, Tweed actually lowered the tax rate)—which, of course, drove up the city's debt, from $43 million in 1867 to $87 million in 1871. When investors learned that their millions had been used not only to fund long-term public works but to pad construction contracts, they refused to purchase municipal bonds, cutting Tweed off from his primary source of funding.[4]

For the next two decades, New Yorkers lived with a remarkable degree of fiscal discipline. Because it took a few years to reveal the full magnitude of the Tweed Ring's borrowing binge, the city's gross debt did not reach a peak until 1878, where it remained fairly steady for a decade, rising only gradually in the late 1880s; the net debt (gross debt less sinking fund) remained steady for nearly twenty years after 1874. Because the population continued to grow, gross debt per capita dropped from a high of $118 in 1878 to a low of $77 in 1893; net debt per capita fell even more dramatically, from $89 in 1874 to $53 in 1894, thanks to a steady increase in the sinking fund, which rose from a low of 23 percent of gross debt in 1874 to a high of 43 percent in 1895 (see graph 1 in the appendix). Even in a strongly deflationary era (except for a short period between 1878 and 1882), this meant that real net debt per capita decreased by one-third ($40) between the 1870s and the 1890s. Even more impressive, the city reduced the ratio of its debt to the assessed value of its real estate—a key indicator of fiscal health, since property taxes were the principal means for paying bondholders. After an increase in the ratio of gross debt to assessed value in the decade after 1868, the figure fell from 11.48 percent in 1878 to 8.15 percent in 1891; the ratio for net debt fell even further, from 8.56 percent in 1878 to 4.77 percent in 1894—that is, compared with its growing ability to carry debt (since, between 1865 and consolidation, assessed values increased in both nominal and real terms), the city became progressively less debt burdened (graph 2 in the appendix). This did not mean that the city was not spending or borrowing money. In fact, between 1871 and 1896, the city incurred $205 million in new debt for additions to the Croton water system, bridges over the Harlem and East Rivers, parks, docks, and public schools. It did mean, however, that borrowing slowed in relation to the growth of wealth and population. (Even this was not enough for a public burdened by Tweed: in 1884, city voters overwhelmingly approved a constitutional amendment restricting public indebtedness to 10 percent of the assessed value of real estate.)[5] During the same period, yearly budget expenditures per capita also fell (as did the tax rate, which dropped from 2.94 percent in 1875 to 1.72 percent in 1895), indicating that public spending overall decreased as a burden on taxpayers. Indeed, after the city addressed its debt problem, it demonstrated an essential feature of late-nineteenth-century fiscal conservatism: as assessed values increased, the tax rate dropped (a pattern that holds true, roughly, between the mid-1870s and mid-1890s). By nearly every measure, therefore, New York cut back in response to Tweed, and these measures had the desired effect of restoring the city's credit: "No class of securities rank higher than municipal bonds," the *New York Times* reported in 1896, "and New York City bonds are regarded by investors here as the best in that class."[6]

Laissez-faire ideology accompanied this period of fiscal conservatism. The threat of the lower classes capturing government and redistributing wealth prompted E. L. Godkin, editor of the *Nation*, to call for a sharp separation of economics and politics.[7] Godkin's arguments for laissez-faire liberalism had an immediate appeal in New York, where the sting of Tweed's excesses remained acute (the city was still paying off the Ring's debt in the 1890s). But Godkin's rejection of an active state also restricted the use of government for promotional purposes, which had always attracted business groups, neighborhood associations, and city boosters (not all of Tweed's spending could be blamed on the immigrants whom the machine occasionally served), and herein lay the real difficulty with fiscal conservatism. Limited government as such meant government unable to invest in subways and unwilling to respond to the increasingly influential engineers, sanitarians, and planning advocates who called for a more vigorous public response to the dirty, unhealthy, immoral, ugly urban environment. The respectable classes rejected corruption and the downward redistribution of wealth, but laissez-faire threw out the baby with the bath water.[8]

The prolonged period of fiscal conservatism therefore generated considerable pressure for spending—in New York and in adjacent communities, where residents looked enviously at the growing wealth of the nation's largest city. The assessed value of Manhattan real estate nearly doubled between 1879 and 1896. Even though officials made some use of that new wealth, their efforts did not go as far as they could have: only one bridge spanned the East River; the water supply remained inadequate and water pressure was dangerously low; steady growth brought suffocating congestion; the port was badly in need of new facilities; and even though the city had spent $14 million on schools between 1884 and 1896 (following a period of restricted school construction after Tweed), the population more than doubled between 1870 and 1898 and showed no signs of slowing (immigration would not peak for another decade).[9]

By the early 1890s, the effects of increasing wealth and population could be seen in almost every one of the city's fiscal indicators: gross and net debt per capita crept upward (in nominal and real terms); debt increased as a percentage of assessed values; and the tax rate and budget stopped their downward trends. However, even though New York had dramatically improved its ability to borrow (by 1896 it had a debt margin of about $55 million), moving out from the shadow of Tweed and tapping the growing private wealth represented by increased property values could quickly overwhelm its ability to pay. Commercial interests, especially, had big plans for the city's borrowing power. Their proposed subway was so expensive, one judge warned in 1896, that building it would make it "impossible for the city to provide for public schools, the improvement of docks, the furnishing of additional water supply, and the establishment of additional parks, and the city would lie helpless, bound hand and foot by this octopus of debt created by these rapid transit contracts."[10] New York had grown, its credit had improved, its borrowing capacity had increased, but it still faced the danger of wanting more than it could afford.

The situation in Brooklyn was even worse. The nation's second largest city was also spending borrowed money on water, parks, schools, and its share of the Brooklyn Bridge, but it did so without the growth experienced in New York City. "Brooklyn men engaged in business and owning real estate in New York, as the most capable of them do, contribute to its prosperity, while that of Brooklyn languishes," the *New York Times* commented in 1895 (the joke was that residents of Brooklyn were called "Brooklynites" because they spent only their nights there). That did not stop the city from borrowing money, however; during the 1880s, the city inflated assessed values to boost its debt limit while maintaining a low tax rate. By 1897, Brooklyn had depleted its borrowing margin to only $485,022, at the very same moment that officials scrambled to deal with a fourth straight year of severe water shortages, among other problems. "Nothing can then save Brooklyn," an increasing segment of the public concluded, "except consolidation."[11]

As a result, the consolidated city began its existence with a debt crisis. In Brooklyn, Long Island City, Kings County, Staten Island, and many other of the ninety-six administrative units incorporated into Greater New York, officials went on a borrowing spree with the idea that their debts would fall on the combined taxpaying strength of the greater city—securing, in effect, a "free ride" for communities in the outer boroughs. By January 1898, Brooklyn had exceeded its debt limit by $17.7 million (its net obligations totaled $74.6 million, while its debt limit was only $56.9 million); Richmond brought $3.6 million in debts to the consolidated city, $1.1 million over its limit; and the communities that composed the Borough of Queens brought $13.3 million in debt to Greater New York, $5.1 million over their limit of $8.2 million (an astonishing 62 percent more than the constitution allowed). In fact, for months after consolidation, officials discovered bonds from school districts and towns incorporated into the outer boroughs that had been issued at the eleventh hour and for which the greater city now stood legally responsible. And even though the former City of New York had some borrowing capacity (not enough, it should be emphasized, even for its own wants and needs), the collective debts of its sister boroughs consumed much of it. "It would appear, therefore, that this big fine city of ours begins its career without the power to borrow one dollar," surmised the *New York Times*. And because of the initial confusion over just how much debt there really was, the greater city's first comptroller, Bird S. Coler, actually halted payments to contractors on public works projects funded by bonds for fear that no borrowing capacity existed at all.[12]

Thus, even without another Tweed Ring and after an extended period of fiscal restraint, on the eve of consolidation the constituent communities of Greater New York had managed to borrow their way to the verge of financial crisis. They still wanted more water, more bridges, a new subway system, and a host of other physical amenities that were rapidly becoming associated with modern metropolitan living. Paying for those improvements required a new approach to public finance that would allow them to expand and then capitalize on their collective wealth while responding to the numerous demands for public borrowing without spending beyond their means.

A New Era of Municipal Finance, 1898–1917

Between 1898 and World War I, a new era of municipal finance dawned in New York City. Until the fiscal crisis precipitated by the war, the city adopted an approach to taxing, spending, and borrowing that was virtually the opposite of the philosophy that had guided its financial management between the fall of Tweed and the mid-1890s. Assisted by experts from the Bureau of Municipal Research and Columbia University, public officials succeeded in gaining greater access to private wealth than any of their nineteenth-century counterparts (including Tweed) ever had and channeling that wealth toward municipal projects of every sort. Although their approach found common ground with traditional public works boosterism, these experts thought in broader terms; they claimed an increasing portion of private wealth as a "social surplus" to be applied to the advancement of public goals whether or not endorsed by real estate developers and business promoters. Thanks to a new philosophy of borrowing—bolstered by new institutional relationships and implemented through new fiscal practices—their combined efforts nearly freed the city from the constraints imposed upon it by late-nineteenth-century fiscal conservatism, even as it generated an alternative set of problems.

To some degree, the city's new approach to public finance can be seen in the relative size of its budget, which grew from $77 million in 1898 to $198 million in 1915 (in 1913 dollars, it grew somewhat less, from $118 million to $197 million). In nominal terms, New Yorkers were paying 71 percent more per year for annual expenditures in 1916 than in 1898 (when corrected for inflation, the increase was not as dramatic: yearly spending per capita rose 31 percent between its nadir during the Low administration and its peak during the Mitchel years). When compared with the growth of real estate values, New Yorkers were also spending more each year: in 1903 the budget amounted to 1.8 percent of the full value of real estate, and by 1916 it had grown to 2.4 percent. The tax rate rose—from 1.49 percent in 1905 to 2.04 percent in 1916—which meant that in nominal terms the tax rate and assessed values tended to move in the same direction, the opposite of the low-tax consensus pattern (although when corrected for inflation, the pattern varies considerably).[13] However, it should be noted that, in constant dollars, New Yorkers paid approximately the same amount per person for yearly budget expenditures in the 1900s and 1910s as in the 1870s and 1880s—about $37 per person. The budget as a percentage of assessed values was also about the same in 1913 as it was in 1875—around 3.3 percent (this figure declined from the 1870s through the 1890s, rising to 3.2 percent in 1899, dipping to 2.7 percent in 1906, and rising thereafter). After a period of belt-tightening in the 1880s and 1890s, in other words, the city increased annual appropriations to levels that had not been seen since the crisis that followed Tweed.

But the most dramatic change in the city's approach to public finance appeared in its borrowing habits. "The city is a stupendous borrower," the city chamberlain—Henry Bruère, a municipal reformer—reported in 1914, and his choice of words was entirely appropriate. Between 1898 and 1914, gross debt exploded from $342 million to $1.1 billion (and in an important sense the growth was actually larger since the debt in 1898 included last-minute, preconsolidation spending by every borough). By 1908, New York's debt was nearly twice as large as the combined debts of the next ten largest cities in the United States (only Boston rivaled Gotham in per capita debt), and by the beginning of World War I, New York City owed just a little less than the whole of the federal government. City-building investments dominated New York's expenditures: water supply, streets and roads, schools and school sites, docks and ferries, bridges, rapid transit—and the largest of these, the water and subway systems, were incomplete in 1914.[14] Part of this growth could occur simply because the borrowing power was there: the value of real estate more than tripled between 1898 and 1914, from $2.5 billion to $8.0 billion (in real terms, assessed values grew by almost two and a half times). But borrowing did more than keep pace with inflation, real estate values, and population. Corrected for inflation, gross debt per capita grew from a low of $159 in 1902 (almost precisely the peak reached in 1878) to a high of $226 in 1915; and net debt grew by 77 percent, from $108 in 1902 to $191 in 1915, while the sinking fund plummeted from 32 percent of gross debt in 1902 to 15 percent in 1916. The ratio of gross debt to the assessed value of real estate increased from 9.87 percent in 1903 to 14.7 percent in 1917; the ratio for net debt nearly doubled, from 6.87 percent in 1903 to 12.4 percent in 1917; and the ratio of net debt to the *full* value of real estate grew from 5.7 percent in 1902 to 11.5 in 1917 (assessed values were a percentage of the full value of real estate—between 60 and 95 percent, depending on the year) (graph 3 in the appendix). In other words, the city increased its borrowing even faster than its overall wealth (as represented in taxable real estate values), thus becoming more debt burdened between 1898 and the war. With the sole exception of budget expenditures per capita, therefore, the city was spending more money than at any time in its past.

This new level of borrowing cannot be attributed to easier access to money. In fact, in order to compete in the capital markets, New York City had to offer its bonds at ever increasing rates of interest. Thanks to late-nineteenth-century fiscal conservatism, the consolidated city had good credit, paying just over 3 percent on its bonds in 1902; but by 1913, it was paying nearly 4.5 percent. The city's credit was just as good, but competition for money was more intense. Western cities such as San Francisco offered bonds at 5 percent, and railroad securities paid 6 percent or better. And New York City itself had saturated the market: from 1904 to 1914, it accounted for between 10 and 20 percent of municipal borrowing in the country each year (the city claimed 11 percent of the country's urban population in 1900 and 1910).[15]

Nor is it possible to attribute the growth of public indebtedness to a cyclical battle between reform and machine rule. In strictly nominal terms, the city's budget increased almost continuously between 1898 and 1917, except for slight decreases during the Van Wyck and Low administrations; compared with the tax base, the budget declined somewhat between 1899 and 1906, but it then rose fairly steadily through the early 1920s; indeed, the Mitchel administration's budgets all exceeded those of any previous Tammany administration. In other words, although there were minor declines in annual spending during a handful of years in the early twentieth century, they do not show a pattern of machine extravagance and reform retrenchment. The growth of public debt shows even less variation by administration. In nominal terms, gross debt increased every year between 1898 and 1916. As the city grew its way out of the preconsolidation borrowing spree, gross debt as a percentage of the full value of real estate declined slightly (from 8.9 percent in 1898 to 8.26 percent in 1900) during the Van Wyck administration and dipped as well during the Low administration (from 8.87 percent in 1902 to 8.71 in 1903), but then grew steadily to 13.8 percent in 1918. Whether adjusted for inflation or measured per capita, the size of the municipal debt tended to increase, with a few exceptions, under reform and machine leadership; debt per capita grew at an annual rate of more than 5 percent during years in the Van Wyck (Tammany), Low (Fusion), McClellan (Tammany), Gaynor (Tammany), and Mitchel (Fusion) administrations. And while the growth of real gross per capita debt was somewhat more pronounced during the McClellan and Gaynor years (with average annual increases of 2.7 and 2.6 percent, respectively, compared with .8 percent under Van Wyck, .5 percent under Low, and 2.4 percent during Mitchel's first two years), borrowing for subways and water supply, which made up a large part of the increase in indebtedness during this period, extended over multiple administrations and can hardly be described as Tammany projects. In short, changes in governing coalitions had little effect on spending and borrowing trends. Rather than a cyclical, reform/machine pattern of borrowing, therefore, what we see in the consolidated city's early history is the tendency to spend as much as possible and borrow up to and then beyond the constitutional limit. That trend began sometime between 1891 and 1894 and was not halted until the extraordinary financial pressures of World War I temporarily disrupted it. Since the city's financial practices were not cyclical, therefore, the reasons for their steady upward trend must lay beyond electoral politics in the realm of larger changes affecting policy making.

Nor does demand alone explain New York's new pattern of borrowing: the desire for spending had been there for years (Brooklyn's water problem was decades old; the rapid transit debate started after the Civil War). Now, however, policymakers reacted to it differently. In an earlier era, the level of indebtedness reached by the consolidated city had caused a crisis, precipitating two decades of retrenchment. What made the increased indebtedness possible for the new metropolis was an approach to financial management involving both private and public sector actors. The city could continue to borrow from capital markets because experts working in municipal government thought about financial problems in the same way as experts working in the investment industry. While running up debt, these experts were also improving the management of the city's tax base and spending practices. Unlike the Tweed-era city, the Progressive Era city made a more conscious (although not entirely planned) decision to expand public claims on private wealth, forming a partnership with lenders rather than hiding from them.

By the eve of consolidation, a chorus of municipal experts had begun to criticize late-nineteenth-century fiscal conservatism and to lay the foundation for this new philosophy of public borrowing, grounding their approach on the notion that the ability and willingness to pay for more public goods and services should go hand in hand. "This is an age of accumulating wealth as well as an age of great and growing cities," municipal reformer Delos Wilcox wrote in 1898; "we are in an era of vast municipal expenditures made necessary by the newness of our city civilization." Large cities generally, and New York particularly, so closely juxtaposed great private wealth and glaring public need that the case for an interventionist government hardly needed elaboration. While the problems of urban life could be attributed in part to the overconcentration of private wealth (in the hands of corporations, for instance), that wealth could also be used by government to fashion a more humane civilization. For those inspired by the marriage of charity and social science that emerged in the late nineteenth century, the resources for that great project of reform spread out before them in the tall buildings and buzzing factories of New York City.[16]

From this vantage point, fiscal conservatism looked like "false economy" rather than enlightened governance. "The real problem is to secure a proper return for money spent, not merely to cut down outlay," Edward Dana Durand noted in his history of the city's finances in 1898. Since there was plenty of money available and plenty of things on which it should be spent, retrenchment led inevitably to penny-wise, pound-foolish public policies. Sensible men knew better. "Colonel Waring in 1896 spent $2,845,000 or $649,000 more than was expended by his predecessor in 1893," wrote subway advocate Milo Maltbie, a future public service commissioner, in 1900; "but no one will deny, no matter what his political affiliations, that the street cleaning department during the Reform administration was more wisely and economically administered than under the Tammany regime." It was clearly wiser to spend more and get the job done right than to claim savings at the expense of crucial public services. "The taxpayer and the citizen is not injured or aggrieved by high governmental expenditures," Maltbie insisted, "provided the return is proportionate, and essential and important functions are efficiently performed."[17] Commentator Henry De Forest Baldwin had taken this point even further in 1897. "It must not be forgotten that to fail to make an expenditure when good business methods require it, is a sign of inefficient and short-sighted government," Baldwin declared. Far from being a measure of good government, miserly public spending cloaked ineptitude, foul play, and mismanagement. "The tax rate is apt to be lowest when the city is being governed worst and is apt to be raised after the overthrow of a corrupt or inefficient administration," he warned. The point was not merely to spend for the sake of spending, but to meet the very real needs (as he conceived them) of the modern city. For Baldwin, the choice was clear: "It is better to have a high tax rate than a high death rate"—a philosophy that proved to be one of the most durable elements of the civic culture of expertise.[18]

This approach was so appealing because it was, in part, a restatement of the old idea that local government should promote growth through public works construction. Any businessman knew that appropriate expenditures for streets and sewers would raise the value of affected real estate, which would increase tax levies, which would in turn allow local government to pay bondholders from whom it borrowed to build the streets and sewers. This argument had even more force with regard to investments like water systems, docks, and subways that generated user fees that (ideally) liquidated their own debt. For this reason, experts argued, investments in revenue-producing projects should not be counted against the debt limit, nor should the city be compelled to set aside so much in its sinking fund as a guarantee of repayment. Properly managed, public debt was really public investment that only increased the city's ability to pay.[19]

Implementing this philosophy required new institutional practices to strengthen the city's relationships with the private financial markets that bought and sold public debt. To sustain a "stupendous" level of indebtedness, the city had to borrow almost exclusively from institutions rather than individuals, and that fact set the course of change for fiscal policy in New York. In 1911, for example, the city offered $60 million in bonds at one sale: 80 percent ($48 million) were bought by investment banking houses, 19.25 percent ($11.5 million) by banks and insurance companies, and a minuscule .25 percent ($500,000) by individual investors. Many individuals made their purchases through banks and investment companies, of course, and the city tried issuing bonds in smaller denominations to attract smaller investors directly. But as Mayor Gaynor wrote to one concerned citizen in 1913, "We have in this city heretofore tried to get 'the people' to come forward and take the city bonds, but we have never succeeded in getting them to do so. When they have to compete and bid [against financial institutions] they have to enter into a fine, informed calculation as to how much they must bid in order to get the bonds." That sort of calculation was the specialty of professional investors. As a result, the only way to satisfy the city's huge borrowing needs was to work with big firms like J. P. Morgan, Kuhn, Loeb, and William A. Read. In large-scale municipal finance, public officials discovered, it was the institutional investors that counted.[20]

To create those necessary relationships with private sector institutions, the city had to put its finances on a "scientific" basis. Toward this end, municipal reformers and public officials enacted accounting and procedural changes to reduce waste and corruption, to render public financial management more transparent, and to insure that taxpayers and bondholders were getting a proper return for their money. And here, the classic elements of the reformers' approach to municipal finance can be seen.

Beginning in earnest during the Low administration and continuing through the Mitchel administration, New York City became the proving ground for cutting-edge fiscal reform as officials increasingly sought advice from the leading lights of scientific government: Frederick Cleveland, William Allen, and Henry Bruère—the intellectual nucleus of the Bureau of Municipal Research. Cleveland, a Midwesterner with an interest in municipal reform, had studied political science at the University of Chicago before receiving a Ph.D. from the Wharton School of Business in 1900. While at Wharton, Cleveland had become close friends with Allen, a fellow Midwesterner, who had been a student of Thorstein Veblen at Chicago and later a protégé of economist Simon Patton. In New York, Cleveland and Allen joined the Association for Improving the Condition of the Poor, where they worked with Bruère (another Chicago graduate and student of Veblen and political scientist Charles Merriam), who was active in Hull House and other reform organizations. When Cleveland and Allen established the Bureau of Municipal Research in 1907, Bruère became its first director. Thanks to a close working relationship between bureau experts and Comptroller Herman Metz, in 1908 the city adopted "the first budget in the United States based on a systematic classification of work being carried on by departments." Through their studies of how budgetary requests were made and how monies were actually spent, experts at the bureau and in the comptroller's office even began wresting control over expenditures from department heads (who exercised almost complete discretion over their budgets), forcing them to prepare detailed estimates that revealed where requests were padded and where savings could be made.[21]

Comptrollers played an especially important role institutionalizing these changes. As one of only three officials elected on a citywide basis (including the mayor and president of the board of aldermen), the comptroller had considerable power to reform municipal financial practices and put the city on good terms with capital markets. Comptroller William Prendergast (1910–17), with the help of his division of expert accounting, made a variety of administrative changes that saved money and increased accountability. By preparing tax bills promptly and collecting taxes biannually, Prendergast reduced short-term borrowing and saved approximately $1.5 million a year in interest charges. He made accounts-payable procedures more rigorous, reforming the "system of petty imposition practiced by clerks" and ending the "general pot approach" to municipal bookkeeping. These and other reforms impressed upon those who dealt with the city that the public treasury operated according to the same principles of financial accountability as the business community.[22] So deeply ingrained were these administrative changes by the end of the Mitchel administration—so thoroughly had the "technology of Government" been altered by successive reform-minded administrations and bureau experts—that even the return to power of Tammany Hall in 1917 could not seriously alter the new financial practices.[23]

Although there can be no question that these reforms (all designed to make the public sector more "businesslike") aimed to limit the downward redistribution of wealth through party politics, they reveal only one aspect of fiscal change in New York City, which in fact was focused as much on spending money as on saving it.[24] Creating a modern city with its multitude of capital projects required vast spending to overcome problems of congestion, disease, and economic conflict. Boston Mayor Nathan Matthews said it best when he noted that "the theory that the affairs of a city should be managed like those of an ordinary business corporation is attractive and widespread; but it is founded on the fallacy of supposing that a municipality is a business corporation. Municipal corporations are organized not to make money, but to spend it; their object is government, not profit."[25] The point of efficiency was not merely to reduce waste, fraud, and abuse but to clear the way for more active government: efficiency facilitated spending, even while it trimmed the fat. By the early twentieth century, proponents of this school of thought clearly had the upper hand in New York; they managed the city's affairs more scientifically to maintain good standing in the financial markets, which in turn allowed them to spend unprecedented amounts on a wide variety of projects.

This approach becomes all the more clear when we look at the other strategies that officials and their allies used in conjunction with their efforts to make public finances more scientific, such as searching for new methods of taxation and increasing assessed values, both designed to provide greater access to the abundance of private wealth in the city. Because real property taxes accounted for between 70 and 80 percent of revenues, spending was constrained by fluctuations in the real estate market and by the wishes of real estate interests.[26] As a consequence, Mayors McClellan, Gaynor, and Mitchel all established commissions to search for new sources of revenue, and these commissions drew on an extensive debate among experts in economics, planning, and municipal reform over the best ways to tax private wealth. For example, Nelson Lewis, chief engineer for the board of estimate, argued for the expansion of special assessments to fund street, sewer, subway, and water projects, even though the city used this approach far less extensively than it had in the nineteenth century. Other experts called for greater use of taxes on personal property, but as one tax commission concluded, "the personal property tax is a farce"—it was burdensome, inequitable, and produced only limited revenues.[27] In 1913, Mayor Gaynor's Commission on New Sources of Revenue recommended a tax on annual increases in real estate values, but only after considering a panoply of taxes and fees ranging from the municipalization of electrical conduits (for which the city could then charge rent) to licensing fees for electric signs and increased use of prison labor.[28] In 1914, Mayor Mitchel established a new committee on taxation to evaluate a modified single tax, in spite of the radical implications of any plan even remotely associated with Henry George. After two years of exhaustive analysis (which included research trips to western Canada to evaluate single-tax experiments in Winnipeg, Regina, Saskatoon, Edmonton, Leduc, Lloydminster, Ponoka, Calgary, Vancouver, Victoria, and Moose Jaw), the committee recommended the adoption of a habitation tax (based on the rental value of residential property), an occupation tax (based on the rental value of business property), and a salaries tax. With the addition of an increment tax on land values, this package was designed to function like a progressive municipal income tax.[29]

No one was more conspicuous in these efforts than Edwin R. A. Seligman, professor of economics at Columbia University, who epitomized the alacrity with which experts took up the cause of finding new ways to tax private wealth. The scion of an assimilated German Jewish family of bankers, between the 1880s and the 1930s Seligman became the nation's foremost authority (along with Henry Carter Adams) on public finance and a leader in a bewildering variety of reform causes. Among his other activities, he helped found the American Economics Association, Greenwich House, the City Club, and the Bureau of Municipal Research; he also organized tenement-reform projects, advocated the living wage and collective bargaining, and shaped numerous state tax-reform measures, federal income tax legislation, and the Transportation Act of 1920. His tireless commitment to reform grew out of a secularized religious impulse, later expressed through his involvement with Felix Adler's Society for Ethical Culture (for which he served as president from 1906 to

1921), and his academic training reinforced that concern for social action. Before receiving his Ph.D. from Columbia in 1885, Seligman studied in Europe, where, along with so many of the founders of American social science, he was influenced by German historicism, particularly the idea that economic theories were "inextricably interwoven with the institutions of the period."[30]

This awareness of the historical and institutional contexts of economic thought made Seligman critical of theories that emphasized the individual while neglecting the paramount role of group processes in modern society; as he noted in his 1903 presidential address to the American Economics Association, "to magnify the individual at the expense of the social group is to close our eyes to the real forces that have elaborated modern liberty and modern democracy, not in the backwoods of a frontier community but in the busy marts of commerce and the complex home of industry." Economic growth still provided the key to social progress, in Seligman's view, but the economy was no longer dominated by individual producers struggling against the natural environment; the modern economy was instead an "artificially created environment" and the product of collective action. To steer a course between quiescent conservatism and the vagaries of radicalism, therefore, Seligman insisted social scientists recognize that "the individual as he exists today has been hammered out by society, that individual ethics is the result of social ethics, and that individual progress is largely the consequence of social progress."[31] A recognition of social interdependence, rather than an outdated belief in economic independence, provided the starting point from which appropriate collective remedies for the problems of poverty and inequality could be developed while preserving individual liberty in the contemporary world.

This perception of economic interdependence had important consequences for tax policy. For Seligman, the ideal of social justice required that "each individual should be held to help the state in proportion to his ability to help himself"—those who had more should pay more.[32] There had been a time when taxes on land approximated this basic principle of progressive taxation, but the nature of wealth in the modern economy had changed; land no longer served as an adequate indicator of ability-to-pay, even in large cities where land values were exorbitant and land ownership was often concentrated in a few hands. As a consequence, "the changing conditions of economic life were making *income* rather than property the proper measure of wealth and of taxable capacity."[33] Although this belief made Seligman an opponent of Henry George's single tax (he publicly debated George in 1890) and thus an object of criticism for those who misinterpreted his opposition as political cowardice (an unwillingness to confront powerful land owners), this only masked his larger commitment to a conception of social justice achieved through forms of taxation more attuned to the conditions of modern life.

Indeed, when coupled with his belief in active government and his interest in incremental social change, the notion that ability-to-pay should serve as the criteria for taxation made Seligman a vigorous advocate of new ways to tap the great private wealth so evident in New York City, and the rapid growth of public expenditures after consolidation provided him with the ideal opportunity to press the case for more modern approaches to tax policy. Seligman watched with great interest as assessed values, yearly expenditures, and tax rates climbed during the McClellan and Gaynor administrations. As a director of the Bureau of Municipal Research, he helped guide efforts to make local government more efficient and to reduce waste even as it spent more. Rather than view rising tax rates as evidence of the need for retrenchment, therefore, Seligman saw them as the logical and necessary (although unintended) outcome of progressive government. "Toward the beginning of the second decade of the present century a new situation developed," Seligman recalled in 1919; "an unlooked-for tendency in the direction of vastly increased outlays on the part of both state and locality now appeared; the final fruits of a democracy which was determined to utilize government to achieve definite social ends disclosed itself."[34]

From Seligman's perspective, modern taxation should provide the resources that would allow local government to continue to pursue those "definite social ends"—from tenement reform to civic beautification. As chair of the executive committee of Mayor Mitchel's committee on taxation, Seligman defended that goal even in the midst of the city's most serious financial crisis since the fall of Tweed. With the real estate market in a downturn and the city under pressure from bankers to reduce spending, Seligman wrote to the mayor in 1915 urging that the city respond to the crisis by exploring new methods of taxation rather than by curtailing outlays. "I do not think that in view of the wealth existing in New York City we are spending too much money," he counseled. "I think on the contrary that the tendency of the present age is to demand a setting aside for public purposes of a continually larger fraction of the social surplus of the community. The trouble in New York has been that outside of real estate the wealth of the city has not even begun to be tapped. The problem is not to reduce desirable social services," Seligman insisted, "but to mobilize the resources of the community so as to make a development of these expenditures possible.[35] The trend of the age, in Seligman's view, was to treat more of private wealth as a public resource to be taxed away and redirected by government toward great civic purposes, even though the city's traditional approach to that goal—taxing real estate—was clearly showing its limitations.[36]

In spite of Seligman's best efforts, however, the mayor's committee was too beset by conflict to endorse new forms of progressive taxation. Real estate interests criticized Seligman's advocacy of new taxes (especially his support of an increment tax on land values), even though they appreciated his opposition to the single tax.[37] Although they, too, knew that municipal investment in streets, sewers, and subways enhanced property values, they were not prepared to accept higher taxes, and in this sense they remained unreconciled to the relationship between private wealth and public spending that Progressive Era government necessitated even while they benefitted from it: faced with the choice between less spending and more taxes, their opposition to enhanced revenues led to greater pressure to borrow.

Even more revealing was the criticism Seligman received from single-tax advocates. While Seligman embraced the greater complexity and interdependence of modern economic life, single-taxers like Benjamin Marsh, the planning advocate who in 1907 founded the Committee on Congestion of Population in New York, hoped that much higher land taxes would bring about a dramatic simplification of government and the economy.[38] For Seligman, institutional interconnectedness—the close relationships among bankers, insurance companies, real estate developers, taxpayers, and the public purse—was the necessary starting point for appropriate public policies, while Marsh and other advocates of the single tax still hoped that the city could be returned to "the ideal of Jeffersonian democracy" (to quote Henry George) by government control of land. Seligman argued for using those new institutional relationships to gain access to the resources necessary to ameliorate overcrowding, disease, and privation, whereas proponents of the single tax adamantly maintained that their plan would do away with those conditions once and for all, thus freeing government from the necessity of doing anything about them.[39]

In the face of these fundamental disagreements, officials were left to raise property tax rates, increase assessed values, or cut spending, rather than take advantage of other forms of taxable wealth.[40] Given both their philosophical commitment to spending and the multiplicity of demands for it, the Mitchel administration raised taxes (from 1.78 percent in 1914 to 2.04 percent in 1916), taxed real estate at a higher percentage of its full value (93 percent in 1917 as opposed to 91 percent in 1913 or 89 percent in 1903), and finally managed to reduce the budget by a measly .9 percent between 1916 and 1917 after three years of growth.

Far more successful than the search for new sources of revenue was the effort to raise the city's debt limit by increasing assessed values. Between 1898 and 1902, the city assessed real estate for the city as a whole at between 66 and 68 percent of its full value (as late as 1930, one-quarter of the cities in the state of New York were still assessing real estate at 60 percent of its full value). But in 1902, New York City raised its assessed valuation limit from 70 to 90 percent and with this one stroke increased its debt margin by $110 million (since the charter allowed it to borrow up to 10 percent of the assessed value of its taxable real estate). Between 1903 and 1928, real estate in New York City was assessed at between 88 and 94 percent of its full value. And between 1913 and 1918, some Manhattan property was assessed at more than 105 percent of its full value, with figures ranging from an average of 106 percent in 1916 to a remarkable 125 percent in 1917.[41]

Although boosting the assessed value of real estate immediately raised the debt limit, even this was not enough. In accordance with their new philosophy of borrowing, officials aggressively sought to exempt so-called revenue-producing investments from the constitutional limit. Water debt was exempted on the grounds that the system would eventually produce enough revenue to pay for itself (especially if a program of universal water metering was ever adopted; even though it was not, the exemption was retained). In 1909, the legislature passed another amendment exempting subway debt (but because of postwar inflation and the popularity of the five-cent fare, the subways were not even close to self-supporting).[42] These amendments, along with increasing assessed values, kept debt nominally within the constitutional limit, while allowing officials to expand the city's total borrowing. By 1913, net debt exceeded the 10 percent limit by $91 million; and by the end of Mayor Mitchel's term in 1917, the net debt exceeded the limit by $195 million. Even with the substantial growth in the tax base, therefore, officials had to tinker with the definition of debt to keep up with the demand for long-term borrowing.

Although the city's extraordinary record of expanding its borrowing capacity was due in large measure to the efforts of key officials (like the comptrollers) and their allies at the Bureau of Municipal Research to reform public spending habits, it would be a mistake to overestimate the control they established. They succeeded in making the budget process more honest and open, they centralized oversight of budget requests in the hands of experts, and they convinced the private sector that public monies were not being wasted. By removing or significantly reducing the threat of corruption, they made it possible for the city to borrow extravagantly, and in this sense the experts created a new problem for which they had no easy solutions.

Separating long-term borrowing from current spending posed one of their most serious challenges. As a matter of theory, a municipality should reserve borrowing for durable investments like school buildings, aqueducts, and bridges, spreading the cost over the useful life of the improvement. In practice, however, officials often used borrowing to fund short-term improvements or current expenses, rather than increasing taxes to pay for such items from current revenues. Under Seth Low, a reform mayor, the city floated forty-year bonds to pay for street-cleaning equipment—horses, harnesses, cans, burlap bags—that lasted nowhere near forty years. The city also used forty-year bonds to fund the repaving of streets, even though pavement usually lasted less than a decade and was considered a yearly maintenance item in most cities. As commissioner of accounts, future mayor John Purroy Mitchel once discovered that proceeds from long-term bonds had been used to purchase carbon paper. And while the docks department did have durable investment needs, in practice it used long-term bonds to pay for maintenance and operating expenses. This method—employed by reform as well as Tammany administrations—hid the current cost of government by shifting it to future taxpayers through the bond market.[43]

Democratic mayors McClellan and Gaynor made great efforts to limit the use of long-term borrowing to underwrite current spending—with only mixed success, however. In 1913, for instance, Comptroller Prendergast had to tell Fire Commissioner Joseph Johnson that the new policy required his department to postpone its corporate stock requests. The commissioner was understandably dismayed: "Permit me to say, and I hope without officiousness," he wrote to Prendergast, "that there may be more important subjects for expenditures of Corporate Stock than the Fire Department but none so urgent." Prendergast was impressed with the Johnson's "energy in securing appropriations for his department" and decided to call in the mayor to put the commissioner in line with the new procedures. Gaynor complied, but not without some discomfort. "I have made him understand," he reported to Prendergast, "and I suppose I will have to make all the heads of departments understand, that they will have to be very careful about asking for corporate stock *for a while.*"[44] A radical effort to reform borrowing practices would put additional pressure on tax rates or assessed values—something no politician wanted to do, especially when both were already creeping upward after 1907.

For all their efforts at reform, therefore, the experts in public finance who played such a large role in city government between consolidation and the war were unable to develop a system for sorting through the multitudinous demands on the public purse they confronted, especially where public borrowing was concerned. Eliminating waste and inefficiency was a relatively easy matter, but accomplishing that worthy goal only exposed the more fundamental problem of public finance: overwhelming, legitimate public demand.

Henry Bruère captured the essence of the problem in 1909 when, after it was revealed that the debt margin was (once again) too small to accommodate a rapid transit contract, he took a close look at what the city had purchased with its borrowed money: "The items range from nearly one hundred and three millions for schools (in order of expenditure, through docks, water, bridges, parks and roads, subway, museums, armories, hospitals, etc.) down to two and one third millions for public baths. What a tale these figures tell," he exclaimed, "described from one point of view as 'progress toward bankruptcy' and from the opposite view of public needs as 'progress toward civilization!' " Bruère did not blame the problem on Boss Murphy or Tammany Hall or the ten of thousands of immigrants streaming into the city. Schools were important, and still there were not enough of them; bridges were necessary for the city's future, as was the subway, which had (incomplete though it was) already "revolutionized" the city's transit problem; and parks were one of the few things that made the increasingly crowded city livable. Indeed, as Bruère so cleverly put it, the road to a new urban civilization seemed to lead simultaneously to bankruptcy. Rather than point the finger at the boodlers, grafters, and ward heelers, therefore, Bruère condemned the woeful lack of planning that characterized the city's spending habits, observing that

> New York has followed no systematic plan of development in the use of its credit resources. For example, where local pressure has been brought to bear, parks have been bought without reference to any city plan. Here and there, in a haphazard way, sites have been acquired, many of them now awaiting development for lack of funds. Each section of the community, each division of the city government, has competed without restraint for authority to embark on improvements payable out of corporate stock.

Local and special interests dominated the borrowing process, just as they had in the preconsolidation spending frenzy, pressuring politicians to distribute each new sliver of debt margin without reference to any citywide plan for using collective wealth to achieve collective goals. "The next most pressing need with reference to the continued growth and increased well-being of the city is the formulation of a *city plan*," Bruère insisted, "based upon a systematic, scientific study of local social and economic conditions, for providing equipment for education, health, recreation, transportation, and commerce." The point was not to stop borrowing (since Progress demanded it), but to go about borrowing in a more systematic way—a way that made "intelligent use of future municipal credit for civic advancement," rather than dividing it up as a political prize for parochial interests.[45] Many of those interests were quite legitimate and would ultimately find their place in a proper course of public investment, in Bruère's view, but the process by which they were satisfied pushed the city toward financial crisis and prevented it from borrowing the huge sums needed to pay for projects like the subway that could, if completed, positively affect the whole city.

George McAneny, New York's foremost advocate of city planning, likewise concluded that the incessant clamor for local projects made it difficult for political leaders to focus on the bigger picture. As the mastermind behind the dual-contract subway deal, McAneny stood foremost among those who called for local government to spend liberally for the good of the city as a whole. "I do not think we need to be afraid to encourage governmental activities of the kind that some good people are pleased to consider 'paternalistic,'" he insisted in 1914. "The city must not only do its share toward supplying fresh air and pure water and recreation and education; it must even bestir itself to promote economic improvements that will make it easier, at least, for willing workers to earn their livelihood." These physical undertakings were of "transcendent importance" in the life of the city and "vastly more extensive and important than they were in past times because of the acceptance of new theories as to the proper functions of government." The modern metropolis was itself an expression of a philosophy that demanded that great private wealth should be directed toward even greater public purposes. However, the projects that embodied this philosophy had to compete with the "pet schemes" of local taxpayers' associations, which appeared with such regularity and in such profusion on the board of estimate's public-improvements calendar that the city would quickly bankrupt itself (and in fact nearly did) if it acceded to all of them. McAneny held out the hope that his fellow board members would be guided by "their responsibility for the city at large" since there were "so many grave problems of city-wide interest always demanding their attention that petty local considerations are apt to have little weight with them." Practically speaking, however, those petty local considerations ate away at the city's limited financial resources in spite of the philosophy that facilitated increased access to taxable values in the name of the city as a whole.[46]

This problem boiled down to the way the city formulated the public interest. Political parties and local officials tended to approach the problem of the public interest as an exercise in interest aggregation—moving from parts to whole, responding primarily to individual, sectional, ward, neighborhood, and borough demands. As long as the city could spend and borrow more, more of these interests could be accommodated, even though, as Bruère observed, their satisfaction did not add up to any systematic response to the needs of the city as a whole (as he conceived them). Alternatively, Bruère and McAneny approached the problem of the public interest in a disaggregative manner, beginning first with a plan for the city as a whole and working from the whole to the parts. Individual projects would be funded when they achieved a predetermined public objective—whether improving transit, water, sewage, parks, schools, or bridges. In the absence of that approach, even the remarkable administrative reforms achieved by the experts from the Bureau of Municipal Research only helped to squeeze more out of a fundamentally flawed approach to resource allocation, leaving the basic process of sorting through public demands beyond expert control. As a result, large needs were still unmet (such as the sewage problem), while millions were spent on half-completed, ill-considered projects.

Thus, it became increasingly clear in the years before World War I that financial reform had succeeded primarily in making it possible for the city to live beyond its means. True enough, the experts had institutionalized a new set of administrative practices, dramatically improving the city's financial management and thereby opening the way for increased spending and borrowing. But the larger problems of finding new sources of revenue and of prioritizing the plethora of demands for public spending remained unresolved. The city did have planning advocates, but not yet a plan that would provide the centralized control that experts like Bruère hoped would bring greater rationality to the process of managing incremental increases in collective wealth. The consequences of this lack of control became more evident every year; because of both the volume of borrowing and increases in interest rates, the city faced mounting levels of debt service in its annual budget: in 1898, 15 percent of the budget was devoted to repaying principal and interest on the debt; by 1910, the percentage increased to more than 26 percent. As one student of the city's finances scolded, "There is something radically wrong in the fact that out of every 100 dollars for current administration, the debt service alone absorbs $26.65." And it grew even more. By 1917, debt service accounted for 29 percent of the budget—with 21 percent devoted to interest alone (and this in an era when current expenditures as a whole were on the rise).[47] This glaring fact, coupled with the stalemate over new sources of revenue, would play an ever greater role in shaping the policies that officials could pursue, thus slowing progress toward a new era of public finance for Greater New York.

Fiscal Crises and the Return of the Pay-as-you-go Philosophy, 1907–1918

If the city could not control its borrowing or tap new sources of revenue, in spite of its improved financial management, then it exposed itself to control by the private-sector institutions that bought and sold public debt. Because the city tended to borrow faster than its debt base grew and because debt service comprised an expanding share of ever larger annual budgets, bond dealers might easily conclude that even the experts from the Bureau of Municipal Research, working with respected city comptrollers, could not restrain the politicians (reform or machine) who so readily parceled out the city's borrowing capacity. Of course, the problem faced by the McClellan, Gaynor, and Mitchel administrations was clearly different from the one created by Tweed; here, there was no large-scale corruption to justify drastic measures against the city. The problem was more intractable and therefore less easily resolved than in Tweed's day. Given the fact that tax rates and assessed values were already on the rise, that the demand for borrowing had not diminished, and that major projects like the subway had only just been started, public officials were left with few choices but to look for new ways to tax the considerable private wealth that existed in New York City, hope for continued growth of the tax base, or risk turning over the management of the city's finances to the private sector.

The first sign that the city might lose control over its finances came during the bank panic of 1907, brought about though it was by factors completely beyond the reach of public officials. Throughout 1907, stock prices had fluctuated under pressure from speculators, dropping precipitously in March and June and again in October. Credit was very tight, due in part to the drain on insurance companies caused by the catastrophic San Francisco earthquake the previous spring. As a result, two city bond sales (one for $29 million in June and another for $15 million in August) had been "flat failures"; the 4 percent offered on New York's long-term notes could simply not compete with 6 percent short-term interest rates or industrial securities selling at 7 percent or better. The city even began paying its creditors in corporate stock rather than cash, as Comptroller Herman Metz traveled to Europe to find buyers for New York debt. The situation took a turn for the worse in mid-October, when F. Augustus Heinze, a speculator who had recently gained control of the Mercantile National Bank, failed to corner the market on United Copper stocks; to cover his obligations, Heinze had to pay out all of his and the bank's assets, rendering him penniless in two days and unable to pay back Mercantile's depositors. From there, the panic spread to the Knickerbocker Trust Company, whose president, Charles T. Barney, had been associated with the United Copper stock scheme. Depositors at other trust companies began withdrawing their money for fear that the speculators had already gambled it away. In the absence of a central bank to provide the cash necessary to stop the bank run, the New York financial community turned to the venerable J. P. Morgan, now seventy years old and semiretired, who made an early departure from an Episcopal convention in Richmond, Virginia, to attempt to forestall a full-scale collapse. For two weeks, Morgan led an ad hoc recovery effort, standing by while the most culpable trust companies closed their doors, persuading banks with greater cash reserves to make available the millions necessary to bail out those he thought it was possible to save. Working with a brains trust composed of James Stillman of National City Bank and George F. Baker of First National Bank (along with a team of the city's leading younger bankers to carry out their orders), Morgan also raised $27 million (in less than fifteen minutes) to prevent the New York Stock Exchange from closing.[48]

It was in this atmosphere of very scarce money and terrible uncertainty that New York City needed to raise $30 million to cover payroll and other expenses. With short-term loan rates at the stock exchange as high as 150 percent, there seemed to be no prospect of the money markets providing the necessary funds at reasonable interest. With the city's credit at stake, Mayor McClellan turned to Morgan. A month earlier, Morgan's company had agreed to bid on the largest bond issue in the city's history—$40 million worth—and that guarantee had so boosted confidence that the sale was oversubscribed (with 960 bidders—800 more than on any previous bond sale). Now, with the markets in a far more desperate condition, Morgan himself arranged for a banking syndicate to purchase the entire $30 million issue at 6 percent interest (which, although more than the city usually paid, was considerably less than the 10 percent Morgan charged the stock exchange, and an absolute gift in view of short-term rates), even though there was almost no chance of selling 6 percent bonds to anyone in the immediate future.[49]

Thus Morgan, with the whole financial community following his lead, saved New York's credit. The city's budget for 1908, which was being adopted at the very moment that Morgan was putting together the bailout package, had to be "cut to the bone"—increasing by a mere $13 million over the previous year—on advice from financiers that municipal bonds would not sell without such austerity; Morgan also required that the syndicate monitor the city's book-keeping practices.[50] However, when compared with the collapse of Knickerbocker Trust and the near collapse of the stock market, Morgan's rescue of New York was just a sideshow in a much bigger picture that had very little to do with the city's financial situation per se. In spite of the high drama of the moment and the temporary budgetary restraint it required, the city's borrowing practices were unaffected by the panic or the bailout. By January, Comptroller Metz was calling for a higher debt limit, and in February a $50 million bond sale—again, the largest in the city's history—was oversubscribed by $250 million. The return of "liquid capital" made the *New York Times* positively giddy, inspiring it to hail the "promise and potency of the subways, Catskill Aqueduct, bridges, bridge approaches, public libraries, piers, and other things" made possible by municipal borrowing. "Later on it may become necessary to preach prudence again," the paper sagely counseled, "but just now judicious expenditure is almost an equal duty."[51] With Barney dead by his own hand, dozens of careers in ruins, and greatly renewed interest in a central banking system, it seems fair to say that the city emerged from the 1907 crisis virtually unscathed, even though it was quite clear that its enormous borrowing needs (both short-term and long-term) made it vulnerable to the very institutions that had saved it that October.

That would become painfully clear in 1914. The fiscal crisis of that year was an altogether different matter, even though, like the panic of 1907, the events precipitating this new crisis were largely unrelated to the city's finances and financial practices per se. In fact, by 1914 the city managed its finances better than ever before; and with the electoral triumph of Fusion mayor John Purroy Mitchel in 1913, the right people were certainly in place to take administrative efficiency and financial reform even further. But the pace of municipal borrowing had not slackened—the city's gross debt had grown from $730 million to $1.124 billion (an increase of 46 percent)—and the recent adoption of the subway dual-contract signaled that there would be even more borrowing to come (see graph 4 in the appendix). Between 1908 and 1913, the budget had grown by $49 million (34 percent), and both tax rates and assessed values were higher than ever before, in spite of the recent downturn in the real estate market. All around there was evidence of municipal progress made possible by monies wisely spent; but as Bruère observed, the city also looked like it was spending its way toward bankruptcy without any significant control over that ominous trend. The crisis of 1914 provided private-sector actors with the opening to impose just such control from the outside; and unlike the crisis of 1907, this time they took full advantage of the opportunity.

As in 1907, the city's need to borrow huge sums of money made it highly dependent on capital markets. Between September 1914 and January 1915, the city had to redeem $77 million worth of short-term notes (fully half of American obligations abroad) that it had sold in London and Paris when interest rates there were lower than in New York (one of the ironies of this crisis was that the city's vulnerability arose from a smart financial practice—selling bonds at the lowest interest rate possible). Normally, the city simply purchased foreign currency to pay its debts, but the outbreak of World War I in July had resulted in a severe shortage of exchange and exorbitant rates as European creditors demanded immediate repayment of their holdings; by early August, the price of the English pound had risen from a prewar rate of $4.87 to as high as $7, making it inconceivable that the city could buy the necessary foreign currency. To make matters worse, the British were now calling for repayment in gold; to raise $77 million in gold at any time would have been difficult, but in August 1914 it seemed like an utter impossibility. Faced with this grim situation, Mayor Mitchel considered the unthinkable: defaulting on the city's obligations. But before taking that drastic step, Comptroller Prendergast turned to J. P. Morgan Jr. (the elder Morgan had died the year before), now head of the company, and he, like his father seven years earlier, "determined to do everything in his power to save the credit of the City of New York." Morgan was aided in that effort by bankers Arthur Anderson, Dwight Morrow, and Henry Davison, who had helped execute the orders of Morgan Sr. during the crisis of

1907. For these men, all of whom were thoroughly familiar with the city's financial condition, saving New York's credit had profound practical and symbolic implications. "The failure of the greatest city in America to meet its obligations would only have added to the world confusion," wrote Thomas Lamont, Davison's biographer and later chairman of Morgan & Company; "moreover, it would have dealt an almost irreparable blow to the credit of New York City, as well as cast discredit upon all the United States." New York bankers were poised to play a bigger role in international finance—a step they could not take with much credibility were they unable to keep their own city afloat, since the aspiring leaders in world finance could not operate from a bankrupt metropolis.[52]

With so much riding on their actions, Morgan, Anderson, Davison, and Morrow put together a syndicate of 126 New York banks to purchase $100 million in short-term notes from the city, for which they would surrender $80 million of their gold reserves, thus allowing the city to cover its foreign debt. All but three banks in the city participated, each paying into the pool in proportion to their gold holdings, with contributions ranging from a mere $3,005 from Tottenville National Bank to an impressive $7,800,980 from City National Bank. Morgan & Company covered a substantial portion of the first installment to British banks in Canada in mid-September with their own gold. The only profit on the deal came from the final transaction's amounting to only $78 million (due to a decline in exchange rates, partly in response to the syndicate's action); the $2 million difference was split between the city and the syndicate, with Morgan receiving a meager $11,121; the banks also refused to charge the city their customary 0.5 percent commission. Saving New York's credit was thus a collective enterprise, led by Morgan and executed largely by Davison, but truly the work of an entire community of banks extending themselves to rescue the city with which they were so closely associated.[53]

Although the Morgan syndicate did not charge much for this enormous transaction, it exacted a heavy price through the other terms of the deal. In exchange for their services, the banking syndicate effected what amounted to a counterrevolution in the city's borrowing practices. As Morgan wrote to William Gibbs McAdoo, then secretary of the treasury, the New York banks would undertake the transaction "only upon the understanding that the finances of the City are to be conducted upon a policy different from that followed during many years; that the new lines should lead to the stopping of debt increase and gradually, by additional taxation, to a reduction of the debt."[54] Going well beyond the temporary cutbacks of 1907, Morgan's syndicate ordered the city to abide by a strict pay-as-you-go policy, issuing long-term bonds only to pay for revenue-producing improvements like the water supply and subway systems; all other improvements (even long-term but non-revenue-producing investments like schools, libraries, roads, and sewers) would have to be defrayed out of current tax revenues—thus ending the "charge-it-and-pay-tomorrow" approach that accounted for so much of the debt.[55]

The bankers' stringent new requirements were an expression of their belief that New York simply could not afford everything it wanted. They acknowledged that the city had made important reforms in its financial management. They also affirmed the value of some of the non-revenue-producing expenditures it had made—like constructing new schools. But other items, though desirable, they considered luxuries—such as public libraries, parks, and the new courthouse. They objected not to the nature of the expenditures themselves—like Bruère, they saw them as progress towards civilization—but to the pattern of spending and borrowing that the expenditures exemplified. This was not like the financial crisis that ended Tweed's reign: the bankers admitted that the city's finances were scientifically managed. But as Arthur Anderson insisted, "we want to see New York conservative, not merely scientific."[56]

Here, then, was the crux of the problem faced by the city and the reason that actors outside the public sector chose to intervene in the city's affairs: in spite of the best efforts of the most knowledgeable minds in budgetary and administrative reform, New York had succeeded only in becoming more expert at managing its money (more efficiency, more honest, more transparent; "clarity, simplicity, and accountability," the three essentials of proper budget making, had been achieved), rather than gaining control over the rising level of spending itself. Indeed, becoming scientific allowed the city to become far more liberal than it ever had been. Since 1903, when the city dramatically raised its borrowing limit, it expended money on a growing array of projects and at a rate faster than the growth of the debt base. The city had devised ways to keep net debt within the constitutional limit, but that hardly fooled the bankers, even as they rushed to buy and sell the city's debt. From an economic standpoint, the city could not continue to borrow at the pace it had kept between 1903 and 1914. Morgan, Anderson, Morrow, and Davison wanted the city to slow its borrowing in order to let the debt base grow, though they and their colleagues in the investment community had played a major role in allowing the city to get as far into debt as it had (such as guaranteeing bond sales to insure there would be enough buyers). To accomplish this, they required the city to shift funding of non-revenue-producing improvements from the bond market to the annual budget. This meant either that taxes would have to be raised to fund improvements, that the improvements would have to be postponed, or that some other current expenditure would have to be reduced to make room for the new improvement.[57] In other words, the city now had to make the hard decisions that increasing financial and administrative efficiency had allowed them to avoid.

The return to the pay-as-you-go philosophy had profound implications for the city's finances, bringing to a halt the borrowing trends that had been building for two decades. Although the city began "applying the brakes vigorously," as the city chamberlain, Bruère, observed, the pay-as-you-go philosophy was a rather clumsy and unsophisticated approach to public financial management, in spite of the fact that it was imposed by some of the most sophisticated bankers in the world. As Mayor Mitchel told Senator Elon Brown, who used the crisis to scold city officials for not adopting the pay-as-you-go approach when Greater New York was first created, "those seventeen years [since consolidation] were the years of the greatest physical growth of the city, and it was necessary to build and enlarge the municipal plant. This could not have been done under the pay-as-you-go policy. If that policy had been in effect, the city should be in splendid financial condition today, but it would not have its present municipal plant." And Mitchel knew full well what it meant to adopt the policy now: the Brooklyn Marginal Railroad would probably not be built; the city could not invest in a sorely needed bridge across the Hudson River; park space could not be purchased; and a multimillion-dollar sewage-treatment system, only recently urged by the Metropolitan Sewerage Commission, would have to be indefinitely postponed.[58] More important, the city would undoubtedly have to raise taxes to keep up with all yearly expenditures. The days of borrowing to cover current costs—and in fact of progress toward a new approach to public finance—were over (or so it seemed).

The Battle against the Pay-as-you-go Plan and the Road to a New Crisis

Although the pay-as-you-go policy was informally followed by the board of estimate in 1914 and formally incorporated into the city's charter in 1916, Mayor Mitchel succeeded in postponing its full implementation until the end of his term; it thus fell to Tammany Mayor John Hylan to face the real challenges of renewed fiscal conservatism. No sooner had the ballots been counted then Hylan's comptroller, Charles Craig (who would very shortly begin a running battle with the mayor over how the city's money was spent), announced that one of the first orders of business for the new administration would be to abolish the pay-as-you-go plan: by 1917 it was clear that fiscal conservatism in an age of increased expectations for municipal spending meant higher taxes. The public still wanted all sorts of non-revenue-producing amenities, but under the pay-as-you-go plan they could be secured only through current spending. Craig feared tax rates as high as 2.5 or even 3 percent, which would be fatal for the city's economy, not to mention its politicians. Not only would Tammany have a much harder time adding to the payrolls (after four long years of Fusion cost cutting) and keeping its campaign promise for new schools, it would also face the prospect of both delivering fewer services and increasing taxes.[59]

Craig and Hylan were right, of course: the pay-as-you-go plan forced them to place a much greater burden on taxpayers than any preceding administration, and this while curtailing capital improvements. Under the plan, gross debt increased by a mere 6 percent and net debt grew by only 4 percent (in nominal terms) between 1918 and 1921, even though over the same period assessed values grew by 20 percent. As a consequence, gross debt dropped from 14.69 percent to 12.96 percent of assessed values. The pay-as-you-go approach thus succeeded in reducing the relative size of the municipal debt, just as the bankers had wanted. Over the same period, the budget grew by 45 percent (in nominal terms), while increasing as a percentage of assessed values: from 2.86 to 3.47—the highest in the city's history. In per capita terms, gross and net debt held nearly constant, while the budget grew from $44 to $61. The tax rate likewise rose to record levels; in 1918, rates ranged from 2.36 percent in Manhattan to 2.46 in Richmond, and by 1921 the rate throughout the city reached 2.77 percent.[60] As Craig had predicted, the mounting tax rate and steady upward pressure on assessed values squelched any prospect of growth. The increased tax burden also induced mortgage foreclosures and tax-lien sales, thereby pushing down property values and creating the very real prospect that the city would exceed its debt limit without ever incurring any new debt. So great was the threat to the city's financial stability that Governor Whitman signed a bill in May 1918 suspending the pay-as-you-go plan until one year after the end of hostilities in Europe (even so, the city was pledged to use its borrowing power only for investments required by the war).[61]

This very temporary reprieve protected Hylan for only a few months, and for the rest of his first term he was under constant pressure to make difficult budgetary choices. Every salary increase for city workers translated into higher taxes, for which Hylan was berated by good-government groups (who viewed any budget increase by a Tammany mayor with deep suspicion, even when they were bona fide attempts to keep up with inflation) and by rival politicians like Fiorello La Guardia (the Republican candidate for president of the board of aldermen in 1919, who delighted in pointing out that while Hylan had campaigned on a promise to reduce spending, he instead had given the taxpayers "the biggest budget in the history of the city, the size of which is without parallel in any city in the world"). At the same time, La Guardia and others chastised the mayor for allowing the school system to languish; with no new buildings constructed during Hylan's first two years (since they would have to be paid for entirely out of current spending), the schools were now so overburdened that some fifty thousand students were taught in classrooms with more than fifty children and a quarter of a million students attended only part time.[62]

It was the emerging crisis in the school system that illustrated just how crudely the pay-as-you-go formula worked in practice. When Comptroller Craig argued, as he did every chance he got, that it was inappropriate to prevent the city from borrowing for durable but non-revenue-producing investments such as school buildings, he was making precisely the same argument that Mayor Mitchel had used to defend the city's record of indebtedness: officials had to build up New York's physical plant to keep up with population growth, to insure the city's competitiveness, and to render the city more livable. Of course, the Hylan administration did not help its cause when it tried to sell corporate stock to pay for current expenses and thereby lower the tax rate; but the larger point—that it made no sense to require roads, libraries, schools, and sewers to be defrayed entirely out of current spending—was still valid.[63] By 1920, Craig's pleas for a more reasonable approach to debt limitation were sounding less like a Tammany plot and more like a necessary modification of an excessively restrictive policy. In the face of mounting pressure for new schools, the city obtained an amendment to the pay-as-you-go policy to begin a $170 million construction program—and as one observer noted, "the door, having been unlocked, was now opened." Other constituencies began to argue for their own exemptions from the pay-as-you-go plan: bonds were granted for a municipal building in Brooklyn, for the American Museum of Natural History, for the Metropolitan Museum of Art, for the Brooklyn Museum of Arts and Sciences, and for street paving and waste-disposal plants; the police commissioner received money for a new garage and repair shop under the Williamsburg Bridge; the fire commissioner secured funding for new stations and motorized fire trucks in Queens; the board of education received money to address fire-code violations in public schools; and the department of public welfare was granted funds for the Cumberland Street Hospital in Brooklyn and a new hospital in the Bronx. By 1923, Craig, who had wanted a more sophisticated approach to holding the line on borrowing, charged Hylan with "frittering away" the city's debt margin. Thanks to pressure for museums, schools, and hospitals from upper- and middle-class groups, for new facilities and updated technology from department heads, and for local infrastructure from borough presidents and neighborhood groups, the pay-as-you-go policy was well on its way to becoming a dead letter.[64]

By the mid-1920s, in fact, prewar trends in spending and borrowing had nearly reasserted themselves. Between 1922 and 1929, gross debt increased by 50 percent, with yearly growth rates only slightly smaller than those during the McClellan and Gaynor years. Real and nominal debt per capita also resumed their upward climb; indeed, per capita debt (gross and net) grew by more between 1920 and 1929 than it had between 1900 and 1914. New Yorkers were again devoting nearly thirty cents out of every tax dollar to debt service, making it by far the single largest item in the budget. The only thing that kept the city from exceeding its debt limit was the phenomenal growth of assessed values, which increased by 67 percent between 1922 and 1929. Even though the city borrowed an additional $650 million during that time, thanks to a boom in the real estate market—this was the period during which so many landmark skyscrapers (e.g., the Chrysler Building; the American Radiator Building; the Paramount Building) were built in Manhattan and new suburban areas sprouted up along subway routes in Queens and Brooklyn—the debt base grew even more, making it appear that the city was not again rushing headlong toward a financial crisis. And because the return of prosperity made it easier for real estate owners to pay their taxes, it no longer seemed so burdensome to live with tax rates that remained near 2.7 percent and budgets that continued to run at 3.3 percent of assessed values.[65] In spite of the resurgence of the trends that compelled J. P. Morgan Jr. to impose the pay-as-you-go plan in 1914, the city appeared to be growing its way out of financial danger.

The city was aided in this effort by the state, thanks to the taxation experts who had continued their search for additional sources of revenue. Although he could not persuade his colleagues on Mayor Mitchel's committee to impose new taxes on the city, Seligman nonetheless continued to push for more modern and effective ways for government to access private wealth. As soon as it became clear that the committee would be deadlocked, Seligman joined forces with State Senator Ogden Mills, then chairman of a legislative committee on taxation, to endorse a state income tax, which, when adopted in 1919, amounted to, in Seligman's words, nothing less than "a real fiscal revolution." By 1926, state taxes comprised more than 8 percent of New York's yearly cash intake, and the state began pouring money into the city for education (state aid for public schools increased more than 800 percent between 1915 and 1926). By 1932, the state contributed $90 million to the city's $715 million annual budget.[66] In this sense, the piecemeal destruction of the pay-as-you-go plan and the introduction of new revenue sources represented an improvement over the philosophy of retrenchment imposed in 1914.

Of course, the experts always admitted that fiscal conservatism did not provide a sufficiently sophisticated approach to managing the city's money, and even bankers began calling for better public investment strategies. By the mid-1920s, a chorus of experts called for countercyclical government-spending policies to ameliorate the effects of the business cycle—ideas that would finally find a receptive national audience in the mid-1930s when the federal government turned to "pump priming" to get the country out of the Great Depression. Along with economists like Wesley Mitchell and Mary Van Kleeck at the National Bureau of Economic Research and the Russell Sage Foundation, they helped make the case that public debt, when properly deployed, was a vital tool for modern urban managers.[67] However, it would be an exaggeration to claim that the passing of the pay-as-you-go plan in New York represented the triumph of expert fiscal management, since the city's borrowing can hardly be described as countercyclical. To be sure, under Mayor Jimmy Walker the administration did try to develop a city plan that could guide capital expenditures, but that effort failed; Walker also tried to channel borrowing toward sewage treatment, only to have the depression call a halt to capital spending. At the same time, Walker presided over a resurgence of the machine, which resulted in an ever larger portion of the budget going toward patronage. But because the city remained within its constitutional debt limit, even the banking community did not protest the return of the city's old spending habits, preferring instead to buy and sell debt without much comment on the haphazard process of allocating each new chunk of borrowing capacity. It was, in the end, the growth of assessed values, rather than a more rational approach to public finance, that postponed the next fiscal crisis until 1932.

Indeed, thanks to the rise of assessed values, New York appeared to have a fundamentally sound fiscal structure, even while it spent its way toward bankruptcy. The budget increased through 1932, with the Walker administration dumping millions into salaries and hoping that revenues would not diminish. By January 1932, however, the depression caught up with real estate in New York, and tax receipts fell precipitously. As the financial situation deteriorated, bankers found it more difficult to sell the city's debt to an increasingly nervous public, at which point they began to press for fiscal restraint. As Columbia University professor Lindsay Rogers noted, "so long as New York City's obligations could be sold the bankers were not greatly concerned over whether the city was living beyond its means." As the market soured, the city's creditors found themselves wondering whether they would ever get their money back; when accumulating tax delinquencies prevented the city from covering its obligations in December, the bankers had the leverage they needed to impose fiscal austerity. In September 1933, the heads of J. P. Morgan, National City Bank, Chase National Bank, and others concluded an agreement with Governor Herbert Lehman, negotiating in loco parentis, that imposed a four-year plan of budget cuts on the city (the chief purpose of which was to make sure that the banks were repaid for their earlier loans) and effectively placed the private sector in charge of the city's finances—for the third time in Greater New York's brief thirty-six-year history.[68]

The Uses and Limits of Collective Wealth

In 1950, the board of estimate authorized a management survey of the City of New York—the sort of administrative and financial analysis that had been pioneered by the Bureau of Municipal Research and had now become standard practice in local government. The financial survey team was directed by the distinguished Robert Murray Haig—Seligman's protégé, chief investigator for Mayor Mitchel's tax committee (it was Haig who traveled to Moose Jaw to research the Canadian experience with the single tax), author of the metropolitan economic survey for the Russell Sage Foundation's *Regional Plan* and a man with four decades of perspective on the city's financial problems. Haig, now in the role his mentor had played during the Progressive Era, noted in 1952 that there had been several important developments in the city's finances between the 1920s and the 1950s. Most obviously, the city had undergone a "radical change" in its pattern of expenditures: New York's "primary function, judging from the amount of money spent, is the care and rehabilitation of those who have been stranded economically through unemployment, illness, physical or mental, or old age," for which the city spent $309 million, or about 31 percent of its yearly budget. Next in line was education, which consumed $269 million. "The traditional core of city functions that is usually thought of when municipal government is mentioned—fire protection, police, sewage disposal, and street construction, maintenance and cleaning—accounted as a group for only $294 million." Partly because of wartime shortages and partly because property values had not increased in two decades, officials had also significantly curtailed capital spending (which gave them another excuse to tell the Interstate Sanitation Commission that the city was still too poor to invest in sewage-treatment plants), while deferring maintenance on existing facilities, particularly the subways. Local government, in other words, had reduced its role as city builder and had become a social service agency.[69]

Thankfully, Greater New York had not taken on the burden of social welfare alone. The city still raised most of its revenue from local sources, but it had diversified its tax base: property taxes accounted for 48 percent of revenue, down from 85 percent in 1922; but now one-fifth of revenues were generated by sales and business taxes, which had not existed thirty years earlier. Even more significant, grants from the state and federal governments provided the city with $308 million, or 28 percent of yearly revenues. Although most were earmarked for education and welfare, the state contributed around $50 million each year for general purposes. Far more than ever before, New York City now relied on intergovernmental transfers to make ends meet.[70]

On the other hand, not enough had changed with regard to the way the city spent its money. "One of the most serious deficiencies in the city's financial administration has been the failure to look ahead, to study present trends and possible future developments and their implications for the city's financial future," Haig observed. "The result is that the city lives from hand to mouth, and each year's budget presents a crisis, the magnitude of which usually turns out to have been largely unforeseen." Capital budgeting had been written into the charter in 1936 to remedy this problem, but the process had not been taken seriously enough. Consequently, Haig recommended the creation of a new bureau of the budget that would work with a new tax-research unit (whose job it would be to find additional sources of revenue) and the city planning commission (which was supposed to be managing land uses and their accompanying infrastructure needs). Or, as Henry Bruère had urged forty-three years earlier, officials should develop a city plan to guide the way they allocated their scarce resources.[71]

The city likewise continued its struggle with the constitutional debt limit. Arguably, New York was better off than it had been in a decade where public borrowing was concerned. Real per capita debt was down by almost half, since net debt had fallen slightly (by about 10 percent, or $250 million), while population had increased and prices had jumped. Although nominal debt per capita was still larger than it had been in the 1920s, no bonds had been sold since 1945, a pay-as-you-go plan had been instituted in 1938 (with numerous exemptions, of course), and certain types of public borrowing were now done by specialized agencies, like the Port Authority and the Triborough Bridge Authority, and were thus not included in the debt-limit calculation. Even so, "New York City has been, and is now, fairly close to the full utilization of its debt limit," Haig noted; as of July 1, 1951, the city was within $260 million of its limit (with gross debt running at just over $3 billion and net debt at $2.3 billion), of which only $18 million had not already been committed to planned projects.[72]

Frederick Bird, who investigated the city's debt problem for Haig, concluded that this tendency to flirt with the debt margin was, unfortunately, nothing new: "This conflict between constitutional borrowing margin and capital requirements has been more or less chronic over the entire period since the consolidation of the greater city. It has been alleviated from time to time by upward adjustment of assessed valuations, constitutional amendments, judicial interpretations, sporadic pay-as-you-go plans, and capital undertakings by special authorities; but this combination of devices has failed to produce a lasting solution to the problem and presently the city is seeking additional restrictions on the incurrence of debt."[73] And yet there was so much more to be done. "The most difficult problems with respect to New York City's debt will develop during the coming decades," Haig warned, "when the city will have to decide whether to undertake substantial capital projects, requiring large amounts of borrowing, for which there is already great pressure." The physical plant that had been started in the early twentieth century was now in need of repair and expansion. Officials had to coordinate capital budgeting, current spending, and city planning and invest in those physical improvements that would increase property values and thus generate the taxable wealth that could be tapped to fund the spending and borrowing to keep up with age and growth, as well as the enormous costs of social welfare that the city had taken on since 1929—a step that they seemed incapable of taking, even though experts had been recommending it for more than four decades.[74]

The root of this financial mess was not machine politics or lack of expertise in government, but New York's civic culture. From the beginning, New Yorkers had proceeded on the expectation that joining the greater city meant there would be more people to foot the bill for local projects (whether for merchants, ward heelers, social service providers, neighborhood associations, or boroughs). To this profusion of local needs, experts and their allies added expensive citywide projects, like the water and subway systems. During the Great Depression, a new generation of social welfare advocates had added yet another expense function to the budget, but one oriented primarily toward consumption rather than investment. Together, the combination of local, citywide, and welfare spending constituted an enormous drain on the tax base, and without any plan to guide the use of resources even the vast private wealth contained in New York City could not support everyone's demands. Back at the turn of the century, city boosters urged government spending on large-scale projects to insure that New York could compete with Boston, Chicago, Philadelphia, and Baltimore. By the second half of the twentieth century, New York would also have to compete with Sunbelt cities and suburban enclaves for middle-class tax dollars. No longer was it merely physical infrastructure—water, transit, docks, and so forth—that made cities competitive; lower tax rates would become increasingly influential as homeowners, retailers, factories, and corporate headquarters were presented with new locational choices. Thus, long before the city faced the challenges of the late twentieth century—the decline of the port; the exodus of corporate headquarters, downtown retail, and the garment industry; skyrocketing energy costs; white flight; a steady stream of postwar immigration; the fight to make local government more responsive to a wider variety of citizens—officials struggled unsuccessfully to sort out the multiple demands for public investment and public services confronting New York City every day.

Instead, the terms of togetherness for the new metropolis negotiated during the opening years of the twentieth century drove the city toward financial crisis year after year. Although experts like Henry Bruère, E. R. A. Seligman, and William Prendergast had successfully argued that the artificial constraints of late-nineteenth-century fiscal conservatism should be discarded to allow the city to build up its public systems, they had not prevailed in the larger effort to institutionalize a more deliberative approach to the management of collective resources. True enough, they had increased public access to private wealth—through new sources of revenue and greater borrowing capacity—but they had not convinced their fellow citizens that municipal debt should be strategically distributed to achieve collective needs (as they saw them) according to a predetermined plan. Without that sort of focused spending and borrowing—which, when properly directed, would actually pay for itself, through increases in property values—the multiplicity of demands unleashed by the experts' successful fight against fiscal conservatism would quickly overwhelm the city's ability to pay. As a result, the needs of the city as a whole remained in constant conflict with local demands for spending, and together they pushed municipal debt to its legal limit and beyond. Unable to conceptualize the city as a whole and to see how each piece contributed to a collective achievement, New Yorkers instead fought for access to public resources, hoping for growth to raise borrowing capacity, calling for outside aid to cover rising costs, tinkering with exemptions and definitions to squeeze a few more million under the constitutional cap, and biding their time until the next crisis imposed fiscal discipline from without.

Part 3 / Urban Planning, Private Rights, and Public Power

CHAPTER FIVE

City Planning versus the Law
Zoning the New Metropolis

In January 1912, the eight-story Equitable Life Assurance Society Building on Broadway burned in a truly spectacular fire, changing the way surrounding property owners looked at skyscrapers (figure 12). The original structure had been one of the city's first modern tall buildings, and its demise permitted a burst of sunshine that few of its neighbors had ever experienced. The owners of the Chase National and Fourth National Banks thought the breath of fresh air and sunlight so improved their buildings that they proposed that the city buy the property to create a park—a move that took on added urgency when they learned of the monstrous skyscraper the Du Pont Company planned to construct on the site (figure 13). Although not the tallest structure in Manhattan, the new Equitable Building (its thirty-six stories topping out at 562 feet) imposed on the skyline far more than the slender towers of the Woolworth or Met Life Buildings (792 and 700 feet, respectively). Covering the entire block bounded by Broadway, Cedar, Nassau, and Pine, this "city within a city" housed fifteen thousand, making it the largest office building in the world; its massive shadow covered seven and a half acres, blocking sunlight from nearby buildings that stood twenty-one, nineteen, and fourteen stories high and dark-

Figure 12. Icy ruins of the old Equitable Building, 1912. The eight-story building burned in January 1912. With temperatures below twenty degrees Fahrenheit and winds gusting as high as sixty miles per hour, "ice seemed to form in the very air," the *New York Times* reported. Ice clogged firefighting equipment, "rooting the pieces to the frozen streets. It settled in cloaks over the men themselves, so that they had to be chopped and thawed out from time to time that they might go on with their work." By nightfall, the structure was covered with a halo of ice, lit by the glowing red timbers within. *Source:* Irving Underhill Collection, Library of Congress, Washington, D.C.

ening the 612-foot Singer Tower up to its twenty-seventh floor. Even as he congratulated Du Pont on its remarkable new structure, Mayor John Purroy Mitchel, who attended the laying of the cornerstone in April 1914, took the opportunity to point out that the Equitable just might be the last huge skyscraper to be built in New York.[1]

In 1914, however, that was an empty threat since the city could do nothing to prevent the construction of more buildings like the Equitable, symptomatic though it was of the overcrowding so evident in lower Manhattan—the problem at the heart of almost every city-building challenge Gotham faced, from

Figure 13. The new Equitable Building, 1927. The imposing, thirty-six-story structure, built by the Du Pont Company in 1914, covered an entire city block. It was so much bigger and bulkier than the old Equitable Building that nearby property owners urged the city to buy the site to create a public park. The prospect of yet more massive skyscrapers encouraged the movement to regulate building height in the city, leading to the passage of the 1916 zoning ordinance. *Source:* Irving Underhill Collection, Library of Congress, Washington, D.C.

the construction of subways to the competition for space at the port and the deplorable condition of its tenement districts. New York had almost no legal authority to regulate building heights or shapes, which meant that property owners could build as big and bulky as they wanted, portending a dark future for this city of skyscrapers.

For the planning advocates who wrestled with congestion and overbuilding in the new metropolis, the need for expanded municipal regulatory power arose from multiple institutional failures. The decisions of real estate owners resulted in ugly, overcrowded streets; corporations built skyscrapers in the wrong places and to excessive heights; and the courts could not see beyond the limits of precedent to prevent the ignorance and self-interest of property owners from damaging the city. Markets and corporations were too individualistic in orientation and therefore insufficiently attuned to the collective consequences of their actions; courts were too preoccupied with individual rights to respond to the new conditions of the metropolis. Individualism, embedded in the city's institutions of collective decision making, produced a city that illustrated on a vast scale its own limitations.

No one understood this problem more clearly than Edward Bassett—lawyer, subway planner, Brooklyn civic activist, public service commissioner, and the father of zoning in the United States. "We must reckon first with the fact that Americans take for granted their right to do on their own property anything they please regardless of their neighbors," Bassett told his compatriots in the zoning movement in 1913—a tendency he blamed on the extraordinary protections granted to individual rights and liberties by the courts.[2] From Bassett's perspective, limitations on the police power—the power of government to regulate liberty and property to protect public health, safety, morals, and welfare—represented the single greatest obstacle to the development of new institutions capable of planning the modern city.

An understanding of New York's landmark 1916 zoning ordinance—one of the city's most significant planning achievements (it was during the campaign for the ordinance that Mayor Mitchel warned that there might be an end to mammoth skyscrapers)—must therefore begin with the recognition that it was the work of experts acting simultaneously as city planners, legal strategists, and state builders. Indeed, creating the zoning ordinance required the aspiration of Daniel Burnham (whose 1909 plan of Chicago represented the first modern attempt to refashion an entire U.S. city), the learning of Edward Coke (the great English jurist who helped bridge the medieval and the modern), and the strategic sense of Sun Tzu (author of the Taoist classic *The Art of War*). The New York zoning experts, although simultaneously involved in an extraordinary effort at distributive policy making (with aspiring planners striking deals with real estate interests throughout the city), were engaged principally in an elaborate legal maneuver against an approach to police-powers jurisprudence that kept state regulatory authority confined within traditional categories wedded to an outdated notion of individualism expressed through inviolable property rights.

Zoning, like Julius Henry Cohen's Port Authority, George Soper's regional sewage commission, and E. R. A. Seligman's income tax, thus represented a modern solution to the problems of the modern city, particularly the need, as perceived by experts, to embrace the consequences of social and economic interdependence. First and foremost, zoning was an attempt to come to terms with the new scale, scope, and interconnectedness of life in the metropolis by partially shifting the locus of land-use control from private to public hands and from judges to bureaucrats. In this way, experts like Bassett hoped to negotiate new terms of togetherness for New York, reconciling private activity and public goals by creating a municipal power to address the needs of the city as a whole—a goal that, as with railroad regulation, underground infrastructure, and municipal finance, they only partially achieved.

Institutional Failure and Motives for Zoning

Zoning in New York proceeded as a subsidiary activity of the board of estimate's committee on the city plan. Formed in January 1914 (three months before the legislature gave the board the power to zone the city), the committee on the city plan was intended as the start of a permanent planning bureaucracy, charged with "formulating a general scheme of improvements with which all local improvements can be coordinated."[3] The committee's experts—attorney Edward Bassett, political scientist Robert Whitten, architect George Ford, taxation specialist Lawson Purdy, and engineer Nelson Lewis (author of *The Planning of the Modern City*, which came out the same year the zoning ordinance was passed)—were deeply influenced by the City Beautiful aesthetic, with its emphasis on unity, harmony, variety, and the city ensemble, even though they knew that the ideal of the low-rise metropolis of wide boulevards and grand public spaces would have to be modified to accommodate the skyscrapers and narrow streets of modern New York. All were preoccupied with congestion in lower Manhattan, which interfered with the proper use of the downtown as "national emporium," to use Bassett's characterization. All were distressed at housing conditions and anxious to promote low-density residential areas away from downtown. And all agreed that government alone could restructure the metropolis to achieve a desirable distribution of people, business, and industry.[4]

The metropolis needed planning, Bassett's team believed, because the institutions involved in the management of land uses and building heights had produced a city of extremes: of wealth and poverty, of skyscrapers and tenements, of congestion and underdevelopment. Manhattan had some of the most crowded residential areas in the world: several sections of the Lower East Side, where tenement building were usually around three to four stories high, were packed in 1905 with more than a thousand people per acre. Conditions in the infamous tenement districts were well-known, thanks to exposés like Jacob Riis's *How the Other Half Lives* (1890) and the proximity of these areas to downtown commercial centers.[5] Within five miles of this extreme congestion, vast tracts of Queens, Brooklyn, and Staten Island remained undeveloped.

New Yorkers lived in dark, stagnant, crowded tenements, the zoning experts believed, because market conditions discouraged building in other sections of the city. As George Ford noted in 1915, "a man takes his savings and buys a plot of land in a residential district and builds a house for his own use. After he has lived there a year or two someone buys the plot next to his and puts up a factory. The man who built the house finds his home spoiled and in desperation tries to sell his place and move out. He discovers that the value of his property has been cut in half." The unpredictability of the market forced potential homeowners "to take a gambler's chance" when building in an outlying area, thus discouraging salutary residential decentralization.[6] Robert Whitten agreed. "A large portion of the land of New York City that is now unimproved or poorly improved is in that condition because the owners feel that the character of the section is changing, is bound to change in the near future, or that the permanent character of the section is unknown," he wrote in 1913. Thoughtful developers rightly hesitated to build because the construction of a tenement or paint factory on adjacent property could transform even a brilliantly conceived residential project into a financial disaster.[7] Bassett saw the housing crisis in New York as a direct result of the inability of private market decisions to create a proper distribution of workplaces. Quite naturally, working people wanted to live within easy reach of their jobs, and the concentration of industrial land uses downtown inevitably led to dreadful residential densities. In his view, it was the "main duty" of local government—"a duty amounting to a sacred trust"—to promote a wider distribution of industry through land-use regulation and subway construction that would encourage the building of "small homes where people can live in the sunlight and near the soil."[8]

The concentration of tall buildings in the central business district similarly demonstrated the failure of markets and corporations to create a desirable distribution of population and land uses. The construction of loft buildings along Fifth Avenue disrupted the character of the city's finest shopping street and created a surplus of office space that tempted building owners to rent upper floors to sweatshops, thus filling the retail district with garment workers.[9] The overbuilding of skyscrapers in the Wall Street district was even worse. Along New Street, the average building height reached 11.59 stories, while Exchange Place had an average building height of 14.1 stories, turning streets into canyons. The shadows cast by bulky, closely built skyscrapers made office space less desirable and forced businesses that needed natural light, like the jewelry trades, to seek new quarters in midtown, where building heights were lower.[10]

The aesthetic impact of skyscrapers posed perhaps the greatest challenge for Bassett's team. The ideal of a horizontal visual unity—embodied in the touchstone of the City Beautiful movement, the Court of Honor at the White City at the Chicago World's Fair of 1893—remained a central component of civic aesthetics during the first years of the twentieth century, and skyscrapers disrupted the low, even cornice line that created that impression along major thoroughfares. But New York was already a city of skyscrapers—so committed to building vertically that it made little sense, either as an aesthetic or economic matter, to enact height limits to encourage uniform cornice lines. Criticism of tall buildings was also significantly complicated by the growing number of aesthetically noteworthy skyscrapers—such as the Woolworth Building, which favorably impressed even so dogged a critic as Montgomery Schuyler. The New York zoning movement reflected this important change in civic aesthetics, with architect Ford leading the effort to create a planned city of skyscrapers, rather than eliminating skyscrapers from the ideal of the planned city. That meant developing a new approach to controlling the size and shape of tall buildings, however, since Ford was persuaded that finding the right combination of height, architectural diversity, and consideration for urban context could not be achieved by market or corporate decisions alone or by a simple uniform height limit.[11]

From the perspective of Bassett's experts, the decisions of developers, mortgage lenders, manufacturers, and property owners disrupted the skyline, changed the character of neighborhoods, contributed to unconscionable crowding, and thereby failed to protect property values, public health, economic efficiency, or any suggestion of a city ensemble.[12] Left to their own devices, markets and corporations would not preserve the civic values that Bassett's team thought should characterize a great metropolis—humane population densities, showcase retail areas, picturesque streets and skylines. Aggregating individual and corporate decisions did not produce desirable collective results, they concluded, and thus the time had come for a new approach to collective decision making to create a planned metropolis, if only the city's regulatory power could be expanded to such purposes.

Toward that end, Bassett's team needed the help of influential groups whose motives for property regulation had very little to do with city planning. Initially, the driving force behind the zoning coalition was the Fifth Avenue Association, formed in 1908 by merchants who wanted building height limits to prevent structures that housed the garment and white-collar workers who flooded the streets of their elite retail district. In 1913, the association joined forces with George McAneny, the Manhattan borough president then wrestling with the subway problem, to form the heights of buildings commission (the HBC—the predecessor of the commission on building districts and restrictions, or CBDR, which drafted the ordinance between 1914 and 1916). By that year—thanks to a slump in the real estate market, news of the giant Equitable Building, and a glut of office space downtown—banks, insurance companies, and mortgage lenders all over the city supported height limits to protect the value of their own properties. Sensing another opportunity to link city planning to clientelistic politics, McAneny composed the HBC "to give representation not only to all the boroughs of the city, but to the various interests that should be properly considered in whatever solutions may be proposed," from real estate dealers, mortgage brokers, architects, and builders to lawyers, labor organizers, and transit experts.[13] Between 1913 and 1916, Bassett and McAneny parceled out "legal" patronage in the form of land-use and height restrictions to garner support from those who had a pecuniary interest in the effects of zoning and those who saw it as a device for maintaining boundaries between ethnic groups.[14]

The experts' efforts to separate retail from industrial activities thus reflected their endorsement of the use of municipal regulatory authority for discriminatory purposes.[15] So, for example, when George Ford became a consultant for Norfolk, Virginia, in the 1920s, he advocated restrictive covenants (not zoning ordinances) to create "negro expansion areas" with proper drainage facilities and paved streets to encourage blacks to move into designated sections of the city.[16] Robert Whitten also helped southern cities enact plans with "colored" areas since this would provide blacks with room to expand their neighborhoods without fear of conflicting with whites, even though he thought that zoning to segregate immigrants was "foolish and unjust."[17] Like Woodrow Wilson, that standard-bearer of Progressivism, whose devotion to segregation illustrated that enlightened governance did not necessarily exclude benighted racial policies, Bassett's team accepted exclusion as a legitimate planning objective, and the use of zoning for such purposes constitutes one of the great blind spots of the experts discussed in this chapter. So anxious were they to expand the planning capacity of government that they failed to make clear distinctions between expert judgment and rationalized prejudice.

Despite their approval of exclusionary zoning, however, the experts who crafted the 1916 ordinance were primarily interested in overcoming the limits of governmental power over vested rights in order to create a planned city. Although Bassett, too, disseminated information on zoning as a tool for segregation, he became less supportive as the legal battle shifted from the states, which occasionally permitted such efforts, to the federal courts, which rejected them. "Segregating races by districts in not within the field of zoning and would be contrary to the fourteenth amendment to the federal constitution," he affirmed in his 1936 treatise.[18] Anything that threatened the constitutionality of zoning worried Bassett, and this preoccupation with legal barriers to planning, far more than their elitism, explains the state-building strategy pursued by his team.

The Problem of the Police Power

For Bassett, crafting a viable police-power defense for zoning constituted the primary problem for his team of aspiring planners. In 1914, he wrote to George McAneny that laying the legal foundation for zoning "is not at all clear to any of us, largely on account of the vagueness of the police powers. I am afraid that if Messrs. Purdy, Whitten, Ford and I do not have our own ideas rather safely grounded, we may all be carried off on some false trail."[19] Although municipalities had ample regulatory authority to enforce building codes, basic powers to protect cities from shoddy construction or flammable materials did virtually nothing to assist planning per se. Indeed, as late as 1915, *American City* magazine editorialized that while many cities had developed plans, "adequate consideration has not ordinarily been given to the very important question of the legal power of the municipality to carry out the physical improvements advocated. As a result the plan has been left somewhat up in the air."[20] Combined with the difficulty of holding together the vast zoning coalition, the legal barriers to expanded municipal regulation determined every step Bassett took in the state-building process.

The zoning team began their encounter with the law at a time when the prevailing approach to police-power decision making, categorical legal reasoning, was perceived as an obstacle to progressive social legislation. The categorical approach placed private rights and public powers in a system of carefully delineated spheres or categories.[21] To determine whether a regulation fell within the sphere of legitimate public action, judges employed several tests, based in part on the due-process and equal-protection clauses of the Fourteenth Amendment. A regulation met the due-process test if it bore a "real, substantial relation" to the protection of public health, safety, or morals.[22] To meet the equal-protection test, the regulation could not impose unusual restrictions on particular groups or persons. The limits of the police power could also be found by citing analogous limits on an individual's right to abate a nuisance.[23] Where police regulations successfully met these tests, private rights stood as no barrier to government regulation.[24] Where police regulations failed the due-process or equal-protection tests, or where they extended too far beyond the nuisance analogy, private rights were superior to the desire of the legislature to regulate property or liberty, regardless of the social or economic necessity or benefits of such measures.

Although legal realists criticized formalism as intellectually ossified and socially uninformed, it nonetheless permitted a very wide array of state regulation within the confines of established categories. So broad was the police power, in fact, that some in the zoning movement saw no serious legal barriers to their plans to zone New York City.[25] Two key court decisions—*Cochran v. Preston* (1908), in which a Maryland court upheld a building-height ordinance that applied to a small, elite neighborhood in Baltimore, and *Welch v. Swasey* (1909), in which the U.S. Supreme Court upheld Massachusetts acts creating a residential zone with a height limit of 80 feet and a business zone with a height limit of 125 feet—seemed to dispose of any of the broad constitutional questions raised by height limits along Fifth Avenue.[26] Although inspired by City Beautiful ideals, these ordinances had been upheld as fire-protection measures. *Welch* was especially important because the court admitted that the Boston height limits were designed to promote a uniform cornice line; but because their *primary* purpose was the protection of health and safety, secondary aesthetic motives did not invalidate them.[27] To a lesser extent, similar legal arguments applied to use districts. In three California cases—*Ex parte Quong Wo* (1911), *Ex Parte Montgomery* (1912), and *Ex Parte Hadacheck* (1913), each involving businesses (a laundry, a lumberyard, and a brickyard) within residential districts created by a 1909 Los Angeles zoning ordinance—the court concluded that the regulations were reasonably designed "for the safety, health, and comfort of society at large."[28]

Not all courts were persuaded that use districting was really motivated by a desire to advance legitimate public purposes, however, and an equally important series of decisions called into question the goal of creating residential neighborhoods free from businesses. In *People ex rel. Lincoln Ice Company vs. Chicago* (1913), the Illinois Supreme Court struck down a Chicago ordinance prohibiting ice houses near churches, hospitals, and schools, with a unanimous court scoffing at the city's attempt to frame the regulation as a public health measure.[29] Later that year, in *People ex rel. Friend v. Chicago,* the same court unanimously struck down another Chicago ordinance prohibiting retail stores in residential areas—one of the most crucial objectives for Bassett's zoning team—concluding that such restrictions bore no "logical connection" (that is, no substantial relation) to health, safety, comfort, or the general welfare.[30] Also in 1913, in *Willison v. Cooke,* the Colorado Supreme Court struck down a Denver ordinance that prevented the construction of stores on residential streets, saying that the "these regulations do not, in the slightest degree, have any relation whatever to the health, safety or general welfare of the public."[31] Two years later, in *Calvo v. City of New Orleans,* the Louisiana Supreme Court struck down a similar New Orleans ordinance, noting that a business on a residential street was neither a nuisance nor a threat to public health.[32] That same year, in *People ex rel. Lankton v. Roberts,* a New York court could find no legal grounds for upholding a Utica ordinance creating residential districts, even though the judge openly expressed his sympathy with the idea of protecting neighborhoods from businesses and industries.[33]

Aside from their forthright rejection of one of the main goals of Bassett's team, these decisions presented a real danger for zoning advocates by branding the protection of residential areas as aesthetically motivated. When judges failed to see the connection between a residential land-use ordinance and public health or safety, they assumed (often rightly) that the real motivation of the law was a desire to beautify a neighborhood by keeping out commercial elements.[34] And while the Supreme Court had allowed aesthetic motives as an "auxiliary consideration" in an otherwise legitimate police regulation in *Welch*, in the eyes of most judges limiting the use of private property to enhance civic beauty bore no clear relationship to established police-power objectives.[35]

Debates at the National Conferences on City Planning heightened the confusion over what these cases meant for the New York effort. Though some zoning advocates insisted that planners "should not be bothered by the constitution," others argued that planning measures had often "come to grief upon the rock of unconstitutionality."[36] After a close reading of the California use-districting cases, the well-known tenement-house reformer Lawrence Veiller announced that "a new use for the police power has been discovered," paving the way for "judge-proof" planning laws.[37] On the other hand, Boston attorney John Walsh counseled planners to "waste no time in experimenting with the police power but adopt measures for the amendment of your constitutions to give your legislatures the power which you now demand."[38] However, respected legal scholar Ernst Freund cautioned anxious planners against constitutional amendments that created planning powers beyond the regulatory limits recognized by conservative American courts.[39] Oklahoma lawyer Phillip Kates aptly summarized the general sentiment when he argued that the Fourteenth Amendment had long caused courts "to flounder about among fine spun theories as to the police power of the states, until no one knew whether a law framed after the most careful deliberation, and to meet a pressing necessity, would withstand even the weakest attack upon it."[40]

Bassett's strategic political concerns likewise shaped his sense of the limits of the police power. He faced the very difficult task of holding together an unwieldy zoning coalition and convincing elected officials that the ordinance would benefit each part of the far-flung city. Looking back on Bassett's efforts, Lawson Purdy recalled in 1926 that the two most important questions the zoning team had struggled with a decade early were, first, whether the ordinance would be upheld in New York state courts, and, second, whether the members of the board of estimate would ever approve the measure—a possibility he considered "very, very remote." As a result, "with everything that we did, cautions of extreme prudence prevailed," especially in light of the time constraints the experts confronted.[41] Zoning's political sponsors were elected to office in 1913 and faced reelection in 1917. Bassett thus had four years to craft an experimental ordinance to cover the entire city that satisfied economic interest groups and planning advocates before a potential change in political leadership. And as it turned out, John Hylan, an urban populist with an active dislike for planning, was elected mayor in 1917, and he promptly did away with the committee on the city plan. Although Bassett could not have foreseen Hylan's victory, as a former congressman from Brooklyn and member of the public service commission (where he played a role in subway matters), Bassett did know how ephemeral electoral victories by reform coalitions could be in New York. Bassett thus had one good opportunity to pull his coalition together behind a legally defensible ordinance, which discouraged him from taking a broad view of the police power: should an overly ambitious ordinance be struck down in court after an electoral change, he would lack the political sponsorship necessary to rewrite the zoning law.

Some of Bassett's key allies also told him that the law could not, and should not, be used to transform the ordinance into a real planning tool. In 1913, Walter Lindner, counsel for Title Guarantee and Trust Company, told the zoning team that the police power did not permit them to make the rather subtle regulatory distinctions they wanted. "If health and sanitation warranted limiting buildings in the suburbs at a low height, then these same grounds could be consistently urged in remedying the congestion in the built-up areas of the city," Lindner insisted. "Legislation cannot be passed to limit buildings at different heights in different districts—it must be directed toward the abolition of unsanitary conditions in all localities."[42] Other members of the coalition simply did not want building-height or land-use regulations, which meant that they would likely challenge the ordinance in court. Developer Alrich Man, who would later serve on the CBDR, felt that government should not deprive owners of the right to build as they saw fit, and he warned that zoning restrictions smacked of "confiscation" and "paternalism."[43] Cyrus Miller, borough president of the Bronx, argued that zoning would require the city to compensate landowners since "any interference with the unrestricted use of private property is to that extent a taking of private property."[44]

Although Bassett overcame this resistance by persuading key figures in the real estate industry to support regulations that suited their pecuniary interests, the ordinance itself remained vulnerable to critics who knew that there was more to it than a narrowly construed attempt to aid Fifth Avenue merchants or suburban developers.[45] Bassett's team wanted to plan the city—to rationalize the use of space, to limit the power of markets and corporations, to bring a greater degree of beauty to the skyline, and to reconcile individual activity with collective effect. That meant shifting the balance of power between the private and public sector in the name of the city as a whole, and that balance stood where it was due largely to the fact that categorical legal reasoning gave property rights significant protection from the regulatory power of the state.

In other words, there seemed to be a basic conflict between the way planners and judges viewed the relationship between private rights and public spaces. For Bassett and his colleagues, the city was a model of interdependence. Everywhere they looked, they saw the lines between supposedly private activities becoming blurred. An absolute right to private property was something of an illusion here, for property did not remain confined within clearly delineated spheres. Skyscrapers blocked light from the streets; businesses gushed steam, smoke, odors, customers, and noise into neighborhoods; private building decisions in outer boroughs contributed to congestion downtown; public systems, such as subways and sewers, were essential to the functioning of private businesses. To view private property in a neatly defined sphere no longer seemed adequate for understanding its actual use in an urban setting. The law, on the other hand, seemed preoccupied with clear boundaries around private property, irrespective of the new conditions of urban life. "The courts have endeavored to apply the same views of police powers to great municipalities that they applied seventy-five years ago to country districts and villages, and of course they do not fit," Bassett wrote in 1915. "Some phraseology should be found that will justify the courts in giving a larger scope to the police powers. I sometimes think that a sentence should be added to the constitution stating that in the exercise of the police powers every intendment should be taken in favor of the people of the state as against the owner of private property."[46]

Bassett's team thus began their work convinced that the police power did not provide an affirmative recognition of community welfare beyond the narrow ground of health and safety. Legal precedent might support some basic form of height and use districting, but would probably not allow them to create single-family residential areas, and it remained uncertain whether the mere presence of aesthetic motives could be the Achilles' heel of the entire project. Because they believed zoning required the subordination of individual property rights to broader community goals embodied by a citywide zoning plan, they saw themselves on a collision course with the U.S. Constitution.[47]

Searching the Law for a State-building Strategy

Bassett's effort to create a "judge-proof" zoning ordinance—one that would signify a change in the balance of power between bureaucrats and judges, as well as a shift in the balance of power between bureaucrats and property owners—began in earnest in 1913 when he became the chairman of the HBC. If he could discern how judges drew the line between individual rights and community welfare, he could expand the boundaries of the police power, giving planners increased control over private property and thus the power to enforce a city plan. By focusing closely on how judges determined the scope of the police power and then framing new regulations with those procedures in mind, Bassett and Robert Whitten, who conducted the initial research on the scope of the police power for the zoning team, could establish a foundation for a new governmental activity on the groundwork of accepted practice.

Paradoxically, Bassett and Whitten had to build that foundation from descriptions of the police power that were so expansive as to be misleading. As defined by judges in a handful of key decisions, the police power seemed boundless—an expression of the unlimited power of sovereignty itself. Bassett and Whitten found such a view in Chief Justice Shaw's now legendary opinion in *Commonwealth v. Alger* (1851), which provided a broad definition of the state's authority to regulate private property so that "it shall not be injurious to the equal enjoyment of others having an equal right to the enjoyment of their property, nor injurious to *the rights of the community*. All property in this Commonwealth is derived directly or indirectly from the government, and held subject to those general regulations which are necessary to the common good and general welfare."[48] This view was echoed in *C. B. & Q. Railway v. Drainage Commissioners* (1906), where Justice Harlan held that the police power "embraces regulations designed to promote the public convenience or the general prosperity" and was not limited strictly to public health and safety.[49] In *Noble State Bank v. Haskell* (1911), Justice Holmes went a step further by declaring that "the police power extends to all the great public needs" and "may be put forth in aid of what is sanctioned by usage, or held by the prevailing morality, or strong and preponderant opinion to be greatly or immediately necessary to the public use."[50] Although these pronouncements seemed to provide compelling evidence that state regulatory authority superceded private rights, Harlan and Holmes were hardly representative of mainstream judicial decision making. As state builders, Bassett and Whitten had to construct an interpretation of police-power jurisprudence that reconciled the apparent breadth of state authority revealed in these decisions with their received understanding of the law as bulwark against state interference with private rights.

Here they were aided by two cases decided during their deliberations that seemed to put teeth into Justice Harlan's community-welfare formulation of the police power by giving local officials wide latitude to create new regulatory schemes. *Reinman v. Little Rock,* decided in April 1915, involved a Little Rock, Arkansas, ordinance declaring livery stables a threat to public health and designating an area in the city where they were henceforth unlawful. In words that seemed to speak directly to Fifth Avenue merchants, the city noted that the area in question "composes the greatest shopping district in the entire State of Arkansas, contains the largest and best hotels in the State, and encompasses the most valuable real estate in the entire State," and thus deserved special protection by the law.[51] The Arkansas Supreme Court agreed, and the U.S. Supreme Court upheld the state court's ruling.

Bassett thought he saw in Justice Pitney's decision for the court a broad donation of discriminatory power to municipalities, and he cited *Reinman* as one of the great decisions underlying the zoning movement. "Every word in the [New York City] zoning resolution was written with this decision in mind," Bassett wrote privately.[52] The decision seemed to vitiate the substantial relation test by allowing cities to declare otherwise lawful businesses unlawful, even so mundane a business as a livery stable. "Granting that it is not a nuisance *per se,*" wrote Justice Pitney, "it is clearly within the police power of the State to regulate the business and to that end to declare that in particular circumstances and in particular locations a livery stable shall be deemed a nuisance in fact and in law, provided this power is not exerted arbitrarily, or with unjust discrimination, so as to infringe upon the rights guaranteed by the Fourteenth Amendment."[53]

The U.S. Supreme Court affirmed Pitney's sweeping application of the police power eight months later in *Hadacheck v. Sebastian,* one of the key California use-districting cases. Hadacheck owned a brickyard in Los Angeles worth $800,000. He had chosen the site, well outside populated areas, because it contained a deep bed of clay ideal for making bricks. Over the years, Hadacheck had excavated large parts of his property to a considerable depth, rendering the site unsuitable for much else besides brick making. The city had grown up around the brickyard, and the now largely residential area had been annexed by Los Angeles. The 1909 zoning ordinance designated the area as a residential district and caused the value of Hadacheck land to plummet to a mere $60,000. Justice McKenna upheld the ordinance, citing *Reinman* as precedent.[54]

Hadacheck brought to the fore the battle between private and public that Bassett saw at the heart of planning. A property owner engaged in lawful business incurred great expense because of a zoning ordinance. The brick maker violated the ordinance by no action of his own; the city had simply grown up around him and now declared his business unlawful. In what became a cherished discussion of the police power for planners, McKenna sided with the community: "A vested interest cannot be asserted against it because of conditions once obtaining. To hold so would preclude development and fix a city forever in its primitive conditions. *There must be progress, and if in its march private interests are in the way they must yield to the good of the community.*"[55] For Bassett, this decision "gave fresh expression to the fact that a city's expansion and growth are superior to the whims of a few capricious landowners who might wish to thwart the greater welfare of the community."[56]

In these cases, Bassett and Whitten found judges expanding the legitimate sphere of government action to accommodate new regulations, especially when dealing with the complex conditions of growing cities. They saw the courts extending the police power most readily where regulations promoted health and safety. They discovered that local circumstances often inspired regulations that seemed necessary and reasonable even though they fell outside the primary purposes of the police power. And they found that judges were often impressed by the role of expert testimony. By building on these observations, Bassett and Whitten could construct a convincing narrative of an expanding scope of municipal regulatory authority that would make it appear as though zoning was the next logical step in the use of the police power. By reading into a series of cases the guidelines that would point the way toward zoning, Bassett's experts could bridge the gap between clearly justifiable regulations (involving public health and safety) and those regulations considered "experimental" (involving general social and economic welfare).[57]

The New York Solution: Comprehensive Zoning

On July 25, 1916, after nearly three years of debate, the board of estimate adopted the nation's first comprehensive zoning ordinance, subjecting every piece of real estate in Greater New York—more than eight billion dollars worth of real property—to use, height, and area regulations. The ordinance divided the city into three use districts—residential, business, and unrestricted. In residential districts, which comprised two-fifths of Manhattan and about two-thirds of the city, tenants could use new buildings as homes, apartments, hotels, clubs, churches, schools, libraries, museums, hospitals, or railroad stations. Buildings in business districts could house all forms of commercial and industrial activity except those generating objectionable odors or byproducts. In unrestricted areas, buildings could be used for any purpose.[58]

The height limits of the ordinance were its most distinctive feature, producing a form of building known as setback architecture. The ordinance created five height districts, each establishing a relationship between street width and building height. In "1½ times" districts, for example, the street wall of a building on a sixty-foot-wide street could rise to a height of ninety feet (figure 14). After that, the structure could rise an additional three feet for every one foot that it was set back from the street wall. The most liberal limits applied to lower Manhattan, business districts in Brooklyn, and waterfront areas in Queens and the Bronx. Lower limits applied to large sections of every borough; Richmond and Queens, where population was still sparse, received the most strict designations to encourage residential decentralization.[59]

The ordinance also created area districts, with restrictions preventing new buildings from covering their entire sites and mandating open spaces at the rear and sides of structures: the taller the building, the more space required on all sides. Although these regulations left plenty of room in the central business district for bulky buildings, certain area districts provided for rather small structures. Area "E" restrictions, applying to interior lots in residential districts, prevented the first floor of structures from occupying more than 50 percent of their sites, with second floors covering only 30 percent (figure 15). These "villa districts" were located primarily in the upscale residential sections of Brooklyn, with some in a few parts of Queens, Staten Island, and the Bronx; here, too, the ordinance encouraged residential decentralization.[60]

Although this ordinance reflected Bassett's conception of the limits of the law, it did break new legal ground, going well beyond existing precedents like *Welch, Reinman,* and *Hadacheck*.[61] The Los Angeles ordinance had created residential and industrial districts; the New York ordinance included residential, business, and industrial districts. The Boston ordinances had created two height districts, whereas the New York ordinance had five (which, because they were calculated based on street widths, created a tremendous variety of specific

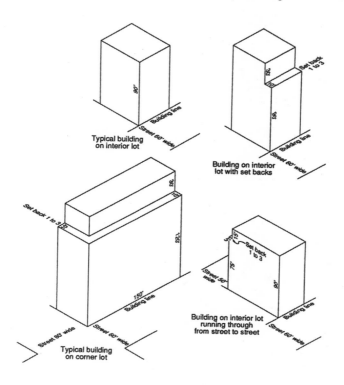

Figure 14. Height limitations of the 1916 zoning ordinance in the "1½ times" districts. To allow more light and air into city streets and onto surrounding structures, to encourage more interesting skyscraper architecture, and to prevent bulky skyscrapers like the Equitable, the 1916 zoning ordinance required the upper portions of new buildings to be set back from the street wall. Building up to the limit of the setback envelope produced ziggurat-like structures. *Source:* Commission on Building Districts and Restrictions, *Final Report* (New York: Board of Estimate and Apportionment, Committee on the City Plan, 1916), 260.

height limits, including some of unlimited height). The New York ordinance also had five area districts, which Bassett and Whitten used to create zones for single-family homes, duplexes, and apartment buildings. Together, these designations allowed seventy-five possible combinations of height, area, and use limits, although the original ordinance employed only thirty-six of them.[62] No court had ruled on so elaborate a regulatory scheme. And while judges had stretched the substantial-relation doctrine to cover near-nuisance zoning (involving brickyards, laundries, and stables), the courts had not yet recognized planning as a legitimate purpose of government.

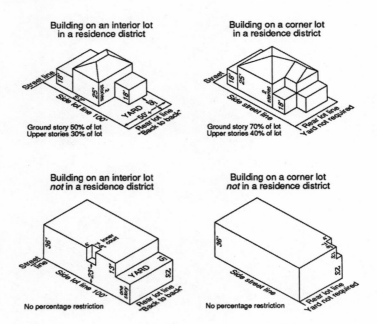

Figure 15. Area limitations of the 1916 zoning ordinance. Edward Bassett's team of planners intended "area" limitations, like these in "E" districts, to restrict residential building to single-family homes. Rather than explicitly creating single-family districts, which they thought could not be justified under existing interpretations of the police power, they hid them in "area" districts, where the limitations could be more easily connected to public health and safety through the provision of light and air. Note the "villa" style at upper left. *Source:* Commission on Building Districts and Restrictions, *Final Report* (New York: Board of Estimate and Apportionment, Committee on the City Plan, 1916), 271.

To give this ambitious ordinance the appearance of legal familiarity, Bassett and Whitten devised a strategy to link the primary and experimental applications of the police power. This meant, first, stating their goals in legally acceptable form since the intention of public officials was a key issue for judging the validity of police regulations. The 1916 ordinance, they claimed, was motivated by three major considerations: public safety (especially the threat of fire), public health (particularly the "importance of *light, air* and the prevention of congestion to health and sanitation"), and the "general welfare" (achieved through the "more adequate provision for *light and air*" to preserve the residential character of neighborhoods and property values.[63] Light and air united health and planning. By this account, planning was essentially a health matter—the one public value that judges seemed to recognize as superior to private rights. Rather bold steps toward the subordination of private property to public direction were thus taken under the auspices of disease prevention, not community rights or planning per se.

The success of this move, not new or innovative in itself, depended on expert testimony showing a "real, substantial" relationship between health and zoning—one of the crucial elements that emerged from Bassett and Whitten's study of successful police-power cases. Bassett's team made this connection everywhere. George Whipple, professor of sanitary engineering at Harvard University and one of the nation's leading public health authorities, played an especially important role in this regard. On July 2, 1913, Whipple measured the density of dust particles in the air from various floors of the Woolworth and Metropolitan Life Buildings. The air around the Woolworth Building contained 221,000 dust particles per cubic foot at ground level, 85,000 at the tenth floor, 41,300 at the fortieth floor, and 27,300 at the fifty-seventh floor. At comparable levels, the air around the Metropolitan Life Building had fewer particles: only 173,000 at ground level, 38,000 at the tenth floor, and 24,000 at the fortieth. "The lower counts at the Metropolitan Life Building," Whipple speculated, "may possibly have been due to the fact that this building is situated near Madison Square and is more isolated than the Woolworth Building which is located down town in a more crowded section." Whipple's data suggested that tall buildings in crowded areas were more likely to harbor the dust that spread tuberculosis.[64] The zoning ordinance thus represented a decisive step toward planning a healthier city, they could claim, providing a low-cost approach to sanitation, since regulation under the police power did not require the city to compensate property owners—an especially important consideration in this era of rising municipal expenditures.[65]

Other experts likewise testified to the health and safety benefits of zoning. Traffic experts agreed with fire chiefs on the value of height restrictions for controlling street congestion. Officials from the sanitation department corroborated the testimony of health department doctors on the connection between population density and disease. A neurologist discussed motion sickness in elevators, emphasizing the need for height limitations. A tuberculosis authority, Dr. Adolphus Knopf, and the city health commissioner, Dr. Haven Emerson, hailed the salutary effects of adequate sunlight and street ventilation.[66] This phalanx of experts secured the relationship between zoning and public health and safety. Bassett knew that the courts often deferred to local authorities; by making experts the most prominent local authorities, he assured judges that they deferred not merely to city politics but to an informed connection between distinctive local conditions and recognized police-power applications.

The real legal innovation of the New York zoning team lay in their handling of the equal-protection test. With a clear connection between health and zoning, it followed that regulations could be enforced under the police power. But how flexibly could municipal authorities apply these regulations without violating the equal protection clause? To what extent did the zoning-health connection make planners prisoners of their own rationale? For instance, zoning advocates cited the massive Equitable Building as a prime example of the threat to public health and safety posed by skyscrapers. Had the Equitable complied with the new height and bulk regulations, it would have lost 483 square feet of space on the second floor, 11,109 square feet on the twenty-third floor, and 36,261 square feet on the thirty-seventh floor.[67] If zoning prevented similar structures because they threatened public health, why should the Equitable be allowed to stand?

To surmount this barrier, Bassett's team emphasized the "reasonableness" of their comprehensive system of regulation—another key element to emerge from their study of accepted uses of the police power. Zoning restrictions, generally justified as health and safety regulations, could not be enforced without reference to their economic effects: "In certain districts suburban conditions of light and air can be maintained with great public advantage and with slight private loss; in other districts such favorable conditions of light and air, while theoretically just as desirable, are entirely impracticable, and any law that attempted to enforce them would be clearly unreasonable and void." Regulation of private rights had to yield proportionate gains in public welfare. If too stringently enforced, zoning regulations would so seriously harm property values "as to render [the ordinance] of doubtful expediency and constitutionality."[68] This "reasonable classification" discriminated against future building owners, but for the greater good of the community.

Emphasizing the reasonableness of the overall classification scheme, so firmly anchored to health and safety, shifted the zoning resolution into the experimental region of the police power. Allowing buildings that violated the new ordinance to stand could not be justified under the primary purposes of the police power, zoning advocates admitted. Nevertheless, zoning was a legitimate use of the police power in a general way. "While a specific regulation taken by itself may not seem to have a very direct relation to the purposes for which the police power may be invoked," Bassett's team of experts acknowledged, "yet when taken as part of a comprehensive plan for the control of building development throughout the city, its relation to such purposes may be unmistakable." Extending this logic, they made the giant leap to the fundamental conclusion of the comprehensiveness rationale: "Grant that a comprehensive system of districting is essential to the health and general welfare of the city and it follows that every specific regulation that is an essential part of such a comprehensive system is justified under the police power."[69]

Bassett and his colleagues thus maintained that zoning was directed toward health and safety in a broad sense, but many of the details of the ordinance were motivated by political expediency, civic beautification, enhancing property values, segregating garment workers from wealthy shoppers, and rationalizing the city's transportation system. These were all worthy ends, but not strictly obtainable under the police power. However, when included as *essential* parts of a *comprehensive* city plan, crafted by experts and directed toward health, safety, and general welfare, they became (hopefully) justifiable as goals of the police power. By carefully balancing incidental private losses against greater public benefits, the meticulous work of the planning bureaucracy—Bassett, Whitten, safety experts, fire marshals, physicians—insured that a comprehensive ordinance was a proper exercise of the police power.[70]

Or, as zoning attorney Alfred Bettman put it in 1924, "comprehensiveness puts the 'reason' into 'reasonableness.'" This bit of legerdemain allowed experts to make two crucial claims. They could insist that zoning was not a radically new type of regulation; it was simply an extension of "sanctioned traditional methods for sanctioned traditional purposes." At the same time, they could proclaim, without batting an eye, that zoning was not confined to traditional legal categories: "Though zoning has the same fundamental basis as the law against nuisances, no greater fallacy could exist than that zoning is restricted to or is identical with nuisance regulation."[71] This logic gave judges familiar police-power justifications, apparently confining the ordinance to the customary categories of health and safety, while asking the courts to endorse the novel bureaucratic process of balancing slight private losses against greater public gains for unfamiliar purposes such as protecting single-family homes and promoting civic beautification. With this strategy (a response to the limits of existing judicial practice and legal discourse as Bassett understood them), the New York zoning team sought to retain the legitimacy of police-power precedent while creating new possibilities for municipal regulation.

Defending the Ordinance: Edward Bassett and the Art of War

Although there is no evidence that Edward Bassett ever read *The Art of War*, his strategy for defending the 1916 zoning ordinance was nonetheless an admirable application of Sun Tzu's famous dictum "To win without fighting is best." Although the comprehensiveness rationale provided a plausible argument for expanding the scope of municipal regulation, constructed as it was from the elements of accepted police-power doctrine, Bassett knew that it remained vulnerable to attack from the certain ground of orthodox constitutional interpretation. Because he believed that neither the politicians nor the real estate interests who had backed the measure were fully committed to its underlying principles of city planning, he was sure his experimental legal logic would quickly come under fire: three years of exceptional political effort would hang in the balance with every court ruling. Rather than fighting the constitutional battle head-on, therefore, Bassett elected to defend the ordinance by disguise, by deflection, and by proxy.

The legal barriers Bassett saw in his reading of police-power cases compelled him to camouflage some of the coalition's key planning objectives, making them appear as commonplace municipal regulations rather than as novel extensions of state authority over private property. For example, Ford, Whitten, Purdy, and Bassett wanted to offer legal protection to residential areas, but Bassett remained convinced that the courts would not recognize the promotion of single-family housing districts (which would exclude apartment buildings and duplexes) as a legitimate use of the police power. As a consequence, the zoning team created one broad residential category and left the distinction between single- and multifamily dwellings to area districts (relating to the bulk of structures). "Fine distinctions of uses would not, we thought, appeal to the courts," Bassett later revealed, "but distinctions in the area regulations having to do directly with light, air, ventilation, etc., would always afford a handle on which the court could hang a favorable decision." Bassett thus buried single-family housing regulations in "D" and "E" area distinctions, where they could be more readily connected to public health through light and air.[72]

Concerns about the police power also encouraged the suppression of the aesthetic debate over skyscrapers that had been so much a part of the City Beautiful milieu of many members of the zoning team. Early in the deliberations of the HBC, builder Otto Eidlitz and architect Burt Fenner urged their colleagues to consider the "artistic value of different skylines and setbacks." But Lawrence Veiller counseled that " 'aesthetic consideration' [should] be struck entirely out of the program, on the ground that even the mention of the term would cause the public and the lawmaking bodies to feel that the Commission was wasting time, as the courts of New York do not recognize aesthetic considerations" as a valid reason for exercising the police power. Purdy, Fenner, and architect C. Grant La Farge responded that "it would be fatal to a comprehensive consideration of the subject not to take into account the aesthetic side" of the problem. Rather than drop their interest in civic beautification, however, the HBC decided to disguise it, referring only to "aesthetic considerations in relation to rentability and the value of land."[73] The next step was avoiding any mention of aesthetics entirely.

Thanks to the efforts of George Ford, however, the ordinance incorporated key elements of the City Beautiful into regulations that encouraged a vertical city. Ford designed setback restrictions to promote visual diversity and architectural experimentation, but clothed these considerations in the language of rentable space, property values, and the medical role of sunlight. He did this to give judges as many familiar legal handles as possible on which to hang a favorable decision and to avoid giving opponents any opening to brand zoning as civic beautification. The legal role of light and air allowed this formula to work since experts could easily connect both to public health, and such connections went a long way toward satisfying the substantial-relation test. The preservation of light and air for reasons of public health became the leitmotif of the zoning ordinance, camouflaging the protection of single-family homes and the sculpting of the skyline. Under the guise of preventing tuberculosis (the leading cause of death in the city), the ordinance thus encouraged architectural experimentation in the downtown and expansion in the suburbs.[74] By this sleight of hand, the authors of the 1916 zoning ordinance reconstituted the language of civic beautification, which they inherited from the City Beautiful, through the language of public health, which the courts recognized as a legitimate reason for infringing on private property. At the same time, they accepted the skyscraper as part of a planned city, taking a crucial step toward reconciling the individualism represented by tall buildings with the civic values comprised by planning. By regulating (rather than outlawing) the vertical city, they opened up new aesthetic possibilities both for planning and for architecture.

Even though he draped the most controversial goals of the zoning team in the language of health and safety, Bassett knew that even the best efforts at camouflage would not prevent litigation. Legal challenges in fact began almost immediately, and by 1922 there were two to five court decisions per month involving the ordinance.[75] Since Bassett anticipated this onslaught, he devised a legal-administrative maze to deflect court battles away from constitutional issues and thus prevent (or at least postpone) a direct test of the comprehensiveness rationale. To set up this maze, Bassett persuaded state legislators to pass an enabling act specifically giving the city the authority to zone; and in response to the onslaught of litigation, he later convinced them to amend the act to create a board of appeals. Bassett knew that even the most painstaking process of drawing zoning boundaries would result in "unnecessary hardships" on some property owners. The board of appeals, as a quasi-judicial body composed of five members appointed by the board of estimate, could "determine and vary" use, height, and bulk restrictions, which gave aggrieved property owners a chance to seek variances or exceptions to the ordinance on a case-by-case basis, without changing whole districts or undoing the entire ordinance. Not all petitioners would be granted variances, of course, and the amended enabling act specified that the board's rulings could be challenged in court by writ of certiorari rather than by writ of mandamus. The distinction was crucial. Under certiorari, the courts examined the record of the board of appeals "in order that errors and irregularities in the proceedings may be corrected." This review process allowed the court to certify whether the board acted within its authority (as specified by the enabling act) and either validate or invalidate the decision to grant or deny a variance. Such cases did not address the larger issue of the constitutionality of zoning, which lurked in the background. Had Bassett not directed disgruntled property owners through the board and into certiorari, they could have challenged the ordinance directly through mandamus proceedings. In mandamus, the courts could command that the city restore to a complainant "the rights or privileges of which he has been illegally deprived." Such a remedy was not a reversal of the board decision on a variance but a declaration that the ordinance itself violated the property owner's rights. The entire zoning measure (or large parts of it) could thus be struck down as an unconstitutional encroachment on basic liberties.[76]

By channeling property owners into the appeals-variance-certiorari maze—

whether those who had a legitimate grievance against a poorly drawn district boundary or those who merely sought to defy the ordinance for pecuniary advantage—Bassett succeeded in deflecting the sort of direct constitutional confrontations he most feared. By the early 1920s, after the revival of the real estate market quickened the pace of building activity and, in the process, expanded the number of complaints against the zoning ordinance, the wisdom of this strategy became increasingly evident. "If variances could not be made, subject to review by the courts in certiorari, there would be at least one case of successful mandamus against a building commissioner each week," Bassett reported to his colleagues. "This would make about fifty-two cases a year, and in a few years the zoning resolution would be on the scrap heap because of court decisions of unconstitutionality."[77] However, "on account of our sensible provision of a board of appeals and certiorari, there has not been a single court decision against the resolution on the grounds of constitutionality since it has been in existence."[78] The board's decision on an individual variance could thus be overruled, but the ordinance itself, along with its cloaked distinctions between single-family and multifamily districts, remained unaffected.

The most dramatic case to illustrate the effectiveness of this deflection strategy involved the Astors and the Morgans.[79] The estate of William Waldorf Astor owned a Madison Avenue building that his executors wanted to convert to a dry goods store. The property was located in a residential use district and the building department refused to provide a permit. Astor's heirs then applied to the board of appeals for a variance, which was granted, on the grounds that most of the neighborhood (although once part of a very exclusive residential area developed in the 1870s) was well along to becoming a business district. This enraged neighboring property owners, including J. P. Morgan Jr., who wanted to preserve as much of the area's residential character as possible. Morgan obtained a writ of certiorari to have the board's decision reviewed by a state court, which concluded that the board of appeals had exceeded its jurisdiction in granting the variance (but left untouched the issue of the constitutionality of zoning, since that was not under review in a certiorari proceeding). The Astors then appealed the court's ruling, which sustained the board of appeal's original decision. Although the Astors eventually got what they wanted, in their appeal they attempted to get the court to consider the constitutional question (hoping that a ruling of unconstitutionality would remove the legal barrier to using the property as a store) and even threatened to appeal to the U.S. Supreme Court on the grounds that the machinery for enforcing the ordinance did not prevent confiscation of their property. But Bassett's maze foiled the effort.[80] He observed later that

> the Sheldon case was in reality between two very wealthy and influential men—Astor on one side and Morgan on the other. The court decided first one way and then another way in their case, but they could never get out of certiorari and could never get the court to consider the question of constitutionality. If it had not been for the rather perfect provisions of the Board of Appeals and certiorari, this case would surely have been adverse to the city.[81]

Bassett thus managed to focus the courts on the exercise of administrative discretion rather than on questions of fundamental rights, and in the process he effectively got the judiciary to endorse the city's use of the police power.

Bassett also prevented cases from ever reaching court from his post as counsel for the Committee to Protect the Zoning Resolution, a voluntary organization funded by private donations. Established on December 7, 1916 (only five months after the board of estimate adopted the ordinance), the committee worked "to sustain the constitutionality of the [zoning] resolution and to defend the substance of that resolution from either judicial or legislative attack"—a job that it performed, primarily through Bassett, until 1942.[82] Because Bassett knew as much about the details of the ordinance and the niceties of the law as any one in the nation, the committee provided him with an ideal position from which to counsel public officials, write amicus briefs, and funnel conflicts through the legal mechanisms he designed to dissipated them. His efforts involved "almost daily conferences in person or by telephone with city officials," including "members of the board of estimate, the chief engineer and his staff, the building department, the fire department, the board of appeals and the corporation counsel," in addition to fielding calls from citizens regarding the proper procedures for the board of appeals.[83] In this way, Bassett educated property owners and public officials about the zoning ordinance he had created, molding public policy in his own image wherever possible and directing problems through the appeals process and away from constitutional issues. "This field work," as he called it, "has assisted in the solution of many problems which otherwise might have gone to court or developed into a weakening of the law."[84] Bassett had his hands full especially during the late 1920s, when the staff of the Regional Plan helped create local planning boards in the New York–New Jersey metropolitan area and advised them on zoning. "The administration is ragged," Bassett admitted in 1929; "little by little, however, we are bringing order out of chaos. At least three times a week I have a chance to set some municipality right."[85]

Bassett also played this role at the national and state levels, insuring the diffusion of the comprehensiveness rationale and the appeals-certiorari technique and establishing legal precedents in other states. Under Bassett's leadership, the Committee to Protect the Zoning Resolution quickly became "the headquarters of the entire country for the collection and dissemination of material on zoning."[86] Bassett believed that "the courts will more readily recognize the enlargement of the police power if employed elsewhere in this state," and to further that goal he coached zoning efforts in other cities to generate successful legal cases (often certiorari decisions) that he could cite whenever the New York ordinance was challenged. Whenever a favorable case was decided, Bassett made sure that New York judges received copies.[87] He also sowed the seeds for successful cases in other states through correspondence and speaking trips. Such proselytizing laid the broader legal groundwork for the ultimate acceptance of zoning. Cases from Illinois, California, Massachusetts, and elsewhere were won with arguments developed by Bassett, thus facilitating the adoption of a consistent constitutional rationale for the extension of the police power. As Bassett told his committee in 1921,

> This national field of zoning work has been thought desirable because if many cities of the country are well zoned it establishes zoning as a proper invocation of the police power and to that extent assures the permanence of the zoning resolution in this city. If for a period of ten years only one city should be zoned, the courts would be inclined to look upon it sooner or later with disfavor. A proper and useful application of the police power tends to become universal.[88]

Not only did this make it seem perfectly natural that the police power should be extended to zoning (since so many other cities—about five hundred by 1926—were already using it), but it virtually guaranteed that a major constitutional battle over zoning would be fought somewhere else in the country. If a constitutional showdown had to occur, Bassett wanted it to happen outside of New York so that the damaging effects of an unfavorable decision could be minimized.

Although this strategy eventually provided the legal context for the acceptance of zoning by the U.S. Supreme Court, the success of the New York ordinance also inspired numerous instances of "freak zoning," as Bassett termed them, which threatened his systematic approach to precedent building.[89]

Whereas Bassett had anchored the New York ordinance with carefully constructed legal arguments, zoning enthusiasts in other cities often looked upon the comprehensiveness rationale as a license to enact any sort of regulation they pleased, which meant that they used it for purposes that Bassett considered beyond the scope of an expanded police power or that they proceeded without the expert commissions, enabling acts, or boards of appeals that guarded the New York ordinance. "There is a tendency," Bassett noted in 1922, "especially in smaller cities, to disregard the fundamental bases of zoning. They forget that zoning must relate to the health, safety, morals, and general welfare of the community. They confuse it with private restrictions. Knowing that the New York City zoning ordinance has been upheld by the courts, they think that any ordinance that may be called a zoning ordinance will be similarly upheld. Sometimes the zoning is piecemeal and preferential. Sometimes it is made to depend on the preference of a majority of frontage owners. All these short cut zoning methods invite court criticisms."[90] Favorable court rulings only made the problem worse. As Lawrence Veiller observed in 1923, "It would be a mistake to assume that because courts have sustained zoning one can 'get away with anything' by simply labeling it zoning. That is, perhaps, the greatest danger the cause of zoning faces today. And that danger comes not only from the zoning enthusiast who longs for 'fresh fields and pastures new' but often from realtors themselves, who wish to make zoning do things it was never intended to do."[91] By the early 1920s, Bassett had shown that a defensible zoning ordinance depended "on the completeness of the state enabling act, the willingness of the municipality to do its zoning within the police power (that is, base the regulations on the health, safety, morals and general welfare of the community), and the alertness and intelligence of the city attorney in forcing the cases into certiorari instead of allowing them to come up on mandamus or injunction."[92] But not everyone followed his advice.

In no state were the dangers of "freak zoning" more clearly illustrated than New Jersey. Although just across the river from the most successful and ambitious zoning experiment in the country, a "deplorable situation" developed in the Garden State when cities attempted to create exclusive residential districts without properly drawn enabling acts. As a result, Bassett regretfully reported, "practically every case has been on mandamus, and practically every case becomes a constitutional case with all the chances against a municipality."[93] New Jersey courts issued writs of mandamus in zoning cases in South Orange, Westfield, East Orange, Nutley, Jersey City, and Newark, declaring it beyond the scope of the police power to create residential districts. In such cases, overeager local officials (often at the behest of property owners anxious to keep apartments or stores out of their neighborhoods) simply assumed that declaring a residential district to be in the interest of health and safety would persuade the courts. Without the context of comprehensiveness protected by an appeals mechanism, residential districts could not meet strict police-power tests and were subject to the same orthodox logic that overturned similar efforts in cases like *Friend, Willison, Calvo,* and *Lankton*. Even Bassett's intervention (with an amicus brief) could not salvage a poorly conceived ordinance.[94] "Almost every day we are doing something intended to help the strengthening of zoning in that neighboring state," Bassett remarked in 1924, but the initial negative precedents made it all but impossible.[95]

It was more than neighborliness that compelled Bassett to help New Jersey, since freak zoning gave rise to a competing line of decisions that disrupted his plan for blanketing the country with defensible ordinances. In spite of his best efforts to prevent cities from proceeding incautiously, unfavorable rulings in state courts pushed the national debate over zoning toward the difficult constitutional issues that the New York zoning team had wrestled with between 1913 and 1916. Guided by nuisance analogies and the substantial-relation test, courts in New Jersey, Missouri, Maryland, Delaware, and Ohio struck down attempts to expand the traditional categories of police-powers regulation to cover the exclusion of businesses from residential districts, while courts in California, Louisiana, Illinois, and Oregon accepted the need to extend municipal regulatory power even where that meant restricting seemingly innocuous uses of private property.

Increasingly, courts arrayed themselves into two camps on a series of key issues in these zoning cases, all of which portended a change in the balance of power between judges and bureaucrats. In one camp were judges who ruled in favor of zoning by acknowledging that the courts played an important role protecting something like the rights of the public; that the conditions of modern urban life clearly required an increased scope for the police power; that the police power could be used to promote the general welfare alone and therefore should not be confined by traditional nuisance analogies; and that the comprehensiveness rationale was sufficiently persuasive for courts to defer to the wisdom of local authorities where the regulation of property rights was concerned. In the other camp were judges who struck down zoning ordinances by arguing that the primary role of the courts was the protection of individual rights against regulatory encroachments by the state; that while cities had certainly changed over the years, the police power should be extended only very, very cautiously; that the general welfare alone could not justify the use of the police power and that nuisance analogies remained absolutely essential to guide such use; that the comprehensiveness rationale was unconvincing, offering very little to justify new restrictions on fundamental rights; and that courts needed to make sure that police-power measures clearly met established constitutional tests before turning local officials loose to regulate private property.

By the mid-1920s, however, Bassett's strategy of fighting the constitutional battle by proxy had, in spite of an unfavorable line of precedents, succeeded in laying the groundwork for the acceptance of zoning. The constitutionality of zoning finally came before the U.S. Supreme Court in an Ohio case, *Euclid v. Ambler*, in 1926 (actually, the first Supreme Court zoning decision involved the 1916 ordinance and was handed down in 1925, but did not raise the issue of constitutionality).[96] In *Euclid*, Chief Justice Sutherland, writing for a divided court, concluded that experts had shown that zoning helped prevent fires, decrease noise, increase the amount of light and air around homes, and provide children with quiet, safe places to play. He thus accepted their portrayal of zoning as an effort to deal with the unsafe, unhealthy conditions of modern urban life and declared that they had provided a sufficient response to the substantial-relation test to give this new form of regulation the presumption of constitutionality. The veneer of health and safety that Bassett so liberally applied persuaded Sutherland to place zoning in the sphere of legitimate public regulation, even though the Euclid ordinance (which designated an obviously industrial area as a residential zone) was itself an example of freak zoning—a fact admitted even by Alfred Bettman, who argued the case for the village.[97]

Fortunately, Bassett's field work gave the court many other examples of zoning that did represent careful applications of the comprehensiveness rationale. As legal scholar Thomas Reed Powell noted of the decision: "It is fortunate for the policy of zoning that the practice had become so widespread before the issue of its constitutionality reached the Supreme Court. There is reason to believe that for a time the issue hung more closely in the balance than appears from the final vote of the court."[98] Indeed, it is remarkable that a judicial conservative like Sutherland would endorse so significant an extension of the police power. But the comprehensiveness rationale—which had made its way into numerous court decisions and legal briefs by that time—persuaded him that stores and apartment buildings in residential districts came "very near to being nuisances." By juxtaposing both traditional police-power concerns (health and safety) with experimental police-power goals (the protection of residential conditions), supported of course by testimony from numerous experts, Sutherland adopted Bassett's logic for closing the gap between clearly accepted and novel regulations.[99] By 1927, Bassett could rightly conclude of his decade of proselytizing: "Probably our work in this direction has helped to bring about the present favorable attitude of [the] courts."[100]

Although it would take the better part of another decade for the courts to sort out the full implications of the comprehensiveness rationale in practice, *Euclid* opened the way for a major change in police-powers jurisprudence. No longer would individualism embodied in law pose so great a barrier to planning advocates in their pursuit of a more orderly, efficient, and attractive urban environment. To be sure, the law was not completely plastic; zoning could not be used for any purpose public officials and real estate developers might dream up—as many subsequent court decisions would show. In fact, as lower courts struggled to apply *Euclid* to the tidal wave of zoning cases that developed after 1926, it became increasingly clear that judges required the experts themselves to adhere to their own claims that the real purpose of zoning was the creation of a citywide plan that balanced private losses against public gains in the interest of public health, safety, and an appropriately expanded conception of the general welfare. In *Women's Kansas City St. Andrew Society v. Kansas City* (1932), for example, a federal court used Bassett's logic (rather than categorical reasoning) to disallow a provision of an ordinance that excluded a philanthropic "old ladies' home" from a single-family residential district on the grounds that the restriction did not reflect "a plan clearly discernible and fixed in the ordinance itself." Because the ordinance did not embody any such effort, the court concluded that "the restriction upon the use of plaintiff's property is *not an essential of the general zoning plan.*" In short, there had been no attempt at comprehensive planning, merely a codification of local preferences that did not rise to the level of experts carefully exercising regulatory power to protect the public against individualism gone awry.[101]

Comprehensiveness thus allowed judges to hold zoning advocates' feet to the fire as they moved away from legal formalism. If comprehensiveness justified greater municipal regulatory authority, then that authority had to be exercised in accordance with its own standards of legitimacy—citywide planning, expert decision making, balancing of private and private rights. Judges adopted Bassett's own reasoning, in large part because comprehensiveness had successfully accommodated an enlarged police power to the idea of constitutional limitations on regulatory authority. That reasoning was based on the role of expertise in local policy making and thus represented something other than judicial deference to local politics. Expert balancing of public and private rights in the zoning process, verified through judicial oversight, provided a set of standards that convinced the judiciary to part with some of its authority, thus establishing the legal foundation for planning the new metropolis.

Freak Zoning in New York City

Even as the law receded as an adversary, localism—that omnipresent tendency to employ municipal powers (to tax, to spend, and to regulate) primarily for the advantage of particular neighborhoods, boroughs, or interest groups—frustrated efforts to use zoning as a planning tool. Bassett and his team supported zoning because they wanted to plan the city as a whole. Persuaded that neither markets, corporations, nor courts could adequately manage the entire city, they succeeded in partly transferring control over land-use decisions to zoning commissions, the board of estimate, and the board of appeals. By making, monitoring, and modifying zoning maps, these institutions could supplant localistic, individualistic decision makers, offering instead a metropolitan planning perspective that recognized the interdependence of land-use decisions in the five boroughs—or so Bassett had hoped. In practice, however, a far less comprehensive vision of the city shaped the use of partially centralized zoning powers.

Because zoning transferred power to a municipal board more susceptible to direct manipulation by political parties than markets, corporations, or courts, corruption posed a threat to the citywide planning perspective. While Bassett had exercised considerable influence over the CBDR, he had no control over the board of estimate, which had ultimate authority over zoning maps, and while he could monitor the board of appeals and give advice to its members and pleaders, he did so in an entirely unofficial capacity. The board of appeals was, as Bassett so delicately put it, "a sensitive spot in the zoning plan," and as the board's long-serving chairman William E. Walsh noted in 1926, "we have had almost insurmountable difficulties in maintaining its integrity."[102] Not surprisingly, it took only a few years before rumors of influence peddling surfaced, though Bassett discounted them at first. "I think there are four or five ambulance chasing lawyers or real estate men who practice before our Board of Appeals, who intentionally allow the word to go forth that they have influence with the Board of Appeals," he told a colleague in 1924; the unscrupulous fabricated such stories mainly for the purpose of "deluding innocent applicants" and drumming up business.[103] Of course, Bassett had his concerns. Since citizen members of the board were political appointees, there was ample reason to be watchful. "One cannot expect that [the board] will be much better than the administration which creates it," Bassett admitted. As late as December 1925, however, he could contend that "we have never been able to locate a case of corruption" at the board of appeals.[104]

But the rumors Bassett dismissed often seemed to be substantiated by the board's actions. In 1926, for example, the board granted a variance to the Milef Realty Corporation for a skyscraper at Broad and Beaver Streets that dramatically exceeded height regulations. The street wall of the proposed building was 248 feet high (without a setback) in a district where the maximum allowable street wall was 145 feet—so gross a violation of the spirit of the ordinance that Lawson Purdy termed it an "unconscionable exercise of power."[105] The culprit in this case, and in many other suspicious variance cases, was one Dr. William F. Doyle, formerly chief veterinarian for the New York City Fire Department, who enjoyed a "most lucrative practice" pleading cases before the board of appeals. In 1930, Doyle was investigated for bribing public officials, including William Walsh, but later acquitted of tax evasion after it was revealed that his practice of "fee splitting"—whereby he received payment from his clients in cash so that he could split the fee with the "higher ups"—was an illusion. The inquiry revealed that no money had ever been paid to a member of the board and that Doyle had capitalized on the idea that the appeals process was corrupt in order to squeeze exorbitant fees from suggestible clients. Ironically, Walsh, who was acquitted of accepting a gratuity for a favorable variance decision, resigned, while Doyle continued his practice.[106]

Although charges of corruption haunted the board during the late 1920s, illicit attempts to influence the variance process posed less of a problem than the efforts of property owners of all sorts to turn regulations intended for the city as a whole to strictly local purposes. In Queens, Brooklyn, and the Bronx, developers attempted to change use districts from "E" (single-family residential) to "C" (multifamily residential) in order to put up apartment buildings.[107] Requests for such changes often came from one or two owners who eventually petitioned their way to the board of estimate, which had the final word on zoning matters. Along Shore Road in south Brooklyn, for example, "zoning bootleggers" purchased property in a single-family home district and then petitioned for a zoning change that would have permitted apartment buildings, hoping to capitalize on a sudden increase in property values because of higher densities; although neighboring property owners objected, the bootlegger had the ear of the borough president. "In an obliging spirit the Board of Estimate has gradually let down the bars until at present all sorts of requests for changes of boundaries are placed on the calendar," complained Bassett; "shall all of Greater New York become gradually Harlemized—in the sense of solid apartment house construction?" During the depression, such attempts to change zoning laws for personal gain (requests for gas stations or parking garages in residential areas were typical) grew more frequent. "Property owners are ingenious in setting forth reasons for variance purposes," Bassett noted, and "administrative officers are prone to let down the bars" because of hard times.[108]

Even more disturbing were attempts by large groups of property owners to use zoning in ways that actually undermined planning. The "preferential" zoning cases from other cities that so worried Bassett cropped up in his own backyard. What his team of planners intended as a means to achieve a more desirable distribution of population and industry within a planned metropolis thus became a tool for institutionalizing idiosyncratic local preferences.

Bassett inadvertently exacerbated this tendency by encouraging local property owners to petition for changes in zoning maps to create exclusive residential districts. In the Rugby area of south Brooklyn, Bassett assisted homeowners who petitioned for the expansion of a residential district to prevent sand miners from using nearby property. Bassett also tried "to convince land owners that they should petition for residence districts which will have a large amount of light and air."[109] And he was gratified that "E" districts were "being created more for people of small means than for rich," especially in Queens.[110] "Tenants of small means are demanding more light and air than formerly," Bassett noted approvingly in 1922. "The second and third generations of immigrants show a decided tendency to live away from congested districts. The 3 percent immigration law may lessen the number of those who are willing to live in dark rooms. Already there are signs that builders recognize the dangers in putting up congested tenements. Loanable funds are not so readily forthcoming for dark buildings as formerly."[111] Key developers also endorsed such restrictions on land use. Walter Stabler, Metropolitan Life comptroller and one of Bassett's closest allies, cited zoning restrictions as one of the factors in Met Life's real estate lending decisions. "We could not make loans in the remote parts of Brooklyn or Queens Borough on dwellings or apartment houses if we did not know that this property was restricted to residential use," Stabler reported in 1919.[112] Bassett interpreted these developments as evidence of a broad and growing constituency for zoning—and that support reduced the likelihood the ordinance would be challenged as an unconstitutional encroachment on private rights. When local property owners and developers backed zoning, in Bassett's view, they pursued their own gain in ways that achieved the goals of the city as a whole; individual and collective interests coincided, since stable patterns of land use benefitted individuals and corporations *and* the planners trying to shape the metropolis as a whole.

By the early 1920s, however, Bassett noticed that New Yorkers had begun "to disregard the principles on which zoning is founded and to use it somewhat to favor considerable sections of homeowners who want to prevent some injurious use."[113] Where aggrieved property owners had capable attorneys, such cases exposed the zoning ordinance to the most dangerous sort of legal challenge. Indeed, the very first zoning case to reach the U.S. Supreme Court, *People ex rel. Rosevale Realty v. Kleinert*, revealed just how idiosyncratic New York zoning could be in practice. The case involved a property in the Midwood Manor section of Brooklyn, purchased by Rosevale Realty while still a "C" district. The company obtained a temporary permit from the superintendent of buildings to construct an apartment, but the same official revoked the permit the next day after hearing that the residents of Midwood Manor had filed a petition with the board of appeals to have their neighborhood designated as an "E" district, thus prohibiting apartments. The Midwood Manor Association had previously attempted to exclude apartments with restrictive covenants; but when these failed to stop the Rosevale project, they turned to zoning. In a long and trenchant brief, counsel for Rosevale argued that "E" districts were a thinly disguised effort to create exclusive residential areas (which they showed by quoting one of Bassett's own promotional pieces on zoning); that excluding apartments could not be justified in the interest of health and safety since the ordinance permitted tall, multifamily buildings in other districts (just as Walter Lindner had argued in 1913); that the zoning change was prompted entirely by the failure of restrictive covenants and thus represented the institutionalization of local prejudice rather than a good-faith effort to prevent disease or congestion (just as the courts had concluded in so many New Jersey zoning cases); and "that the real objection made to the proposed construction has no relation to the public health, peace, safety, or welfare, and is solely on aesthetic grounds"—all points for which they had good evidence backed up by ample precedent. Fortunately for the city, the case worked its way through New York courts with Rosevale contending that its rights had been violated when the area designation had been switched from "C" to "E." On appeal to the U.S. Supreme Court, the company focused on the provisions of "E" district itself, in effect, changing the question from one of the validity of an administrative procedure (altering a zoning designation) to one of the constitutionality of zoning itself (and thus the issue of fundamental rights). The Supreme Court concluded that it had no cause to overturn the state rulings on the grounds of constitutionality since that question had not been raised previously. Thus, thanks to a technicality, the flawed practice of New York zoning narrowly escaped the sort of judicial rebuke that led to its ruin elsewhere.[114]

Nevertheless, by the middle of the decade it had became apparent that the residents of Midwood Manor were quite typical of New Yorkers and that many homeowners (the very people whom Bassett was encouraging to embrace zoning) had no interest in planning at all. Motivated by localistic concerns, they actually used zoning to thwart the experts' planning efforts. Along the Long Island Railroad in Queens, for example, the zoning experts had designated vacant areas for industrial use, since they knew that future residents would need work and that future factories would benefit from convenient rail access. But as residential developments sprang up near those areas, "petitions are circulated to have the whole changed into a residence district, thus precluding future industry. Sometimes petitions come from residence districts outside of the industrial district because the property owners do not want industries for neighbors." Arthur Tuttle, chief engineer of the board of estimate and one of Bassett's allies in the zoning fight, explained how detrimental these changes would be to the economic development of the borough. But as Bassett discovered, "in many of these cases the board of estimate and especially the borough president of that borough disregarded the advice of the chief engineer and followed instead the preferences of constituents and voted in favor of the change." Where the CBDR had intended zoning maps to direct the growth of the city, local property owners reached beyond the borders of their neighborhood to prevent industrial uses in nearby districts, and the city's senior elected officials tended to go along with them. "Experience has shown," Bassett concluded in 1925, "that the board of estimate is not greatly influenced by considerations of fundamental city planning."[115]

Bassett had always conceived of residential zoning as but one element in the creation of a planned city, and he looked contemptuously on unsystematic efforts in other cities to exclude business and industry in favor of residences. Such attempts at "strong-arm" zoning posed a serious threat of adverse court decisions and many of the pre- and post-*Euclid* cases involved ill-conceived efforts to create exclusive residential areas apart from broader considerations of city planning. These efforts violated the whole spirit of zoning, which in Bassett's view was a tool for assuring that work and residence could be located in convenient, rational patterns, thus avoiding the overcrowding, inappropriate mixing of uses, and subsequent instability of real estate values that was so apparent in lower Manhattan. And while he discovered a constituency of property owners in favor of restricting the use of private property, their motives sabotaged the goals that in his mind (and indeed in the legal arguments he created) justified such restrictions for the courts.

Zoning as Institutional Success and Cultural Failure

The 1916 zoning ordinance was an expression of a civic culture of expertise that conceptualized the new metropolis in terms of physical and economic interdependence, and the movement that produced it partook of a wider debate over the need for legal and institutional change in a society that had outgrown its individualistic roots. Although Bassett and McAneny brought together hundreds of interest groups who supported the ordinance for rather narrow purposes, zoning in New York was driven by concerns that transcended deals involving homeowners, real estate developers, and merchant groups.

Most important for Edward Bassett, Nelson Lewis, George Ford, Robert Whitten, George McAneny, and Lawson Purdy, New York was not the city they wanted it to be. The conditions of urban life they witnessed every day made it abundantly clear that the techniques employed to make decisions affecting the shared environment were inadequate to the task at hand. Myriad private decisions made by myriad private property owners, coupled with the decisions made by powerful corporations, did not take into consideration essential public values, as the experts defined them. Private activity alone resulted in misuse of space, rather than a rational, healthy, aesthetically pleasing city. Markets and corporations as approaches to collective decision making were, in the eyes of the experts, manifestly insufficient to manage a metropolis with scores of skyscrapers, hundreds of factories, thousands of tenements, and millions of people. Good cities did not happen by accident or thanks to the invisible hand of self-interest, and New York proved that. The experts could imagine a better city, but the existing limits of municipal regulatory power prevented them from realizing that vision.

The state-building process they initiated to deal with this problem involved a dispute among judges, property owners, and zoning proponents over the relationship between the institutional role of the judiciary and the social philosophy adduced through constitutional interpretation. When Justice Kephart of the Pennsylvania Supreme Court declared that "we are living in an era of regulation" in a 1927 zoning case, he identified the key issue in that debate.[116] Ubiquitous regulation, a sign of the transition from individualism to collectivism, was necessary because of the increasing complexity and interconnectedness of urban life, reformers argued. Modern urban conditions blurred the lines between seemingly private activities, producing a world starkly at odds with conceptions of the boundaries between private and public apparent in most police-power cases. "In the city, economic independence does not exist," noted a 1924 text on municipal government approvingly quoted by a zoning advocate, "except as a fiction in the minds of people whom it does not please to admit that the interdependence of modern city dwellers has forever vitiated the old individualistic conception of governmental functions."[117]

If this was the social philosophy behind zoning—if accepting zoning meant discarding the ideal of individualism expressed through rights of property protected by constitutional guarantees as a quaint, archaic holdover from a simpler era—then expanding the police power did indeed represent a serious institutional challenge for the courts and for property owners. Advocates of new regulatory measures based on that philosophy portrayed judges as guardians of an outdated individualism embodied in legal formalism, bound by traditional police-power categories, anchored in common law precedent, and blind to the social and economic realities of modern life. Establishing new regulatory agencies with authority over private rights and liberties threatened to displace the courts from the central position they had occupied in the American state during the nineteenth century, while subjecting property owners to much greater control by municipal authorities. Institutional structure and social philosophy went hand in hand. And there could be little doubt as to the significance of the change proposed by zoning enthusiasts: as counsel for Rosevale Realty argued before the Supreme Court in 1925, "Are we prepared thus to arrest the natural growth of our cities; to substitute for the law of supply and demand, which it has been our experience is highly beneficial in

Figure 16. Setback skyscrapers in midtown Manhattan, 1931. Landmark skyscrapers like the Chrysler Building (center right) showed just how imaginative architects could be working within the zoning envelope. However, even less imaginative, workaday structures that were built right up to the limits of the zoning ordinance showed a certain flair, thanks to the setback requirements, which transformed them into modernist pyramids. *Source:* William Frange Collection, Library of Congress, Washington, D.C.

most human activities, and which the courts have sought to protect, for one of paternalism, restriction and regulation by the State?"[118] Judges who wrestled with the comprehensiveness rationale developed by Bassett thus sought to protect an interpretation of the Constitution, their institutional legitimacy vis-à-vis other branches of government, and a set of ideas about the relationship between individuals and their government.

By the early 1920s, New York City (and a goodly portion of urban America) demonstrated just how successful Bassett's efforts had been in this battle for institutional change. The new setback skyscrapers that graced midtown were the products of individual architects sponsored by powerful corporations, but they were also the result of a collective decision to limit private rights and force private property to conform to public specifications (figure 16). Vast stretches of suburban homes in Queens and Brooklyn illustrated that the hidden residential restrictions in the ordinance had in fact helped to disperse population by maintaining stable patterns of land use. The board of appeals had withstood attempts to corrupt it, and by the early 1930s Bassett could say proudly that "the mechanics for zoning in this city is as good as in any city of this country, and that the zoning administration here is as good as any and far better than most."[119] Hundreds of cities around the country had followed New York's lead, making zoning a basic feature of local regulatory power. Hundreds of zoning cases embodied the arguments Bassett's team had developed to support the 1916 ordinance. The scope of state regulatory authority, the judiciary's approach to decision making, indeed the very meaning of constitutional guarantees of property and liberty had been modified to suit the experts' vision of metropolitan interdependence. In very palpable ways, therefore, Bassett and his team had succeeded in altering the structure of power and the use of space in the city.

Though zoning was, by almost any measure, an institutional success, it could also be interpreted as a cultural failure. "It is remarkable to what extent the zoning plan becomes what the property owners of each district want it to be," Bassett observed in 1925. "Arguments about what will be for the benefit of the future city do not often seem to prevail."[120] The property owners who supported zoning were not very different from those who created the ordinances struck down in *Friend, Willison, Calvo,* and *Lankton.* They had no real interest in planning, either before or after Bassett found ways to persuade the courts that very similar efforts at land-use restriction were justified because their real aim was to create rational, healthy, efficient cities. For them, zoning was not a recognition of metropolitan interdependence or the limits of individualism in the modern world, but a tool for distancing themselves from "undesirable" land uses, even those that made sense from a planning perspective, such as locating industrial areas next to residential districts to shorten the distance between work and home. They embraced the letter, but not the spirit, of the legal changes achieved by Bassett, and their parochial motives for new regulations proved the point that the old restrictions on government power over liberty and property had been there for very good reasons. It remained to be seen whether the experts could organize other public institutions to employ this newly created power to plan, as opposed merely to zone, the metropolis.

CHAPTER SIX

"They shall splash at a ten-league canvas with brushes of comets' hair"
Regional Planning and the Metropolitan Dilemma

In May 1929, Father Knickerbocker, Gotham's mythic founder, returned to New York City from the Great Beyond to offer his descendants some sage advice in the pages of an anonymous essay in *American City* magazine. Two months earlier, the legislature had failed to pass a bill authored by Edward Bassett at the behest of Mayor Jimmy Walker that would have created an official planning commission for the city. Without a planning commission, there would be no city plan, and, as Father Knickerbocker insisted, it was lack of planning that caused the overbuilding and congestion so evident in midtown. History was repeating itself, he noted, reading from his *History of New York*. In 1665, the fatuous burgomasters of New Amsterdam had debated the need for a city plan; but they bickered among themselves and failed to do anything about it: "meanwhile, the town took care of itself, and like a sturdy brat which is suffered to run about wild, increased so rapidly in strength and magnitude, that before the honest burgomasters had determined upon a plan, it was too late to put it in execution." The problem had only grown worse, of course. Knickerbocker's progeny now faced overcrowding on a distinctly modern scale, with population stuffed into skyscrapers and subways "urgently needed where there is hardly room for a crowbar between already existing underground structures." In spite of these pressing problems, the legislature had cost New York another crucial opportunity to change its ways. "What our cities need is not unguided confusion on the one hand, or straitjackets on the other," he counseled, "but city and regional planning commissions constantly applying to changing civic needs the highest possible degrees of constructive imagination, technical skill and unselfish cooperation in community development."[1]

Forty-five years later, in 1974, when New York, then on the verge of its fourth fiscal crisis of the twentieth century, epitomized all that was wrong with sprawling, fragmented American cities, Robert Caro laid the blame for Gotham's decline squarely at the feet of Robert Moses: "For decades, to advance his own purposes, he systematically defeated every attempt to create the master plan that might have enabled the city to develop on a rational, logical, unified pattern—defeated it until, when it was finally adopted, it was too late for it to do much good." In a biography that credited Moses with so much (bridges, beaches, playgrounds, parks, housing, highways—that extensive list of public works that has rightly made him famous), that burden is perhaps understandable. Caro based his conclusions in part on what he heard from city planning officials who detailed how Moses, then at the height of his power, had thwarted their efforts in the late 1940s and early 1950s to develop a master plan.[2] By that time, Moses had established himself as a doer, rather than a planner—a term of opprobrium, in his jaundiced view, since planners never got anything accomplished: the ultimate sin in a city that needed so much.

As hostile as Moses was to city planners, however, New York's failure to develop according to a more "rational, logical, unified pattern" can hardly be attributed primarily to the master builder's disagreements with the city planning commission after World War II, important though that conflict was. Indeed, well before Moses concluded that planners were long-haired radicals and master planning was equivalent (borrowing from Rudyard Kipling) to "splashing at a ten-league canvas with brushes of comets' hair," New York had tried and failed more than once to save itself through city planning—as Father Knickerbocker's return in 1929 illustrates.[3] If the "Fall of New York" was the result of missed opportunities to plan, as Caro implies, then the city's decline began well before Moses ever had enough power to bear the blame.

This chapter situates Moses and his hostility to planning within the larger drama of ongoing attempts to give institutional form to the idea of the city as a whole. That notion, so much a part of the civic culture of expertise, animated the drive for planning in the new metropolis and underlay its most noteworthy institutional successes. The next step—establishing a bona fide city planning agency—required both a vision of a cohesive greater city within an interdependent region and the authority to coordinate city-building powers (from budgeting to zoning variances) to pursue that vision. In the two decades between 1918 and 1938, planning advocates fought and eventually won this battle to create a centralized planning authority with a metropolitan perspective. But that victory so exposed the conflicts between the dream of regional planning and the reality of a stubborn localism embedded in the city's institutions of collective decision making that it represented the high-water mark of the experts' efforts to institutionalize their approach to the public interest. Indeed, Gotham's most decisive move toward urban planning during this period (the creation of the city planning commission in 1938) could be interpreted as both a perfectly reasonable step in the fulfillment of Andrew Green's dream of a greater city and a planners' coup d'état that nearly revolutionized the basic structure of municipal government.[4]

Experts waged this battle on two interrelated fronts: the first to substantiate metropolitan interdependence and the second to set up institutions capable of thinking regionally and acting locally. The sheer size and diversity of the area for which the experts claimed interdependence made it extremely difficult even for them to conceptualize the regional metropolis in a politically compelling way to justify their control over city-building policies. Planning advocates splashed at something larger than a ten-league canvas: by the late 1920s, the city itself covered 298 square miles and contained more than six million residents; and the region (as defined by the Committee on the Regional Plan) comprised 5,528 square miles and included about ten million people—a population larger than every state in the union except New York. With residential densities ranging from 45 people per square mile in Putnam County (north of Westchester) to 103,823 people per square mile in Manhattan, there was hardly a more unlikely area for which to claim some essential unity.[5] Nevertheless, even before they could make a convincing case for it, planning advocates argued that the New York region constituted one community of metropolitan growth and, therefore, that policymakers had to act from a recognition of that interdependence for the city to escape its history of overcrowding, fiscal crisis, ugliness, and inefficiency. To overcome the tenacious forces of localism so evident in the city, however, the experts first had to show that they really knew what they were talking about when they insisted on a regional perspective.

Giving this perspective institutional form posed an even greater challenge, for New York typified the metropolitan dilemma: multiple governments in charge of a single economic and social community. The metropolitan solution required one government for that one community.[6] But in the absence of regional government (something not even the Committee on the Regional Plan advocated), state and municipal officials would have to implement a regional plan, aided by those few institutions, like the Port Authority, that looked beyond established political boundaries. This made for an intergovernmental project of staggering proportions. By the mid-1920s, the region included three states and 436 local governments, each with elected officials, department heads, and ward bosses, all dealing with a profusion of interests divided along geographic, ethnic, party, and economic lines and pursuing their own conceptions of their communities' futures under differing legal, financial, and administrative constraints.[7]

New York City itself offered the most apparent and problematic instance of this institutional fragmentation coupled with an adamant localism. Even into the early 1930s, the Committee on the Regional Plan described New York as "really a federation of about a hundred separate communities" that "still retain their local patriotism." Regional planning advocates freely admitted that "in the boroughs themselves neighborhood leadership is stronger than borough leadership, and there are stronger affiliations between the citizens and some definite local unit than between them and their borough or the city as a whole." They also perceived a growing recognition of the "one great community" of consolidated New York, but, as a practical matter, borough boundaries, and boundaries within boroughs, marked distinct communities with little devotion to the larger dimensions of collective life that so concerned the experts.[8] Although they had established a number of institutions that operated from a citywide perspective (at least in theory), a single planning authority coordinating all those public functions involved in city building did not exist. Creating that sort of coordinating agency required a dramatic reconfiguration of power within the public sector since such an authority would necessarily take power out of the hands of those officials already in charge of building roads, bridges, docks, and sewers, crafting zoning maps and granting variances, and parceling out chunks of the city's borrowing capacity. For New Yorkers to take planning seriously thus implied a degree of centralization far greater even than that represented by the creation of the Port Authority.

This debate over a centralized planning authority capped a decades-long struggle to change municipal officials' approach to the pursuit of the public interest through city-building policies. Experts had long argued that the city's institutions of collective decision making—courts, parties, corporations, and markets—could not address a broader conception of the public good, oriented as they were toward private, competitive, individual, or local interests. With their call for centralized planning, the experts questioned the formal structure of local government itself, directly challenging the ability of the board of estimate and borough presidents in particular to make policies that would safeguard the metropolis from ruin.

Moses watched this process of institutional warfare during his own rise to power and concluded that planners asked too much and accomplished too little. Although he thought of the city in terms similar to those used by the experts at whom he spewed his poetic invective, he chose to pursue his vision of the future by means that were more attuned to Gotham's dominant civic culture, and he succeeded in turning existing political institutions to his own advantage when planning advocates chose to fight them. Ironically, by the time Moses was using Kipling to disparage city planners, he could claim to be more "democratic" in his methods than they.

Discovering the Region: Congestion, Decentralization, and Interdependence

In November 1917, New York took a decisive step away from city planning—one of several such steps during the twentieth century. A mere sixteen months after the adoption of the zoning ordinance, voters swept Mayor Mitchel and other reform candidates out of office.[9] The new mayor, Democrat John Hylan, said that he had no use for "art artists" (a cryptic reference to City Beautiful planners), and on February 1, 1918, the new board of estimate abolished the city planning committee, fired its staff, and ceased municipal sponsorship of urban planning.[10] Had it not been for a popular outcry in support of land-use restrictions from Hylan's own Brooklyn neighbors, the new administration would probably have repealed the zoning ordinance as well.[11] Abruptly removed from power, the experts who hoped to translate zoning into planning had the next eight years to make a convincing argument that they should be given greater control over city-building policies.

As a result of the Hylan housecleaning, in the crucial years between 1918 and 1926 planning advocates could only watch as New York "took care of itself"—a modern city, with skyscrapers and zoning laws, suburbs and subways, emerging from one fiscal crisis and sinking into another, growing by a million and a half residents, and struggling to cope with the new dimensions of an old problem: congestion in lower Manhattan. On a typical workday in the mid-1920s, the equivalent of half of Gotham's six million residents crammed into Manhattan below Fifty-ninth Street, almost as many as had been living in the entire city in 1898. After the workday ended, only slightly more than a million remained in that part of the city as permanent residents—about half the total living in all of Manhattan. Nearly two million commuters crowded into the central business district everyday, along with more than two hundred thousand automobiles, trucks, taxis, and buses.[12] It was as though all of the inhabitants of Mississippi or Philadelphia moved downtown five times a week, creating a city within a city as populous as Chicago or Wisconsin. And this became the experts' central concern through the end of the 1920s.

During this exceptional period of boom and bust, planning advocates refined their sense of the city's predicament and concluded that uncontrolled decentralization, fueled by persistent congestion in lower Manhattan and abetted by misdirected public policies, constituted the fundamental problem facing the new metropolis. As an alternative to this future of overcrowding, blight, and sprawl, they proposed a vision of multiple, diversified, interconnected urban centers—a vision that required both a regional perspective and administrative centralization and that had its most detailed expression in the *Regional Plan of New York and Its Environs*.

The Regional Plan was the brainchild of Chicago business executive and financial expert Charles Dyer Norton, whose remarkable career brought him into contact with so many of the leading figures and projects of city planning that he seemed destined for the job of reshaping the nation's largest metropolis. With the help of consulting engineer Frederic A. Delano—a relative of Franklin Roosevelt who had risen through the ranks of the Chicago, Burlington, and Quincy Railroad and worked with Norton to fund Daniel Burnham's famous Plan of Chicago in 1906—Norton persuaded the Russell Sage Foundation to continue the work begun by the defunct city plan committee but on a much larger scale that would allow him to "find the man to plan for New York as L'Enfant planned for Washington or Burnham for Chicago."[13]

The Regional Plan that emerged from this unofficial but very influential effort bridged an important gap between planning traditions in New York and Chicago. Chicago had Burnham, but his plan remained largely a work on paper since that city had very few policy tools to implement his vision.[14] New

York became the national leader in municipal control of building height and land use, but never had a Burnham, in spite of Norton's best efforts to find one.[15] Planning advocates in both cities lacked an understanding of the organisms they tried to reshape, and here the Committee on the Regional Plan made its greatest contribution. Barred access to the levers of government power by Hylan's hostility to planning, Norton's group could devote their full time to refining their understanding of city-building processes. As a result, the *Regional Plan* was built upon the *Regional Survey*, which admirably embodied the distinctive ethos of reform of the Sage Foundation, whose aim was to discover the root causes of specific social problems so that regulators could effect real change.[16] As Frederick Law Olmsted Jr., one of the leading lights of the Regional Plan, put it in 1923: "Before attempting in any large way to exercise deliberate control over what New York physically is to become, we ought to have as clear an understanding as we can reasonably get of the answer to the question, never completely to be answered, 'Why is New York?' "[17] That meant discerning the patterns behind the seemingly uncontrolled growth that pushed the city upward, downward, and outward. With that goal in mind, Norton, Delano, and Olmsted "decided that they must have a *regional* plan, for it was useless to just go on with only a *city* plan," recalled George McAneny.[18] To prepare themselves to take the reins of power when Hylan departed, the experts thus attempted to see the city's problems more comprehensively, as manifestations of larger patterns that encompassed the entire region.

Discerning those interconnections between the region and its constituent parts required a conceptual leap forward—a leap based, at least initially, on faith rather than facts. Predisposed to think in terms of the whole region, Norton's experts had already completed three years of work before they seriously debated what would appear to have been the most basic question of all: "Are we justified in saying that there is a region for which there can be a regional plan?"[19] Admittedly, every municipality faced the same basic problems, such as providing citizens with water, recreation, and health services.[20] But these generic challenges had not yet encouraged a regional perspective; indeed, on those subjects that seemed most obviously regional in nature (like harbor pollution), officials had consistently avoided thinking beyond the edge of their jurisdictions. Transportation, too, linked cities, but the fact that railroads and highways cut across political boundaries had never spurred cooperation. If these factors alone held the region together, then the Regional Plan was as quixotic as George Soper's dream of a citywide sewage authority.

Scottish planner Thomas Adams, who became director of research for the plan, set a higher standard for regional interdependence. "A region is an area that functions as a unit—industrially, economically and socially, in spite of political sub-divisions," he insisted—a definition very close to those employed by port planners like Calvin Tomkins and Julius Henry Cohen. Not surprisingly, Adams admitted that "the main thing that ties the region together is the Port of New York." Though important, the role of the port did almost nothing to explain why local officials should think regionally to respond to overcrowding in Manhattan or suburban sprawl in Queens. As Adams himself pointed out, aspiring planners had to contend with the fact that "the region can be a structure in respect to certain social features and yet a series of parts in respect of other features."[21] The experts would have to make the case for intimate linkages among some elements of the region—elements requiring a regional plan—while admitting the independence of others, which could happily survive without officials looking beyond their borders.

Because the integrity of the region was more easily assumed than substantiated, the experts' offered their initial answers to the fundamental question of regional unity in such schematic terms that it cast the very notion of regional planning (along with their claims to authority over city-building policies) into doubt. As Delano confided to one of the planners on his staff during this debate in 1924: "It would take a man of boldness and prestige to do something which would convince the public that we have known what we are talking about. The man or men who will help us most may or may not be an architect, may or may not be an engineer, but it will take some real seer, whatever may be his calling."[22] Even the experts had a hard time seeing the relationships between whole and parts amid the profusion of detail confronting the regional planning team. "My thought is that some man like you," Delano wrote to engineer William Wilgus that same year, "must climb to an Olympian height and surveying the field from afar make up his mind what the salient and essential things are."[23] As late as the middle of the decade, in other words, planning advocates had not yet fleshed out the key connections between center and periphery that would allow them to advance a regional vision in the face of far more palpable local interests and thus justify their access to public power.

Three interrelated insights into the process of industrial and residential decentralization then taking place in New York provided the conceptual breakthrough that resolved Delano's doubts and gave regionalism real credibility. The first emerged from the experts' historical sketch of cities in the regional periphery, which revealed that specialized industrial towns and residential suburbs had diversified as a result of decentralization. In their formative years, the survey showed, Danbury had specialized in hats, Perth Amboy in clay products, and Paterson in silk production. But a variety of other businesses had lately moved into these areas, creating "rounded communities rather than one type towns." "A fairly definitely functionalized industrial region had evolved," concluded survey researcher Wayne Heydecker. "Such an evolution presupposes *an organism*"—an interdependent regional system of economic development subject to changes affecting all of its subcomponents. "The entire area is in a state of flux," he observed. "Industry is scattering. It is being pushed out from New York and Newark by high rentals, and it is seeking communities in which lower priced land is available, where workers can be found."[24]

That process of industrial decentralization and diversification resulted in local planning problems. To accommodate new industrial growth, for example, Roselle, New Jersey, designated its eastern border as an unrestricted land-use zone; but that put incoming industries right next to one of the best residential districts of neighboring Elizabeth. Try as they might to see themselves as distinctive communities pursuing their separate futures, in other words, Roselle and Elizabeth, Danbury and Perth Amboy in fact responded to the same larger trends emanating from lower Manhattan, and their reactions to those trends often had a direct bearing on the quality of life they could offer their citizens. Heydecker considered this a hopeful sign. Previously, he noted, "each community thought itself unique, and felt pride in its isolation. Today, struggling against problems beyond their control, these same communities are beginning to see that their security lies in cooperation." Understanding that bigger picture would lead "ultimately to a regional consciousness from which of sheer necessity will come the recognition of a need for a Regional plan"—or so he hoped.[25]

Columbia University economist Robert Murray Haig—E. R. A. Seligman's protégé and the director of the economic and industrial survey for the Regional Plan—led a team of researchers who provided the second and most important insight into regional interdependence: a convincing explanation of industrial decentralization. Their three-volume study constituted the most direct effort "to explain the city itself" by ascertaining "what is doing the crowding and what is being crowded" in lower Manhattan—the engine driving the process of economic change for communities throughout the region.[26] Haig's analysis showed that intense competition had "crowded out" many of the manufacturing activities in lower Manhattan.[27] But this was no mass exodus: some stayed while others left, as individual business owners balanced the advantages of downtown (proximity to transportation, finance, buyers, and labor) against the advantages of decentralization (cheaper rents and more space).[28] As a result, cigar factories, which were concentrated below Fourteenth Street in 1900, had shifted first to Jersey City and Newark, then to Perth Amboy and New Brunswick, driven out by the high cost of land and their inability "to stand the competition of other industries for the labor supply of New York."[29] Wood manufacturing had likewise shifted from downtown to outlying areas, especially Queens and Brooklyn, during the early twentieth century; but many small factories continued to thrive at the core—those that produced very-high-grade cabinets or furniture requiring close connections between marketing and manufacturing.[30] Between 1900 and 1922, foundry work "practically disappeared from Manhattan," with the heavy and bulky branches of the industry relocating to Brooklyn, Queens, and New Jersey, where lower land values allowed factories to achieve economies of scale; but market-sensitive products continued to be manufactured downtown: producers of design-grade metals remained in lower Manhattan close to architectural firms, their primary users; high-end jewelry manufacturers, who made to special order and distinctive customer preferences, also clung safely to central locations.[31]

The garment industry, which retained its hold on lower Manhattan in spite of the decentralization of some of its component trades, displayed these same patterns. Employment in garment making had risen substantially between 1900 and 1922, especially in lower Manhattan. Because tastes in women's clothing varied so greatly and so quickly, the industry had developed a "marvelously flexible organization" that "could scarcely exist anywhere else than in a great city." Styles changed; buyers ordered; jobbers responded; small factories mushroomed and then disappeared; and the cycle began anew when the season changed. Manufacturers needed to stay close to retailers to remain in the loop, but the work depended so much on labor rather than specialized equipment or facilities that the industry could thrive in obsolete buildings (those constructed for other purposes but abandoned by their original tenants, like loft buildings on Fifth Avenue), allowing it to maintain a strong presence downtown. The men's wear industry, less sensitive to stylistic changes, showed a greater tendency to decentralize, even though it, too, maintained a strong presence downtown. Survey researchers predicted that sales offices for men's wear would stay in lower Manhattan where they could respond quickly to market demands, but fabrication would increasingly take place in Brooklyn and other outlying sections of the city. Men's shirts and hats already showed a more pronounced movement away from Manhattan, with Queens, Brooklyn, parts of New Jersey, and even Connecticut emerging as locations where these easily standardized goods could be produced.[32]

The nuanced picture that emerged from the work of Haig's survey team revealed that an array of industrial and nonindustrial activities prospered in the crowded, expensive conditions at the region's economic core. Even as the financial district fulfilled its destiny "to be the home of the large national and international corporations that direct our industrial and financial life," lower Manhattan would retain a remarkably diversified manufacturing economy. The fastest growing occupations, like investment bankers, lawyers, and accountants, served the needs of big business, but photoengraving, dressmaking, newspaper printing, and technical-instrument production had good reason to survive well into the future. And even though retail was decentralizing, as shopping centers in the suburbs took over sales of standardized goods that could easily be made outside of downtown, "New York City's growing importance as a world city and a center of culture points to an increase in its trade in articles of luxury—the unstandardized, the beautiful, the rare and the artistic"—which meant that many specialized small manufacturers would continue to prosper downtown where they could respond most quickly to specialized buyers.[33]

The evidence Haig assembled did not support a knee-jerk reaction against industry in the commercial core. It showed instead that certain industries "will continue to cling tenaciously to sites in the center of the metropolis"—and appropriately so, since the city served as an ideal environment for those manufacturing activities that required no specialized buildings, that served time- or taste-sensitive markets, that relied on specialized labor, and that were relatively small in scale. "Manhattan will remain the center for such activities as benefit greatly from a high degree of accessibility and proximity to each other," the regional planning team agreed in December 1926, "including marketing, banking, shipping, *as well as many kinds of fabricating.*" Indeed, Haig specifically warned against attempts (based on a "hasty and superficial" conception of industrial location) to prohibit activities like garment work through zoning since many of them acted as "chink-fillers"—using obsolete buildings temporarily and providing productive life for structures in transition.[34] Other industries—those requiring large or specialized buildings, those generating waste, fumes, or noise, and those for which time, service, or fashion were unimportant—would naturally continue to "abandon central sites." This did not mean that industries randomly decamped to the suburbs. Decentralization proceeded logically, with seasonal industries drifting eastward into Brooklyn and Queens and those requiring large quantities of raw materials and those serving continental markets gravitating toward New Jersey.[35] The regional core would become more corporate and commercial as large-scale industry moved to the periphery, but an equally compelling economic rationale portended the continued presence—the continued growth—of a variety of industries downtown.

Haig concluded that although the process of competition for prime locations would, on its own, "approximate the ideal layout" of industries throughout the region (in spite of the various inefficiencies, natural obstacles, and "market peculiarities" disrupting the smooth functioning of this system of "territorial specialization"), even when decentralization worked perfectly it increased "friction of space" (the distance between industries and between work and residence), a term Haig coined and that subsequently became a major conceptual element of the plan.[36] By regulating land use and providing transportation, planners could eliminate or reduce the costs of "friction"—the costs of moving goods and people over growing distances—thus redirecting the interdependent processes of concentration and dispersal toward more desirable ends.[37]

Haig even thought it possible that planners could one day specify exactly "where things belong" in the vast metropolitan region, since his analysis provided "a scientific basis for zoning."[38] Equipped with that understanding, planners could exercise "social control" over individual choices and market processes and thereby mold economic and social conditions for the betterment of the entire region. Imperfect markets would no longer have to suffice as institutions of collective decision making; planners would no longer have to rely on allies in the real estate business for advice on how to draw zoning maps or where to put subway lines. Haig's study gave planners the knowledge to treat the distribution of land uses in the city with the same purposeful direction that the great industrialists used when laying out a modern factory.[39] The implication of this claim was not lost on Adams, who knew precisely what Haig's research findings meant: the factors affecting metropolitan growth and arrangement "are subject to government control."[40]

A simultaneous process of residential decentralization—the third key insight developed by the survey—made that public reordering of land uses all the more urgently needed. By the mid-1920s, census data showed that Manhattan, for all its congestion, had been losing residents for at least a decade. Its population peaked around 1910 at 2,331,542, declining slightly to 2,284,103 in 1920, and slipping to 1,945,029 five years later. The state census of 1925 confirmed what students of demographic trends in the city suspected—that Brooklyn, with 2,203,991 residents, had surpassed Manhattan. In fact, every borough other than Manhattan showed steady growth—especially Queens, which swelled from 469,042 residents in 1920 to 713,891 in 1925. These developments had radically changed the distribution of population throughout the city. In 1905, 53.76 percent of New Yorkers lived within a four-mile radius of City Hall, with 35.90 percent living between four and eight miles out and only 8.55 percent within eight to twelve miles; by 1925, the percentage of city residents within the four-mile zone had dropped to 29.45 percent, while increasing to 44.88 percent in the four-to-eight-mile zone and leaping to 22.13 percent in the eight-to-twelve-mile zone. Nassau County to the east of the city, Westchester County to the north, Bergen, Essex, Hudson, Middlesex, Monmouth, Morris, Passaic, Somerset, and Union Counties in New Jersey, and even Fairfield County in Connecticut had all grown while Manhattan lost residents. An outward shift of population was sweeping the region, driven by the process of "crowding out" taking place at the urban core.[41]

Although survey researchers saw value in residential dispersal, much to their dismay they discovered that it reproduced the crowded, unhealthy conditions typical of the worst tenement districts in lower Manhattan.[42] "Deplorable conditions are growing up in the suburbs of New York City," they concluded, and "new foundations are being laid for future slum areas" in the outer boroughs, making "the remedy [for congestion] no better than the disease." Brooklyn, formerly known as a city of homes, "is now seeking to emulate Manhattan in the density of much of its building," with almost fifteen thousand more new-law tenements by 1927 than the city's formerly most populous borough.[43] Low-density development presented problems as well, since it often led to the "wasteful scattering of buildings" in suburban areas. "Next to the problem of congestion the most defective condition is the too wide dispersal of houses on premature subdivisions in the semi-rural areas," Adams emphasized.[44] As Mayor Jimmy Walker noted in 1928, "people in their anxiety to live in Queens have rushed ahead of civilization itself," with unpaved and unsewered streets commonplace.⁴⁵ The city's "policy of non-interference with the place and manner in which private enterprise engages in housing" allowed speculators to cut up property "prematurely and injudiciously and the unfortunate consequences retard or even blight what should have been a healthy and orderly growth." Residential decentralization in its present form, concluded the survey, would only reproduce the most undesirable conditions of lower Manhattan, increasing the cost of getting back and forth to work and doing business and fragmenting the region into "unrelated areas of very rich and very poor."⁴⁶

Here, then, was a compelling rationale for thinking regionally. Taken together, these three crucial insights—industrial relocation to the periphery, a process of "crowding out" at the center, and unplanned residential decentralization, all contributing to increasing "friction of space" in the region—made plain what the experts assembled by Norton, Delano, and Adams had sensed from the beginning: that the millions of citizens living within the region belonged to one system of metropolitan growth. The connections among them were not merely skeletal—not merely flows of wastewater or ribbons of rail. Nor was it simply the port that held them together. The ties of regional interdependence remained invisible to local officials, developers, and homeowners, since so many "present themselves only when one thinks of the totality rather than the parts." To see them, one had "to recognize that the great whole is a living thing, with a certain spirit of its own, a sort of anatomy, and something like a functional physiology."⁴⁷ Every community in the region was caught up, to a greater or lesser extent, in the same overarching processes in that organism; they all belonged to "New York's planetary system" and thus fell within the central city's "gravitational influence"—whether they recognized it or not. "If this study does nothing else," the survey emphasized, "it proves that the entire New York region needs to be planned, irrespective of political boundaries, if its problems are to be solved"—one community, one government.⁴⁸ No one could deny that now, at least in the eyes of survey researchers, regardless of how local communities tried to distance themselves from one another: the evidence clearly showed that they were not dealing with isolated, local problems—how to find laborers for the cigar factory in Jersey City, how to provide drainage for new homeowners in Queens, how to prevent the garment workers from blocking the streets in midtown—but with the myriad manifestations of one common phenomenon caused by the push and pull of concentration and dispersal surging away at the center of the region.

Rather than fighting these trends, the experts of the Regional Plan devised a strategy of "diffuse recentralization"—a three-pronged effort to channel existing patterns of change along more desirable lines.[49] The plan would foster the continued movement of industry to the periphery, both to lessen congestion at the center and to create diversified subcenters in suburban areas. Residential decentralization would likewise proceed, but into "compact" neighborhoods "integrated with industrial sections so as to reduce distances between homes and places of work" (i.e., friction of space). And to make these emerging communities complete, the plan would encourage the "sub-centralization of business so arranged as to provide the maximum of convenience to residents," rather than allowing it to spread itself "in straggling lines along every main highway," as it did so often in unplanned areas of the metropolitan periphery. In each case—whether their plans for widely spaced, setback skyscrapers along broad boulevards downtown or their vision of dense, mixed-use industrial suburbs with ample green space in New Jersey—the experts attempted to use their superior understanding of regional growth to salvage the virtues of density from the perils of congestion.[50]

This outline for regional growth reflected both the experts' preoccupation with congestion and their embrace of centralization—a combination that only *appears* contradictory. "All communities must have centers," they stated in the survey, so matter-of-factly that it seemed as self-evident as a census statistic; "what is wanted is not decentralization but the reform of centralization." Manhattan would continue to be the "mother city" for the region, surrounded by interrelated but economically and socially self-sufficient communities. This was not a rejection of urban life (with its messy proximity) in favor of homogenized suburbs free of industry, nor elaborate window dressing for an effort to remove factories and immigrants from lower Manhattan to make way for a shining, sanitized corporate-commercial downtown. Instead, the plan sought to recreate the dense, mixed-use character of downtown in largely self-contained communities on the periphery, recognizing both regional interdependence and local independence, and preserving the values found only in closely built areas while encouraging the comforts of well-designed suburbs and small towns.[51]

To achieve this vision, the experts insisted that transportation and land-use planning proceed interdependently, and this constituted the most significant operational recommendation to emerge from the plan.[52] Such coordination precluded relying on the railroads, even though the plan contained recommendations for rail beltways to move freight around the congested center.[53] Because they worked closely with the Port Authority, the regional planners knew that railroad planning was "outside the control of local authorities" and largely the product of battles between federal and state governments and the railroads themselves. "The placing of the railroad and transit proposals on the Graphic Plan does not mean, therefore, that there is any illusion on the part of the Regional Plan staff, as to the absence of power to give effect to these proposals as part of a city plan," they unabashedly admitted.[54]

Instead, they offered a vision of an automobile city, since public authorities had more power to decide when and where to build public roads. Of course, private motor transportation could easily aggravate the problems the plan sought to avoid: the experts recognized that automobiles had "already begun to separate the city into detached neighborhoods between the great radial arteries that spread out from the center of the city."[55] With this in mind, they presented not merely the outline of a highway network, as they did for railroads, but "a completely coordinated and classified system of highways" integrated with land uses for the entire region. Outer beltways would direct traffic away from downtown and among growing communities on the periphery; inner loops would follow the working shorelines of the port, facilitating the movement of cargo to ships and onto radial roadways extending outward from the core. In this sense, the highway system mirrored the rail system, creating a parallel and largely redundant transportation network, but one built with public monies and under public control. Indeed, the success of highways obviated the need for a rail system: "It is admitted that if a bold and comprehensive treatment of the highway system is pursued," the plan noted, "a great deal of the expense in railroad and transit line construction may be avoided"—as long as planners coordinated transportation with land use.[56]

Properly coordinated by local officials, transportation and land-use plans would redirect existing trends of concentration and dispersal into more desirable patterns of development, reducing the "friction of space" that Haig had identified as the chief problem created by decentralization. Because those trends were driven by processes at the urban core that did have an economic logic, planners could capitalize on the most positive features of the urban evolution taking place before their very eyes, while avoiding the repetition of mistakes in newly developing areas and gradually reconstructing downtown as advances in municipal regulatory and borrowing power permitted. By taking into consideration the movement of heavy industry to New Jersey, the flight of tenement dwellers to the suburbs, the decentralizing force of the automobile, and the aesthetic possibilities of the skyscraper, planners could coax that evolution in the right direction, allowing for proximity without congestion, decentralization without sprawl, and local pride without fragmentation—and that combination justified their power over city-building policies.

Critic Lewis Mumford lambasted this vision as a concession to the plutocrats who profited from those urban forms and processes that produced the overcrowding so characteristic of the megalopolis, but he never quite understood the battle the regional planners faced.[57] "Any effort to relieve or reconstitute the metropolis must go *against* the basic pattern," Mumford insisted—precisely the opposite of the plan's strategy.[58] Adams answered this attack in similarly strong terms, describing Mumford as "an esthete-sociologist" with lofty but unworkable ideals who had little conception of what it took to plan a region in a "democratic country."[59] And while the Mumford-Adams conflict was in part a battle between a visionary and a pragmatist, that tension suffused debates among the Regional Plan staff early on, with Adams trying "to steer a safe course between the boldness which is based on getting results of any kind so long as the results are obtained, and the idealism which is indifferent to results."[60]

In the end, the plan exhibited a greater willingness to imagine possibilities beyond the scope of existing laws and heaped more criticism on the limitations imposed by property rights than Mumford acknowledged.[61] But Mumford wanted a more explicit and vigorous condemnation of capitalist urbanization from the plan, and Adams knew that hardly provided a workable theme for building a coalition to implement the vision. Adams understood, as Mumford did not, that the real difficulty of regional planning was convincing local officials to think regionally and use the legal and administrative tools they already had at their disposal for that purpose, rather than getting them to assist in the liberation of technics from capitalism.[62]

Implementation had to play a central role in this process because, as Adams observed in 1925, "we have no power to execute any part of the Plan."[63] The Committee on the Regional Plan had no power to float bonds to build bridges or even erect a park bench, not even the power (which the Interstate Sanitation Commission would soon use to good effect) to bring suit against municipalities—rendering it even less formidable than the "toothless" Port Authority. Planners therefore had to cultivate relationships with local officials to get them to see the same big picture and acknowledge common problems requiring a coordinated solution.[64] In part, that involved founding new institutions to do the job where none existed before, and the staff of the Regional Plan devoted a good deal of their time "to the establishment of official agencies to carry out parts of the Plan."[65] It also involved consulting extensively with local officials, both to educate and to persuade. "Our plan has been prepared with almost unanimous consent and approval of the engineers worth consulting throughout the Region," Adams declared in 1930. "We could have made the plan in half the time if we had not tried to get agreement on all the major proposals with the representatives of the local authorities."[66] Laborious process though it was, Adams's staff took that time to avoid creating the impression "that the Regional Plan had been put forward over the heads of local officials and authorities," and they were "frankly astonished" at the level of agreement their efforts produced. In this sense, the Regional Plan of New York differed from similar efforts in Chicago and Philadelphia. Where those cities built regional plans from the bits and pieces of local plans (the whole being the sum of the parts), New York emphasized the *regional* aspects of the plan, encouraging local officials to think about their role within a larger system of development (the whole being greater than the sum of the parts).[67]

But agreeing on the plan and carrying it out were two different things. Maintaining that crucial regional outlook proved difficult when it came time to spend money and regulate property, since everyone wanted to make adjustments to suit local tastes. And while Adams always characterized the plan as an "elastic outline," little changes could quickly undermine its regional orientation.[68] "What I think requires to be avoided is unduly emphasizing the fact that the plan is merely a preliminary skeleton which requires to be revised merely to satisfy some local point of view that may be unduly stressed by special interests," he cautioned.[69] Localism, the omnipresent enemy of regionalism, would eat away at the vision unless the experts coaxed the actions of localities toward the fulfillment of that bigger picture. Some of the most inspired—and, in a sense, desperate—sections of the plan hailed an emerging regional consciousness that would guide the appropriate expression of local interests. "People take pride in local places within the city and in many cases the unit for social and civic endeavor is a street rather than a district," the plan affirmed. "Neighborhood unity is a good thing, but the satisfaction to be obtained from it depends on qualities of environment and facilities for communication that can only be secured by working through and with the whole city or Region."[70] Adams acknowledged "the value of that degree of disunity in social and political organizations that is essential to vigorous local community life." But in the end, "the whole is greater than the part," even though "it is the strength of each part that makes for the greatness of the whole"—a tricky balancing act, to say the least, requiring officials who could see both dimensions.[71]

The experts working on the plan thus sought to connect the whole and the parts of the new metropolis through a quasi-scientific voluntarism built around a shared regional vision grounded on a social scientific understanding of trends in industrial and residential decentralization. By 1929, they could say with certainty that they understood New York better even than railroad executives, real estate developers, mortgage lenders, or the most seasoned local official. They also knew what it would take to save the metropolis from ruin: the coordination of land use and transportation. They had documented the ties that made the region one community, and now they had the perspective to govern it as one community. In the absence of regional government, however, they would attempt to coordinate existing institutional resources (creating a few of their own whenever possible) by persuading officials to think regionally and act locally. If they could convince them that larger trends shaped their fate—and it was a big *if*—they could go a long way toward enacting the plan. A central question remained: would New York City, the largest and most important municipality in the region and the source of most of its strengths and weaknesses, organize to act on the experts' vision?

Institutionalizing Planning, 1926–1936: The Battle against the Boroughs

"Today, New York has no plan," declared George Ford, general director of the Regional Plan Association (the organization established to oversee the implementation of the plan) in 1930.[72] True enough, the *Regional Plan* served as a brilliant beginning—a vital source of direction in a city where multiplicity could easily overwhelm one's sense of the whole. New York likewise possessed "very substantial power to carry out plans"—vast spending capacity, the best zoning administration in the country, the world's longest subway system, and expertise beyond measure.[73] But it had no official organization (other than the board of estimate, which oversaw everything in government) to coordinate the use of those powers toward that lofty goal. In spite of this lack of coordination, the city constructed roads and sewers and public buildings and regulated land uses more vigorously than any other place in the nation—but not as steps in some larger plan and certainly not with the goal of reducing "friction of space." Instead, borough presidents directed much of the city building that went on, responding to pressures from developers, neighborhood associations, or political operatives and fashioning their own empires with public works and zoning maps. New York had not just one mayor, the saying went, but five little mayors, who voted on their own appropriations and zoning changes through their positions on the board of estimate. Planning New York City—coordinating the multitude of existing city-building activities to reshape the interdependent processes of concentration and dispersal—thus required the centralization of power jealously guarded by the borough presidents and the board of estimate. Changing that configuration took ten years and a sweeping charter revision and pitted citywide planning (and the hopes of regional planning that relied upon it) against a fractious localism embodied in the very structure of city government.

Borough autonomy had troubled the city since consolidation and was itself an outgrowth of the conflicts that had frustrated Andrew Green, the father of the greater city, during the late nineteenth century. The communities that became the outer boroughs, especially Brooklyn, had resisted consolidation for fear that Manhattan would dominate them; Green, on the other hand, envisioned a central government capable of addressing the needs of the city as a whole. The charter of 1898 attempted to strike a balance between those competing concerns—unification and borough autonomy—with specific structural compromises: district boards in each borough determined some local public works priorities, but the municipal assembly controlled appropriations; a public improvements board composed of the mayor and department heads made policies with regard to water supply, highways, street cleaning, bridges, public lighting, and sewers, with borough presidents acting as ex-officio, non-voting members; citywide departments, with executives appointed by the mayor, managed city-building programs, but with branches in each borough to maintain familiarity with local conditions—all to insure that public works proceeded on a citywide basis, since, in the opinion of the charter commission, borough control of physical improvements would lead to "disintegration and not consolidation."[74]

Almost immediately, local officials protested this system, and in 1901 a new charter reorganized city government, striking an alternative balance between centralization and autonomy that clearly strengthened the boroughs.[75] Police, fire, finance, docks, and taxation remained centralized, but almost all key public works became the exclusive domain of borough presidents; henceforth, they had charge of the construction and maintenance of highways, sewers, and public buildings (in Queens and Richmond, they also controlled street cleaning and garbage disposal) and were responsible for making and updating city maps—the starting point for all physical improvements. Borough presidents also appointed their own commissioners of public works and superintendents of buildings and thereby controlled the legions of municipal workers involved in those activities. The 1901 charter revision also gave borough presidents a greatly enlarged role on the board of estimate, which became the most important administrative *and* legislative body in government, managing the public purse and deciding what each borough could spend. In the new scheme of things, borough presidents could vote on all funding decisions, whether citywide or exclusively for their borough, thus transforming them into "commissioners of public works" with the power to determine their own spending—a dangerous combination of executive and legislative authority, warned Mayor Robert Van Wyck (no reformer he!), last possessed by Boss Tweed, who served as commissioner of public works and effectively controlled the officials who voted on his budget. Andrew Green, seeing his handiwork undone, correctly anticipated that the new charter would usher in an era of extravagant spending and chaotic growth as the boroughs pulled in their own directions to the detriment of any citywide plan.[76] And while subsequent charter commissions confirmed that "the division of planning among the five borough presidents had resulted in the complete absence of a city plan," borough autonomy remained a basic fact of governmental structure in New York for the next thirty years.[77]

Into this mix stepped New York's one-hundredth mayor, James J. Walker, who snatched the office from John Hylan in 1926. It is a peculiar fact in Gotham's history that Gentleman Jimmy Walker, a machine politician who looked up to no one in public life more than Charles Francis Murphy, became one of the best friends planning advocates ever had (if not their most effective ally). For reasons that have eluded his biographers, Walker supported city planning from his very first moment in office. "What the city needs," trumpeted the new mayor, "is a scientific distribution of its population," which required a reformation of its city-building policies.[78] "During my campaign for election I stated that in the past we have provided most of our improvements with a view to the benefit of some particular locality, that we have not looked upon the city as a whole, and have not tried in any intelligent, farseeing way to plan our improvements with a view to the best development of the city as a whole," he said in June 1926 as he inaugurated his Committee on Plan and Survey. Walker intended his planning committee to see the thing as a whole—housing, zoning, port facilities, traffic congestion, streets and highways, harbor pollution, bridges, parks, the Regional Plan, the Port Authority, finance and budget—all would be carefully considered in order to provide "decent living conditions for every man, woman, and child within our limits." The speech might as well have been written by Henry Bruère, Edward Bassett, or Thomas Adams.[79]

The mayor's committee came as close to participatory planning as any such effort in New York before the 1960s. Composed, at its peak, of 507 members, the committee included citizens from every borough and seemingly every civic, professional, philanthropic, and business association in the city (from John Savarese, executive secretary of the Italian American Society in Brooklyn to Mary Simkovitch, director of Greenwich House), along with political leaders (the reluctant Tammany boss George Olvany and Belle Moskowitz) and government officials (Long Island State Park Commissioner Robert Moses), more than a few bankers and lawyers, and a handful of the usual suspects from planning circles (Bassett, Purdy, McAneny, Veiller, Delano, Wilgus, and architect Harvey Wiley Corbett, who probably exaggerated when he called it "the most significant thing in municipal development that has occurred in the United States"). Not surprisingly, the mammoth organization of notables proved unwieldy.[80] With so many people to manage and no staff to provide and analyze data, even the subcommittees with members who were, in many cases, the leading experts in the city on key subjects could not do much more than make recommendations that might somehow, someday be pieced together into a workable plan.[81]

What the mayor's committee lacked in depth it made up for in decisiveness. Although the final report admittedly did not amount to a comprehensive city plan, as its principal and unanimous recommendation it called for "the establishment of an official city planning board as an integral and permanent part of the city government" that could produce such a plan. The board of estimate, structured as it was and preoccupied with the multifarious business of the city, could not give "to general far-ahead planning anything like the adequate attention it requires," which resulted in public works construction on a piecemeal and local basis to the detriment of the city as a whole. "The city and borough departments acting independently do not develop plans much ahead of actual construction," which meant that public borrowing occurred whenever someone felt the urge to build something, rather than according to some broader list of public priorities. Though not in charge of actual construction, a bona fide city planning board with control over zoning, a capital budget, and a unified city map would solve all of these problems. So empowered, the board could create a comprehensive "Master" plan for the city and, working with the Regional Plan and neighboring municipal planning boards, achieve a more scientific distribution of population in the five boroughs. In particular (and here the voices of the Regional Plan contingent made themselves heard), the board would "encourage the spreading of certain types of industries over wider areas and thereby lessen the distances between homes and places of employment," using, among other devices, "a great loop highway" that would assist in relieving the dreadful overcrowding downtown and, in conjunction with zoning and housing regulations, "prevent the recurrence of these evils in suburban areas."[82] In other words, a planning board would direct New York's city-building policies toward the fulfillment of the metropolitan vision set forth in the Regional Plan; with the departure of John Hylan, the planning efforts begun under the Mitchel administration could now continue.

Although it took another five months, Walker showed just how seriously he took this project when he hired Edward Bassett to prepare a legislative charter for the city's first planning agency.[83] Perhaps because the mayor's committee, with all its representative citizens, had already completed its work, perhaps because planning now posed somewhat less of a legal problem than zoning, or perhaps because favorable press for the Regional Plan had created such a positive atmosphere for planning, Bassett showed far less prudence and considerably more ambition in his design for the city planning board than he had with zoning.[84] Building on a similar plan by George McAneny tucked inconspicuously into an appendix of the final report of the committee on plan and survey, Bassett proposed a three-member planning board—five members might have suggested borough representation, and there would be *no* borough representation on this board—one of whom would be the chief engineer of the board of estimate. Well paid, appointed by the mayor to four-year terms, and equipped with their own expert staff, board members could focus on master planning (leaving construction to existing departments), which meant controlling the city map (explicitly based on the police power) and exercising authority over almost every physical feature the city had, including roads, bridges, tunnels, viaducts, parks, playgrounds, public buildings, waterways, ferries, sewers, water, and zoning districts. As a first order of business, the board would draw up a new zoning ordinance along the lines presented by the committee on plan and survey (which is to say, by Bassett). The board would also have the authority to withhold building permits in order to enforce the street plan, which developers in the outer boroughs regularly ignored by constructing houses in the middle of mapped streets and making it very difficult to control suburban growth or coordinate transportation infrastructure or sewer lines. Most important of all, the board would not languish in an advisory capacity (since that virtually guaranteed that elected officials would completely ignore it), because Bassett recommended that the board of estimate could change the planning board's decisions only by three-quarters vote. Bassett noted that this provision made the planning board "something more than a mere advisory body and something less than an independent agency with autocratic power," since the board of estimate had the last word on everything. This was no "supergovernment," he insisted—but it came very close.[85]

Certainly it came too close for the borough presidents, who immediately saw its effect on their positions. After Brooklyn Democrat Phillip Kleinfeld introduced the bill in Albany in February 1929, they rallied in opposition, particularly George U. Harvey, Republican of Queens (just then in the process of cleaning up from the sewer graft scandal) and Democrat James Byrne of Brooklyn. Walker, with Bassett at his side, fought back, decrying the logrolling so characteristic of the board of estimate and maintaining that borough representation would only relocate pork-barrel politics to the new planning board to the detriment of the city as a whole. Although the Bassett bill briefly appeared assured of passage, by the end of March forces in Albany "slaughtered" the city's entire legislative package, including the mayor's plan for a consolidated sewage department, a unified subway authority, and centralized control of bridges and tunnels—thanks, it was rumored, to Brooklyn Democratic boss John McCooey, who never had good relations with Tammany.[86]

Walker did not let city planning die there, however, although it might have been just as well if he had. The mayor attempted to get the Bassett bill passed again in 1930, only to have it bottled up in committee. He then submitted to the board of estimate a vastly scaled-down version of the measure that did little more than set up a planning department pledged to a policy of noninterference with the public works in borough hands.[87] Borough presidents again objected that the measure invaded their autonomy, but McCooey relented, and on July 17, 1930, New York finally created a city planning department. Distinguished mechanical engineer John F. Sullivan became planning commissioner and a year later took charge of a staff of forty-four pieced together from other departments. Admittedly, the new department solved none of the problems that had inspired it, since weak authority over the borough presidents constituted its principal political virtue.[88] It did not last long, however. By the end of 1932, the city's fiscal crisis led to severe budget cuts, and the city planning department was one of the first things to go.[89]

That Walker would devote as much time and effort as he did to the cause of city planning in the face of such spirited and effective opposition suggests just how influential the principles of the civic culture of expertise had become. To be sure, other political considerations might have motivated the mayor and no doubt reinforced his support of planning. Political revenge against the borough presidents of Queens, Brooklyn, and Staten Island, who withheld their support in his primary bid against Hylan, might have moved him, but there were more direct routes to punishing them than establishing a planning commission. George Harvey argued that Walker supported city planning because "Tammany is breaking up and going to pieces as the residents of Manhattan move to other boroughs," necessitating a centralization of power over public works to maintain political control of the city.[90] But that accusation came from the lone Republican borough president in a plea to Republicans in the state legislature, and it does not explain why Walker supported the planning department in 1930, which provided but one patronage job to John Sullivan, who hardly qualified as a party man in spite of his name. The borough presidents themselves would certainly have suffered from a loss of patronage and spending power—public works in the boroughs had always been about such things—but even without those jobs under the mayor's direct control, Tammany managed to find thousands of sinecures for its loyalists during the Walker administration, as the looming fiscal crisis and Seabury investigation would show.[91]

More likely, Walker agreed with the underlying arguments for city planning (as did Al Smith, who had some influence over the mayor). The mayor was known as a quick study with an ability to grasp complex subjects, and he abhorred both the inhumane conditions of the most crowded residential areas of lower Manhattan and the haphazard expansion of suburban districts in places like Queens. By the late 1920s, planning advocates in the region unanimously recommended coordinating the efforts of borough and city departments as the only possible solution to those problems, and anyone who dealt with the daily business of the city could see that no existing municipal organization could take that broader view, most certainly not the board of estimate.

Indeed, the difficulties facing the board of estimate made the strongest case for a new city planning organization, particularly in light of the growing persuasiveness of experts on the need for seeing the city as a whole. A typical meeting of the board of estimate had an agenda that ran to three hundred pages and included something like a thousand items, about two-thirds of which related to the physical development of the city, and many of them "disputatious" in the extreme. This "mass of detail" overwhelmed the board and made it impossible to see how the abundance of projects and zoning requests contributed to long-term, large-scale changes sweeping the city and region. Especially when those appeals for funding or variances originated in the boroughs, as a great many of them did, citywide considerations quickly fell by the wayside.[92] Even when projects had implications beyond a single borough—the replanning of the west side of Manhattan, the extension of Riverside Drive north of the Harlem River, or the extension of Linden Boulevard from Brooklyn through Queens—no agency existed to determine how they related to any other city-building activity. As the mayor noted in his support of the planning department: "No one, not even the Borough Presidents, could deny that these, although they might be local improvements, were major improvements relating to the City of New York plan which should be reviewed by a competent department which could take into account the relation of these matters to other pending developments such as the development of the waterfront by the Dock Department, the laying out of park areas by the Park Department, and the location of subway stations and routes by the board of transportation." The city did so much that someone had to start thinking full time about what it all meant when added together.[93] Without that bigger picture, the city would spend its way into bankruptcy without solving any of its problems. Such coordination would eventually require "direct control over the city map and a certain veto power over proposals for public improvements," affirmed George Ford in 1930, but any degree of centralization would be a step in the right direction, and Walker seems to have agreed—hence his support for a planning department that did nothing for Tammany patronage or control over outer boroughs.[94] The drive for city planning and the restructuring of borough government thus came together as essential aspects of the solution to the city's problems, even though this message was carried by Jimmy Walker, whose administration in other respects represented the most compelling argument for reform since Tweed.

Not surprisingly, this combination of urban planning and structural reform resurfaced in the charter-revision effort in 1934. In the wake of the fiscal crisis of 1933 and the exposure of the excesses of the Walker years by Samuel Seabury, Tammany Hall once again became the principal target of reform in New York. In his final report, Judge Seabury vehemently urged the city to change its electoral system to break Tammany's apparent stranglehold on government. The bankers who had just halted the city's borrowing spree agreed. Civic groups also rallied behind the idea, and in 1934 the legislature reluctantly convened a charter commission with twenty-eight members—led by Seabury and Al Smith, seemingly the perfect combination of good-government crusader and reform-minded political insider.[95]

Although electoral change got charter reform started, the future of borough government derailed it. From the first, borough presidents called for further decentralization of authority, claiming that they alone among New York's senior elected officials had close contact with citizens: "The people come to us for everything," said James J. Lyons of the Bronx (unwittingly revealing the basic dynamic of distributive politics in the city), "and if we had jurisdiction over everything in the borough we could give them service, instead of the runaround." Al Smith, in particular, mercilessly denigrated that claim and the whole of borough government along with it. Lack of centralization represented the "basic evil in our present charter," he declared, and whatever lingering sentiment in favor of local self-government that survived had been "twisted and warped by small local politicians and spokesmen for special selfish groups and interests, who are deaf, dumb, and blind to the welfare of the city as a whole." With considerable authority, Smith could say that he knew the city and the diversity of its neighborhoods, but borough government hardly represented that complexity, and more frequently it got in the way of a necessary citywide perspective. "Borough government is district leader government," said the old district leader, "and what is primarily wrong with both major political parties in this city is control by district leaders who are incapable of thinking of the entire city. I cannot believe that my old neighbors from the East Side of Manhattan, who have moved into Brooklyn, Queens, and the Bronx, have suddenly become madly excited about borough autonomy." Coming from a dyed-in-the-wool local politician, this should have been persuasive; but Smith could see the bigger picture, and most members of the charter commission could not. After bitter public disagreement, the commission voted fifteen to thirteen to retain borough government, causing Seabury and Smith to resign in protest.[96]

This failure left charter revision in the hands of Fiorello La Guardia, who embodied reform and popular government and supported centralized executive authority and city planning. La Guardia had praised the Regional Plan in his spectacularly unsuccessful campaign against Walker in 1929, and as one of his first orders of business while mayor-elect he vowed to carry out the "general comprehensive city plan" it provided; by April 1934, he had established his own committee on city planning, announcing that "I know of nothing of greater importance and necessity to the city than city planning."[97] The argument that Charles Dyer Norton had made so frequently—"the money which will carry out the plan of New York is the money which New York will spend in any event, whether it has a plan or not"—now seemed particularly relevant as the new mayor took charge of a city with huge debts and no borrowing capacity and the obvious need to build even more infrastructure to keep pace with growth in the outer boroughs and address blight in Manhattan.[98] Government restructuring, the fight against Tammany, fiscal reform, and the need for planning thus came together in January 1935 when La Guardia appointed the nine-member Thatcher Commission—named after its chairman, Judge Thomas D. Thatcher, who had been solicitor-general in the Hoover administration—to pick up where the Seabury-Smith effort had failed.[99]

Although only one member of the Thatcher Commission had any noteworthy ties to planning—Forest Hills developer Charles Meyer was also a member of the Regional Plan Association—they made it a priority, since it went to the heart of the problem of centralization, fiscal management, and city building with which they struggled. By December 1935, the authors of the new charter had embraced the idea of a city planning commission modeled on the one proposed in the Bassett bill seven years earlier. The mayor would appoint its six members for overlapping eight-year terms, specifically to remove them "as far as possible from political control" as they took up their principal task of preparing a master plan. To undertake that great responsibility, the revised charter vested them with control of the city map: no physical improvement over $10,000 could be made without their approval. The board of estimate could change the commission's recommendations only by three-quarters vote—which would appear to have given the planners almost despotic powers. However, the Thatcher Commission went to great lengths to point out that none of the planning commission's recommendations had any effect whatsoever unless the board of estimate approved them by majority vote: the planning commission might recommend new roads, bridges, and sewage treatment plants all over the city, but nothing happened unless the board of estimate voted to build them. The planning commission could initiate zoning changes and had to approve all changes proposed by property owners or city officials, with its recommendations taking effect automatically within thirty days unless the board of estimate modified or disapproved them by three-quarters vote. The new charter also gave the planning commission authority to prepare a six-year capital budget—which incensed La Guardia (he thought he should have the power to "do all the dreaming and advanced planning of the city"), who supported it nevertheless. Borough presidents would retain control over strictly local improvements paid for by special assessment, but not over any project paid for by the city. By directing all major capital projects through the commission, the "union of borough presidents" could no longer engage in the "log-rolling responsible for costly and wasteful expenditures now frozen into the city's debt service," Thatcher noted. "Thus the special and local interests urging particular projects before the board will be confronted by the responsible report and recommendation of an independent commission advocating the city's interest." Attempting to make the same case that Bassett had in 1929, the judge insisted that this arrangement retained the principle of borough representation, tempered by the paramount need to see the city as a whole, while preserving ultimate authority in the elected officials on the board of estimate.[100]

That argument went over about as well as it had in 1929, the borough presidents again rising in opposition. George Harvey said that the charter revisions created a planning commission with more to do than the board of estimate.[101] Manhattan Borough President Samuel Levy declared that the commission was a waste of money, and Bronx Borough President James Lyons suggested that "super-planners" had moved to New York from Washington, D.C. (where planning had become one of the key themes of radical New Dealers) in hopes of getting on the commission's payroll. Brooklyn Democratic leader Frank V. Kelly warned that the commission destroyed borough autonomy and constituted a "super-power" beyond the control of voters.[102] Outspoken though they were, neither the borough presidents nor Tammany Hall, which had the most to lose from the new charter, could sway the Thatcher Commission as they had the Seabury-Smith commission. Tammany then, in August of 1936, took their fight to the courts, declaring that the Thatcher Commission represented an unconstitutional delegation of legislative power; the Wigwam won the first round, but the New York Court of Appeals unanimously upheld the constitutionality of the charter revision, thus assuring a public referendum on the matter.[103]

On November 3, 1936, the citizens of New York City decisively approved the charter: the vote was 952,186 to 603,072, with only three of sixty-five assembly districts (one of Manhattan's and both of Staten Island's) siding with the opposition.[104] The overwhelming victory resulted in part from La Guardia's typically vigorous campaigning. Although the mayor would have preferred greater centralization of powers away from the board of estimate and into his own hands, he supported the revised charter and defended the paramount need for city planning, which included excoriating the "real estate manipulators" and borough presidents who opposed it. Franklin Roosevelt very likely helped as well, garnering the largest vote and plurality in the presidential race (held concurrently with local races) of any election in the history of the state.[105] But the voters did not accept the charter blindly. The Citizen's Charter Campaign Commission, headed by former Tammany sachem Morgan J. O'Brien (the judge who had chaired Mayor Walker's committee on plan and survey ten years earlier), conducted a very aggressive educational effort that included 108 radio broadcasts, 630 meetings with voters, distribution of 2,000,000 pieces of campaign literature, and almost universal support from local newspapers. Although the charter had only lukewarm support in the summer of 1936, this skilled effort changed public opinion during the autumn before the election. Tammany relied too much on their legal challenge (which failed in early October) and had only a month to organize an opposition campaign. Procharter forces so outmatched them, in fact, that just days before the election a Tammany spokesman denouncing the charter at a party rally at Madison Square Garden was booed and then silenced when the crowd began chanting "We want the Charter."[106]

The creation of the city's planning commission thus hardly constituted a coup d'état even though it represented a radical change in the distribution of municipal power. The elevation of planning in a "non-political, full-time body whose decisions cannot be lightly overridden" represented the conclusion to an old debate over the need for a government capable of treating the city as a whole, rather than as a federation of loosely related parts. The revised charter shifted the balance between centralization and autonomy back to a point closer to that envisioned by Andrew Green, decisively favoring the greater city over less comprehensive expressions of the public interest. Those other interests, the authors of the charter affirmed, "should be controlled by publicly confronting [their] representatives with the interests of the public at large. Too often such interest finds no advocacy because the local political or special interest is organized and the general interest is not."[107] The planning commission institutionalized that confrontation between general and particular interests, and that, ultimately, was the role experts had always envisioned for planning.

New York had never organized itself as a truly consolidated city capable of acting on a general public interest. The movement to create a centralized planning body—like the battle against individualism embedded in legal formalism or the fight against privatism embodied in fiscal conservatism or the struggle against localism inherent in separate sewage policies or the clash with corporate voluntarism in port planning—represented an attempt to institutionalize a more comprehensive perspective on urban problems. The *Regional Survey* had shown that the large-scale trends affecting land use in the metropolis respected no political boundary; therefore, to make the city (or any part of it) livable, healthy, beautiful, and financially stable in the long term, local officials had to think beyond their jurisdictions. New York officials rarely did that. Local interests had vigorous (although very uneven) representation; special interests (real estate, labor, banks) frequently organized to represent themselves effectively in key public decisions. The planning commission, by its very structure, looked at the whole city and could therefore advocate for the general interest, taking those larger trends into consideration when arguing for the city's future. The new charter thus configured local government to think regionally and act locally—a goal that visionaries like Andrew Green had pursued for half a century.

Implementing the General Interest: Robert Moses versus the Planners

In early December 1937, a month before the revised charter took effect, Fiorello La Guardia offered the chairmanship of the new planning commission to Robert Moses. No one could imagine a better choice. Moses already had his hands full, of course. He was city park commissioner, president of the Long Island State Park Commission, chairman of the Triborough Bridge Authority, executive officer of the World's Fair Commission, and alone constituted both the Henry Hudson and Marine Parkway Authorities. Since he already did nearly everything the city planning commission would do, it seemed perfectly natural that he would make it official and fill this most important post in the new era of municipal government. But Moses declined.[108] The *New York Times*, in a moment of unusual insight, thought the refusal entirely appropriate. The city needed *planners* and *doers*, and it was important not to confuse the two, the paper editorialized: "A man who likes to dig, as Mr. Moses does, would have been miscast drawing blueprints."[109] But the two roles coexisted less happily than the *Times* implied, at least where Moses was concerned. Indeed, by the time the city had a genuine planning organization, Moses had decided that doers like himself did not need planners at all. More than a matter of style, this distinction between planning and doing led to bitter conflicts over how to implement the general interest in the modern metropolis.

Like Norton, Moses seemed to participate in almost every step in the emergence of city planning in New York, but the master builder reacted against planning, rather than embracing it. When Mayor Hylan cleaned house in 1918, ridding his administration of the planning advocates and municipal experts who had populated City Hall during the Mitchel years, the young Robert Moses was among those who had to start looking for work. During that lean time, Moses had the good fortune to find Belle Moskowitz and Al Smith, who taught him how things really got accomplished in New York. Thanks to Smith, Moses had his first real taste of power as Long Island Park Commissioner, and he used it with the zeal that would characterize the rest of his career. Moses resigned from Mayor Walker's committee on plan and survey only six months into the work and moved on to other projects where he could actually build something.[110] Although officials had discussed the need for a triborough bridge for years, Moses reenergized the moribund authority and solved the problems that frustrated completion. Exasperated at the slow pace of sewage-treatment efforts, Moses built swimming pools. Disgusted at the endless controversies over how to deal with the west side of Manhattan, Moses built the Henry Hudson Parkway. After his "metamorphosis" from "idealist, theorist, and reformer into a political realist," the good government work that first drew him into public service seemed pointless. Looking back on it, Moses thought that little was lost in the Hylan purge: "I know of nothing actually accomplished by the sterile Mitchel Administration excepting Mr. Bassett's launching of the zoning system," and even that "was loaded down with all kinds of junk."[111]

As Moses became more and more the accomplished pragmatist, he grew to hate the planners, whom he associated with the ineffectual good-government efforts of his idealistic youth. Planning, he concluded, accomplished nothing. Everyone had plans: sanitation experts had decades-old plans for a comprehensive sewage-treatment system; hydraulic engineers had plans for universal water metering; the Port Authority had plans for unified rail and terminal facilities; zoning advocates had plans for stricter density control; transit experts had plans for a unified subway system; traffic experts had plans for reducing congestion in midtown; the Regional Plan Association had magnificent plans for almost everything; architects had plans for the skyscraper city of the future; railroad companies had plans for tunnels under the upper bay. New York City did not lack for plans. Plans it had in abundance—a few of them even made sense; a few of them might have worked. But it also had pie-in-the-sky plans. "That's my criticism of the regional plan," Moses remarked in 1931, "too many plans, too many billions of dollars. There's a distinction between the fancy things that irresponsible people put on paper, and don't have to do anything about, and worthy public improvements."[112]

A decade later, Moses sounded very much like Friedrich Hayek, whose *Road to Serfdom* (1944) equated planning with socialism. "In municipal planning we must decide between revolution and common sense," Moses wrote, "between the subsidized lamas in their remote mountain temples and those who must work in the market place." Mumford, Saarinen, Gropius, Wright, Tugwell— those "itinerant carpetbag experts"—did not like or understand the communities they sought to rebuild in their vainglorious images. "The man who loves his city," Moses insisted (referring most assuredly to himself), "will recognize its faults and shortcomings, but he must be loyal to the institutions and to the local scene in which his lot is cast."[113] And Moses did just that. As Caro has argued, he became a "ward boss" who exceeded anything Tammany Hall ever managed; although personally honest, Moses used the "benefits that could be derived with legality from a public works project" to unite the key political and economic factions in the city and state behind his projects, spreading the money around to the right people just the way the ward heelers did.[114] But where machine politicians had a knack for thinking small, Moses used their methods to work for the city as a whole, at least as he saw it.

Indeed, Moses's inspired attacks on the leading figures of planning hid just how much he shared important aspects of their approach. His thirst for power represented, in part, an effort to achieve the kind of coordination among disparate city-building projects that the *Regional Plan* had identified as the key to saving the metropolis. The Triborough Bridge, for example, allowed him to pull together all of the threads essential for the solution of the downtown traffic problem that had troubled real estate owners, zoning advocates, and public officials for thirty years. He also believed that residents would return to lower Manhattan, with its mix of businesses, industries, and residences, once slum clearance removed the decaying structures that made the suburbs look so good: "An immense number of people like city life much as it is, and at most want neighborhood improvements," which Moses would happily provide. Like the regional planners, he thought the core would remain a center of urban life and not merely a business district surrounded by bedroom communities.[115]

Always the political strategist, however, Moses believed too much in the utility of limited objectives to announce his vision the way the planners did. "What purpose is served by putting [projects] on an official map of the city and thereby serving notice on innumerable interested parties that there is some sanction back of the proposal?" he asked incredulously.[116] It only gave too many people time to get in your way. Cleveland Rodgers—one of New York's first planning commissioners and Moses's greatest fan—put his finger on the master builder's modus operandi: "His method is similar to that of the military leader who concentrates on limited objectives and makes definite gains in given sectors without disclosing larger purposes or strategy." Moses had a vision, but he knew too much about getting things done to broadcast it—since only a fool would disclose his plan to his enemies.[117]

It was his small-minded fellow citizens against whom Moses fought in the battle for the general interest as he saw it. Moses exhibited that most important characteristic of the civic culture of expertise, the tendency to think of the city as a whole. "He loves the public," Roosevelt's labor secretary, Frances Perkins, famously said of Moses, "but not as people. The public is just *the* public."[118] He saw aggregate needs and formulated a conception of the city from a general point of view, but had no faith that anyone would understand that the whole sometimes had to take precedence over the parts—and this was the ultimate meaning of his realism. The borough presidents certainly did not listen to reason, nor railroad executives, nor real estate developers. The great mass of New Yorkers, from the bottom to the top of the social ladder, lived their lives without reference to the general interest, and it made no sense to ask them to see the bigger picture: better to buy them off or bulldoze them or win them over the way Al Smith always had. The public as people would never agree on a general interest. When planners set themselves up to represent it, they only picked a fight that they could not win.

More important still, they really did not deserve to win—a point illustrated most clearly for Moses by the work of the city planning commission. When Moses refused the chairmanship, La Guardia offered it to Adolphe A. Berle Jr., (who stayed only until March 1938), and then to Rexford Tugwell, undersecretary of agriculture and head of the Resettlement Administration under Roosevelt—not merely *one* of the radical New Dealers whom Borough President Lyons worried about, but *the* radical New Dealer. Although the commission lacked adequate staff and remained the target of the borough presidents, Tugwell had big plans for it.[119]

To a large extent, Tugwell merely picked up on the problems and solutions presented a decade before in the *Regional Plan*, focusing on the increasing "friction of space" evident in the sprawling city and advocating a program of "recentralization" to keep work and home together. He took these ideas a giant step further in the commission's first master plan in December 1940, however. Like Mumford (who equated centralization with exploitation), Tugwell wanted to spread population around the city in more uniform densities, and this involved dramatic changes in land use (for Manhattan especially) that would shift both residents and businesses to new centers around the city. In place of unchecked suburban growth, the plan called for the creation of a series of greenbelts in the outer boroughs—an idea Tugwell had developed at the Resettlement Administration—concentrating work and residence in focused centers and tripling open space to a whopping one-third of the city's area by 1990. That represented a far more radical program of recentralization than the one envisioned by the regional planners, and it looked to Moses like "the program of socialistic, planned economy, whose aim is to reconstruct the entire city and with it our economic and political systems." Tugwell, armed with his fantastic notion of planning as the "Fourth Power" of government, had lodged himself in the ill-conceived planning commission with perilous consequences for everyone's personal freedom. As Moses warned, under the new charter, the master plan "is not simply a picture or suggestion. It is not advisory. When adopted, it is controlling." And he knew, as someone working in the trenches on more or less the same problem that Tugwell wanted to solve, that very few suburban landowners would take kindly to having their property subsumed in a greenbelt and very few businessmen wanted to be pushed out of Manhattan just to lower densities to suit Tugwell's conception of nonexploitative development.[120]

Tugwell argued that the planning commission represented the general interest in the face of these lesser expressions of private or political interest. In the fall of 1940, he published "Implementing the General Interest" in the first volume of *Public Administration Review*, setting forth his philosophy of the commission's role as custodian of the interests of the city as a whole. The city planner's method, Tugwell claimed, "attempts to grasp the whole before it considers the parts, since these have only a derived, a contributory significance and none taken alone." In his view, planning was a disaggregative procedure—with the planner proceeding from the larger vision to its contributory elements—not an aggregative procedure in which the parts somehow added up to a coherent whole. "An ideal urban life"—and that was the real goal—was an "emergent" (having a character greater than the mere sum of its parts), rather than an "additive." "It is this whole, this emergent, this relative and conjectural interest which, it seems, the planning commission was intended to represent."[121]

Tugwell recounted the struggle to institutionalize city planning against the corrosive localism of real estate interests and borough presidents; the city had established the planning commission to break the hold of those self-interested parties and to represent the general interest that they had always prevented from emerging, he insisted, but the attacks had not stopped. "As it came closer and closer to representing the true central interest of the people of New York," he revealed of his own efforts to create a master plan, "there was less and less available understanding and support from citizens" ("he loves the public . . . ," as Frances Perkins said of Moses, Tugwell's nemesis).[122] Like Cohen at the Port Authority, Tugwell at the planning commission quickly developed a Veblenian approach to the relationship between the urban system and its constituent parts; but with planning, real estate developers and borough presidents, rather than railroad executives, resisted the assertion of the general interest by the expert who claimed (with complete sincerity) to represent it against their selfish demands. Like Cohen, Tugwell could see the whole almost without reference to the parts.

Tugwell's approach confirmed the worst fears about a centralized planning agency. Resistance to his master plan came not merely from real estate interests and borough officials but from supporters of planning who agreed with the commission's larger purpose. "The fundamental purpose of a city is centralization," the Citizens' Budget Committee insisted, and the commission's proposals threatened to undo the dense network of economic relationships that had long characterized lower Manhattan to make way for a "radically different economic and social order." Tugwell's extremism proved that a disaggregative approach to planning had lapsed into an aspiring despotism. Had he adopted the *Regional Plan*'s approach—connecting the future to the past by redirecting current trends—rather than attempting to change the city entirely, he probably could have achieved a better marriage between whole and parts. But the whole he envisioned seemed unconnected to the present or the past and therefore appeared extreme even to planning advocates. Tugwell thus managed to alienate potential allies who embraced the great urban qualities that accompanied centralization, even as they hoped centralized planning would do away with congestion. Not surprisingly, only two years into the commission's existence, the Real Estate Board of New York, the Brooklyn Real Estate Board, the West Central Park Association, and the Electrical Workers Association, among others, agitated for new limits on its power. Tugwell got the message. In July 1941, he resigned from the commission. Four months later, Robert Moses, who would consistently resist future master planning efforts, took his place.[123]

Moses, of course, believed that he had little in common with Tugwell, but the difference had more to do with strategy than outlook. Moses thought his own methods more democratic since he had to fight to get each project accepted—a procedure more in line with the practices of existing institutions than Tugwell's use of the power given to him under the charter that so many New Yorkers had approved just four and a half years earlier. Moses had to slug it out in the "market place"—working within existing institutions, gathering power, competing for money, and building, always building, whenever he could—rather than having the privilege of making plans that citizens *had* to accept by virtue of his position (eventually, of course, to get the job done Moses garnered so much power that he had more of it unofficially than the city planning commission had officially—which arguably made him even less "democratic" than that "socialistic" body). He believed that he served the "great middle class"—that "inherently, fundamentally, and incurably conservative" body of New Yorkers—who did not buy radical theories but did approve of his own efforts to give them the amenities they wanted.[124] That, in his view, rendered his work a more acceptable formulation of the general interest than the one bestowed on an unwilling city by the ivory-tower planners.

This did not solve the larger problem of city planning, however. Building the various elements of a planned metropolis, as Moses had a genius for doing, could not provide the coordination of land use and transportation that alone could save New York from wasting resources and repeating old mistakes. The city always had building, although not enough and only occasionally of the right sort; the borough presidents built things all the time; real estate developers did as well. Building was not the problem. Coordination was the problem, and Moses could provide only so much of that, for as the *Regional Plan* had shown, the city developed within a system that included New Jersey and Connecticut, which were beyond even Moses's reach. Previously, the city's private, local, voluntary institutions made coordination impossible; now, the agency created to solve that problem conflicted with those very institutions in ways that not only seemed undemocratic but also appeared to show that planners could not articulate a meaningful general interest. Thus, having established a bona fide planning agency, having organized to represent the interests of the consolidated city, having finally created a power capable of seeing and acting on the larger trends affecting the metropolis, New Yorkers now had to face the likelihood that they had something they did not really want.

Planning the Fragmented Metropolis

New York City did not have its first master plan until December 1940. By then, more than a decade had passed since the *Regional Plan* set forth the only coherent vision the metropolis ever had. Unfortunately, the forces identified by the *Regional Survey* had expressed themselves more decisively as the city tried to organize to do something about them. In a span of just ten years, the problems that so concerned Norton, Delano, and Adams—the problems of the congested industrial city that had inspired every planning effort in the previous century—had given way to a new set of difficulties that would characterize the urban crisis of the 1950s and 1960s: blight, sprawl, and the abandonment of center cities by industry and the middle class.

Although the authors of the *Regional Survey* had detected the first signs of those emerging patterns in the mid-1920s, nothing they saw in their detailed study could sway them from the belief that Manhattan would retain its role as "mother city"—but that changed in the early 1930s. At the height of the Great Depression, Manhattan landlords lost tenants at the rate of forty thousand per year. This exodus from the city became a major concern during La Guardia's first year in office since flight to the suburbs meant that landlords could not pay their taxes, further straining the city's credit, then in the hands of bankers demanding their money back from the spending spree of the preceding decade.[125] By 1935, single-family homes dominated the city, occupying more than half of the land used for residential purposes.[126]

By the time Tugwell joined the planning commission, the situation had reached a crisis point. Six months after he took office, he told La Guardia that the city had to put a halt to housing developments in the outer boroughs (no wonder, then, that real estate interests opposed him) because they only further depopulated the urban center while costing millions for streets, sewers, schools, and recreational facilities to make the new areas livable. "Present activity is creating slums faster than any agency can eliminate them," he declared. The city had to do something to attract middle-income groups back to the deteriorating core (a project he shared with Moses), and any future construction had to take place in the blighted areas abandoned by the new suburbanites.[127] By 1940, the commission concluded that subsidized migration to the suburbs constituted "the basic problem of New York City." Meanwhile, the blighted districts of Manhattan had higher rates of tax delinquencies and foreclosures, putting pressure on other taxpayers to make up for the loss—a burden that suburbanites would ultimately reject. To make matters worse, the city now had to fight the state and federal governments, which backed the borough officials, real estate developers, and mortgage lenders who promoted sprawl. "It should be perfectly clear," Tugwell pointed out, "that the lending institutions which map large sections of the city, and by agreement refuse to renew loans or make new ones in these [blighted] sections, are contributing to the rapid deterioration and to the spread of slums [while] lending large sums to develop land in outlying sections, the effect of which is to draw population from the older sections" of the city. New York would never solve its fiscal crisis, he warned, unless it started to rebuild deteriorated areas downtown.[128] Land-use policy (and the city's fiscal health) had thus become entangled in the same web of local, regional, and national institutions that had frustrated railroad regulation at the port for decades—and with similar consequences.

In other words, by the time the city got around to taking its first stab at a master plan, it had become even more difficult to generate that sense of shared problems that Thomas Adams and the experts at the Regional Plan held out as the only true foundation for comprehensive city planning. The outer boroughs no longer played the subordinate role they had for the first forty years of consolidated government. Planners had made a clear distinction between decentralization and sprawl, but lack of control had given sprawl the upper hand. Increasingly, development on the periphery took place without the intimate connections to downtown that the *Regional Survey* documented and the *Regional Plan* endorsed; the city created satellite suburbs not as a way to preserve the competitive standing of downtown areas (as Adams had hoped), but as a way to avoid them entirely. As the city had pulled further apart, the cultural outlook of suburban communities differed markedly from the cultural outlook of central city communities, and their visions of the ideal urban life diverged. Manhattan had a vested interest in reducing sprawl, but the suburbs that surrounded it now had a vested interest in promoting it. Suburbanites would soon begin to argue that they did not need the center city at all. As with sewage policy, later forced on localities, coordinating land-use policies to serve some greater good concocted by the experts involved more costs than benefits for those outlying communities that suffered no direct consequences from isolating themselves from the city as a whole. The tendencies that had long made consolidation difficult, therefore, now made master planning virtually impossible.

The attempt to get the city to develop along some more "rational, logical, unified pattern" had come to grief many times before Moses thwarted the next major master planning effort in 1949. Comprehensive planning—the coordination of land-use regulation and infrastructure—could not occur before consolidation; it became impossible under the new charter in 1901; Hylan destroyed any chance of it in 1918; borough resistance prevented it in 1929 and 1934; Tugwell's radicalism undermined it in 1941. Moses played a role only in this last episode, and while his resistance to planning arguably aggravated the city's problems in the 1950s, the basic pattern of blight and sprawl had emerged while the city struggled to organize itself to think and act comprehensively.

In this sense, the creation of a city planning commission represented a Pyrrhic victory for the experts who had advocated a citywide perspective for the first forty years of Greater New York's existence. On the one hand, it represented the fulfillment of their approach to the public interest and the necessary institutional form for addressing the new scale and scope of urban life. On the other hand, it aroused such opposition from groups who saw the city differently or whose activities seemed to subvert the master plan that the commission never took on the role the reformers fought so hard to give it.[129] New York's experience seemed to prove what political scientist Pendleton Herring had observed in 1936: "To advocate a planned society under our system of capitalism and democracy is to urge a leopard to change his spots."[130] But that merely highlighted the cultural dilemma the experts faced, since everything they knew about the city—its institutional structure, its patterns of growth, its fiscal limitations—suggested that it could be saved only by planning. The one community of Greater New York required one government—or at least one set of coordinated policies—to preserve its future; but the realities of localism generated multiple governments as the region further fragmented into multiple communities.

Moving the system ideal to the public sector with regard to land-use planning thus ran into the problem that had always frustrated it: connecting the whole and parts in a way that satisfied the parts, which wanted to go their own separate ways in spite of their interdependence. The experts acknowledged the difficulty, but could not resolve it, even with their superior understanding of metropolitan interdependence. As Harold Lewis, who contributed to the *Regional Plan*, noted in 1926: "The eventual plan is necessarily a compromise between conflicting proposals and ambitions of the various communities affected. While in its final form *it may not seem best for an individual community*, it probably will prove to be so, as only a well-rounded and balanced plan is practicable."[131] The planners always thought they could get over that problem because the end result would prove better for everyone in general. But the system of distributive politics that retained its hold on New York City was built on the notion that those individual communities—whether neighborhoods, boroughs, real estate groups, banks, or railroads—could appeal their dissatisfaction to other institutions and separate themselves from the eventual plan, however well-conceived it seemed to the planners. A "well-rounded and balanced plan" was not necessary, as the city's growth after 1940 certainly proved.

CONCLUSION

"An almost mystical unity"
Interdependence and the Public Interest in the Modern Metropolis

> The history of municipal politics shows in most cases a flare-up of intense interest followed by a period of indifference. Results come home to the masses of the people. But the very size, heterogeneity, and mobility of urban populations, the vast capital required, the technical character of the engineering problems involved, soon tire the attention of the average voter. The ramification of the issues before the public is so wide and intricate, the technical matters involved are so specialized, the details are so many and so shifting, that the public cannot for any length of time identify and hold itself. It is not that there is no public, no large body of persons having a common interest in the consequences of social transactions. There is too much public, a public too diffused and scattered and too intricate in composition. And there are too many publics, for conjoint actions which have indirect, serious, and enduring consequences are multitudinous beyond comparison, and each one of them crosses the others and generates its own group of persons especially affected with little to hold these different publics together in an integrated whole.
>
> —JOHN DEWEY, 1927

Cleveland Rodgers, who served on the New York City Planning Commission from 1938 to 1951, appreciated the apparent contradiction posed by the modern metropolis. "Anyone interested in promoting greater order and unity in New York should begin by studying multiplicity in all its manifestations," he recommended. "The metropolis is the epitome of multiplicity; the paradox of the phenomenon is that the city is made possible by an almost mystical unity."[1] Finding a way to institutionalize something "almost mystical" posed a considerable challenge to those engaged in building Gotham. Relating multiplicity to unity, parts to whole, in some specific institutional form occupied the attentions of experts in engineering, sanitation, architecture, public finance, and law who attempted to improve the city's commuter and freight railroad systems, fund and build vital public works, and regulate private land uses to create a planned city. Rodgers's "mystical unity" represented their durable belief that Greater New York, in spite of its inexorable tendency to subdivide into distinct communities, did possess common interests that rose above mere collections of special or group interests. New York functioned as an interdependent social organism in many important ways, in addition to serving as a platform for more or less autonomous individuals and groups to conduct their lives unconcerned with each other or with any larger, shared project. When existing institutions of collective decision making failed to act on those common interests—when courts, parties, corporations, markets, and local governments bogged down in so many parochial motives and narrow perspectives—experts built new institutions that could respond to metropolitan interdependence as they saw it.

This question of how to institutionalize the pursuit of common interests in an exceptionally diverse city appeared in mundane guises but stimulated extraordinary changes. Moving goods and people across crowded rivers and through busy streets, digging tunnels underneath skyscrapers, managing the city's borrowing power, and looking for loopholes in obscure legal cases involved the banalities of technical discourses, but these tasks comprised the workaday details of much broader questions of mutual obligation and collective interests. Embedded in those solutions to practical problems were new terms of togetherness; to propose them meant confronting old notions of limited government, private property, and voluntary association.

This world of problem defining and problem solving nurtured the idea of the city as a system and made it seem as though the proper application of expertise could reconcile public and private interests. New Yorkers needed better ways to get across the Hudson: entrepreneur William Gibbs McAdoo and engineers John Vipond Davies and Charles M. Jacobs completed the first tunnels under the river, which to this day provide commuters with a faster, safer method of crossing between Manhattan and New Jersey. The city needed to overcome its long-time isolation from continental rail systems: The PRR's vice president, Samuel Rea, and some of the best engineering talent in the world directed a small army of laborers in the construction of Penn Station, the Hell Gate Bridge, and tunnels under the Hudson and East Rivers, creating the rail route that Amtrak still uses to transport passengers up and down the eastern seaboard. Experts, nurtured by private industry, believed that bringing the principles of system building to the public sector would provide the solution to many of the city's problems—physical, economic, and political.

The application of expertise in its many forms to the practical difficulties of living in New York appeared to confirm that a properly configured and empowered metropolitan government could successfully manage the new scale and scope of urban life. In response to the expensive, unreliable procedures for moving freight at the port, attorney Julius Henry Cohen coaxed hundreds of public officials and railroad executives into supporting the establishment of the nation's first port authority, the bistate agency that built the Goethals Bridge, Outerbridge Crossing, Bayonne Bridge, and George Washington Bridge, among other projects, within ten years of its creation and served as a model for the Tennessee Valley Authority. To forestall water famine, multitalented engineer John Ripley Freeman helped empower a board of water supply to construct a tunnel and aqueduct system 160 miles long, extending 752 feet under the city. Dismayed with growing levels of pollution, sanitary engineer George Soper led the fight that produced an Interstate Sanitation Commission to coerce localities to take responsibility for cleaning up the rivers and the bay. The ubiquitous George McAneny—civil service reformer, Manhattan borough president, zoning sponsor, planning advocate—overcame the logjam between the traction moguls and politicians and added five hundred miles to the subway system, partially relieving the crush of traffic downtown and allowing the growth of new residential areas in Brooklyn, Queens, and the Bronx. Henry Bruère, a leading figure at the Bureau of Municipal Research, worked alongside economist E. R. A. Seligman and Comptrollers Herman Metz and William Prendergast to enlarge the city's borrowing capacity, expand its ability to tax, and thus increase its claims on private wealth, permitting billions of dollars in infrastructure spending during an unprecedented period of improvement in the city's physical plant. Attorney Edward Bassett, political scientist Robert Whitten, and architect George Ford articulated a new rationale for restricting the rights of property owners and convinced the courts to allow zoning of land uses and building heights; their achievement not only reshaped the New York skyline, it also gave planners an indispensable tool for controlling the development of cities and contributed to a major change in constitutional jurisprudence. Chicago businessmen Charles Dyer Norton and Frederic Delano, convinced that the greater city made sense only in a larger geographical context, launched the Committee on the Regional Plan that set a national standard for the analysis and vision required to plan a metropolis. Thanks to their sponsorship, economist Robert Murray Haig and planner Thomas Adams documented the system of growth driving the dual process of overcrowding and decentralization in the region and thereby laid the groundwork for the establishment of planning commissions in New York, New Jersey, and Connecticut. Their work raised awareness of the need for public control of urban development, providing the rationale for the creation of the New York City Planning Commission in 1938, which fulfilled (at least on paper) the vision of coordinated city-building policies for the interdependent regional metropolis that motivated the work of this varied group of experts.

While the civic culture of expertise clearly had a profound impact on the physical and institutional structure of the city, it did not achieve the broader goals of formalizing a new approach to the public interest or making the city as a whole the primary focus of public policy. Although the railroads thought of the city as a system and reacted to the same larger patterns of growth as municipal authorities, freight planning remained trapped within many different institutional contexts with incompatible approaches to interdependence. The Port Authority could never consolidate its power over the private freight systems at the harbor and had to abandon its rail plans to concentrate on building bridges and tunnels for trucks and automobiles. Thanks to the Catskill aqueduct system, the city had abundant water (at least for a time), but engineers could never convince the public or elected officials to address the problem of leakage; New York continued to waste millions of gallons of water even after spending millions of dollars and reaching far upstate to get it. A comprehensive sewage-treatment program took almost a century to bring to fruition, with localities resisting every effort to get them to prevent downstream problems. Hatred of traction companies starved the subway system of needed revenues and restricted its continued growth. Experts in public finance did significantly expand the city's taxing and borrowing power, but never persuaded decision makers to plan the use of precious debt-incurring capacity; they empowered New York to spend beyond its means and left it perpetually teetering near fiscal collapse. Zoning changed the legal and physical landscape in New York and beyond, but it tended to reflect narrow local preferences rather than citywide planning objectives. The *Regional Plan*, though influential, did not quite inspire the sense of metropolitan patriotism that would encourage elected officials to think regionally and act locally, and Greater New York tended to sprawl rather than cohere in spite of the plan's national reputation. How the experts saw the city, it turned out, had more to do with their approach to problem solving than with the interests of the groups with whom they formed temporary partnerships, and the institutions they created received only provisional support from their conditional allies.

Most disappointing of all, while the experts intended the planning commission as a truly centralized city-building authority, it never did play the role of coordinating public works, capital budgeting, and land-use regulation in pursuit of common interests as they had hoped. Instead, it became yet another locus of fragmented, partially centralized public authority, among hundreds of others. The heir apparent to this legacy of institutional change, Robert Moses, showed nothing but contempt for planning, even though he looked at the city the same way planners did. He believed in the centralization of power—in his own hands, at least—but used that power increasingly to make decisions about the city's welfare cynically and unilaterally. Moses believed that New Yorkers had collective interests, but gave up on the notion that the public would ever agree to them.

This mixed record of achievement seemed to undermine the idea of common interests that inspired it, in spite of ample evidence of urban and regional interdependence. The sense of optimism that accompanied the growing recognition of the city as a whole and the belief in the concurrence of individual and collective welfare nurtured by the civic discourse of experts in public and private roles gave way to more pessimistic conceptions of New York's chances in the face of the compromises and failures of state building. The city institutionalized hundreds of different formulations and approaches to interests, common and special, rather than coordinating city-building policies toward the fulfillment of widely shared goals. The prospect of unified development gave way to blight, sprawl, fiscal stress, and divided public authority.

It would be easy to conclude from this very mixed record of Progressive Era state building that the public has no common interests at all or that it has none that governing institutions can meaningfully separate in practice from special or local interests. Other studies of urban policy making seem to suggest as much. For example, the defenders of Chicago's "segmented" system of government in the mid-nineteenth century argued that "there was no such thing as a public interest that city government could pursue citywide" for the simple reason that "when a government's constituency became sufficiently complex, the sum of all 'local' interests no longer added to a 'public' interest," historian Robin L. Einhorn has observed. The reformers who attempted to remove that barrier to a more active role for local government insisted that the city did have "public" interests, but reserved to themselves the right to define what those were.[2] The fiction of public interest rhetoric, historian Harold L. Platt concluded of late-nineteenth-century city building in Houston, Texas, delivered government into the hands of commercial-civic elites who served their own interests under a veil of common concerns. "Metropolitans," led by planning experts representing progrowth interests, emphasized investment in public works that boosted Houston as a regional oil and commercial center, rather than building up services in residential areas as "parochials" wanted—resulting in blatant inequalities between downtown and neighborhoods, white and black, rich and poor, that set the stage for the urban crisis of the second half of the century.[3] Edward C. Banfield and James Q. Wilson, borrowing from historian Richard Hofstadter, maintained that the notion of the "city as a whole" was so completely an outgrowth of middle-class Protestant moralizing that it was "fundamentally incompatible" with the style of politics practiced by immigrants in big cities.[4] Especially for a city as large and diverse as New York, one could argue, the concept of the city as a whole might be so hopelessly abstract that it lulls us into believing that there are applications of it that rise above particular interests.

In our rush to deconstruct the cultural inventions that experts employed to justify the expansion of municipal authority a century ago, we run the risk of missing just how powerful the idea of common interests created by interdependence seemed in the fight to establish new methods, philosophies, and institutions to manage mass society and large cities. Outdated notions of individualism and local autonomy, contradictory claims on government power, and persistent tendencies toward parochial organization made it difficult to respond to the shared problems that emerged from divisive large-scale economic and social transformations, and a belief in common interests served as a vital bulwark against despair in the face of seemingly uncontrollable changes.

In *The Public and Its Problems* (1927), for example, philosopher John Dewey tried to identify the conditions under which a new "Public" could organize itself in an age of interdependence and thus liberate the notion of common interests from the limits of inherited social philosophy. Dewey argued that the technical and organizational transformations of the industrial age had created a "Great Society"—a world of "intricate and interdependent economic relations" that infiltrated small-scale community life and undermined the old verities of individualism. Farmers operated within global markets far beyond their reach or understanding; networks of power and transportation crisscrossed cities and regions; giant corporations produced for national and international consumption—all with profound consequences for individuals and communities that still did not recognize their interdependence. "Indirect, extensive, enduring and serious consequences of conjoint and interacting behavior call a public into existence having a common interest in controlling these consequences," Dewey observed. "But the machine age has so enormously expanded, multiplied, intensified and complicated the scope of the indirect consequences, [which] have formed such immense and consolidated unions in action, on an impersonal rather than a community basis, that the resultant public cannot identify and distinguish itself." Dewey thus saw on a national scale the same linkages that experts perceived in the New York region—connections that reached under the city and throughout the metropolitan area, uniting skyscrapers and suburbs, rich and poor, congestion and sprawl, crowded lower Manhattan and bucolic Putnam County, even while local officials, corporate executives, judges, and the public remained unaware of or unmoved by them.[5]

For Dewey, a new understanding of the public interest had to emerge from a recognition of those ties of interdependence since the new forms of association themselves did not create a "Great Community" capable of harnessing them and endowing them with moral purpose. "The Great Society created by steam and electricity may be a society, but it is no community," he insisted, since "no amount of aggregated collective action of itself constitutes a community"—the central problem facing experts in New York City who attempted to build new institutional forms and encourage regional patriotism to act on the realities of physical and economic interconnection.[6]

Dewey did not believe in turning away from those new forms of aggregation for answers even though he never gave up on the idea that face-to-face relations would always provide the "deepest and richest" sense of community. Reestablishing the fiction of individual autonomy—salvaging the absurd "image of a residual individual who is not a member of any association at all"— only distracted attention from the real question: how to comprehend and choose among the consequences of different forms and configurations of interdependent relationships in an effort to provide everyone with a "fuller and deeper experience" of life. Instead, the Great Society had to expose those interconnections and make them plain to the public since "such perception creates a common interest. Then there exists something truly social and not merely associative." To accomplish that, the experts who did understand (albeit imperfectly) the powerful but invisible linkages of cause and effect in the modern world had to disseminate that knowledge through new methods of "debate, discussion and persuasion." Only "when free social inquiry is indissolubly wedded to the art of full and moving communication" would the Great Society become a Great Community.[7] Along these lines, the rhetoric and spirit of the *Regional Plan* found common ground with the great philosopher of American democracy.

Where Dewey stressed communication to clarify that interdependence created common interests, political scientist Pendleton Herring argued in *Public Administration and the Public Interest* (1936) that bureaucracies had to establish terms of togetherness in the face of contradictory claims on government power. Focusing on the growth of the federal government, Herring concluded that "the purpose of the democratic state is the free reconciliation of group interests and that the attainment of this end necessitates the development of a great administrative machine." The conflicts of industrial society created winners and losers who competed for state assistance, each demanding that the work of government advance their own special interest and each conceiving of the general interest as merely an extension of their own concerns. "Groups have demanded special consideration from the federal government for themselves while condemning the general encroachment of the state into private affairs," Herring observed, echoing Bryce. Thanks to those interests, a large but uncoordinated bureaucracy already existed, but "a collection of federal bureaus created at the behest of aggressive minority groups cannot envisage the general welfare"—a problem experienced, in modified form, by Cohen with the port, McAneny with the subways, Soper with the sewage system, and Bassett, Adams, and Tugwell with planning.[8]

Herein lay the great task of bureaucracy in a democratic society and the fundamental challenge of state building in New York City. In spite of divergent social and economic interests, citizens did share "a basic community of purpose." Expert administrators had to allow special interests and organized minorities a voice in government, "since it is their concerns that provide the substance out of which the public welfare is formulated," but they could not reduce that basic community of purpose to the goals of those groups. Here they faced the essential dilemma of representative bureaucracy. In a democratic society, expert administrators had to determine the proper relationship between community of purpose and group ends by reference to the "public interest," and this put the bureaucrat in the position of Rousseau's citizen: trying to distinguish between the General Will and particularistic interests in the guise of public interests. To do that, the bureaucracy had to see the work of government comprehensively and attempt to coordinate its multifarious interventions into economic and social relations toward clear collective objectives—just as planning advocates in New York City had long urged. Herring, who thought planning went too far, wanted to avoid an approach to common interests that collapsed into a directionless struggle between interest groups (as occurred in distributive politics) or one that verged on the threat of tyranny inherent in a purely disaggregative approach to policy making. And the stakes were high: "This Gordian knot in some countries has been cut by the sword of dictatorship."[9] Only by reference to some eventual resolution of these conflicts—the elusive, mystical common welfare—could bureaucratic institutions, freed of the constraints of limited government, resolve the dilemma within an American framework.

For political scientist E. E. Schattschneider, political parties, rather than bureaucracies, provided the best means to establish those linkages between whole and parts, even though they remained locked in a constant battle against parochial approaches to larger organizational problems. Like Dewey and Herring, Schattschneider remained fully convinced of the existence of common interests in spite of the conflicts of modern industrial society. In *Party Government* (1942), he argued that "the raw materials of politics are not all antisocial. Alongside of Madison's statement that differences in wealth are the most durable causes of faction there should be placed a corollary that the common possessions of the people are the most durable causes of unity." Without a recognition of those common interests, politics became the most cynical of games. "To assume that people have merely conflicting interests and nothing else is to invent a political nightmare that has only a superficial relation to reality," Schattschneider warned. "The body of agreement underlying the conflicts of a modern society ought to be sufficient to sustain the social order provided only that the common interests supporting this unity are mobilized." In his view, "public policy could never be the mere sum of the demands of organized special interests," and he insisted that political parties, in their role as mobilizers of majorities, "are never mere aggregates of special interests." Parties had to build majorities around common interests, and their success in that effort would allow them to resist parochialism.[10]

In practice, however, parties often became the creatures of special interests, sectional coalitions, and political bosses, none of which provided a lasting basis for governance. No society had resources enough to respond to all of these interests—something the consolidated city discovered when it took collective responsibility for the public borrowing of its constituent communities in 1898. Neither could those factions organize to act on "vital common interests," preoccupied as they were with their own ends. "Local bosses," because of their proprietary attitude toward government power, "are hardly conscious of the fact that there is a problem of planning, integration, and overall management of public affairs for the protection of the great interests of the nation"—as Boss Murphy showed during the PRR franchise battles.[11] Someone had to think about discriminating between claims on government power to make sure that common interests would be addressed. Although Schattschneider believed that parties could best perform that task while experts in New York City had lost their faith in that approach to collective decision making, the problem of sorting through competing claims to identify and promote common interests remained the same as the one confronting Bruère and Seligman.

These devices, whether social inquiry wedded to democratic communication, independent but responsive bureaucracies, or truly majoritarian parties, stemmed from the belief that common purposes did exist in modern society. For Dewey, Herring, and Schattschneider, the public interest was not merely a ruse or disguise; nor did it emerge from pluralistic bargaining; nor was it reducible to configurations of private interests or the interests of groups who claimed it from time to time. Obsolete conceptions of individual autonomy hampered social inquiry, special interests captured the bureaucracies that tried to regulate them, and local and sectional concerns dominated the political parties that attempted to rise above them, but such difficulties did not mean that common interests did not exist or that those institutions should abandon the pursuit of them. The difficulty of finding a satisfactory answer to the question of how to organize to act on the general interest in a diverse and conflict-ridden world did not invalidate the search for meaningful expressions of common concerns.

Large cities like New York, where the failure of new and old approaches to social integration had become increasingly evident by midcentury, faced this problem in its most acute and tangible form. In this regard, perhaps the most surprising use of the idea of common interests appeared in Jane Jacobs's *The Death and Life of Great American Cities* (1961), a book that is usually thought of as a defense of the smaller-scale of urban life. Jacobs uncovered the intricate dance of street life that modern city planners had not appreciated and thus destroyed by their actions, but she also understood that cities operated on other levels of self-government—as the city as whole, as street neighborhoods, and as districts linking the two. "It is impossible to say that one is more important than the others," she noted; "all three are necessary." In this scheme, the city as a whole was more than an abstraction—more than a convenient way to characterize collections of street neighborhoods. "We must never forget or minimize this parent community while thinking of the city's smaller parts," Jacobs emphasized. "A city's very wholeness in bringing together people with communities of interest is one of its greatest assets, possibly the greatest." That broader focus clearly did not substitute for an appreciation of community at less comprehensive levels, and that is where planners made their fateful mistake. "Planners like to think they deal in grand terms with the city as a whole, and that their value is great because they 'grasp the whole picture.' But the notion that they are needed to deal with their city 'as a whole' is principally a delusion," she maintained. "Aside from highway planning and the almost purely budgetary responsibility for rationalizing and allocating the sum of capital improvement expenditures presented in tentative budgets, the work of city planning commissions and their staffs seldom deals, in truth, with a big city as a total organism."[12]

Of course, Jacobs wrote during a time when the city had a capital budget and a planning commission (although not very good ones). Sixty years earlier—even twenty-five years earlier—those institutions did not exist and a vital level of social integration remained neglected. Jacobs believed that planners had to change their thinking to connect whole and parts effectively, but the previous generation of planning advocates faced the same problem without any institutional forms capable of addressing the city as a whole. In this sense, *Death and Life* has more in common with the Regional Plan, and Jane Jacobs is more similar to Thomas Adams, than it would first appear.

The experts in this book felt the absence of those citywide institutions most keenly because the problems they addressed in their professional lives convinced them that the interdependence of the modern metropolis served as an obvious and essential starting point for effective public policy. Because of the city's size and complexity, because of the natural obstacles to its physical coherence, because its problems had regional, national, and even international implications (especially port and financial management), because its power structure was so divided, and because consolidation explicitly raised issues of institutional fragmentation (unlike Chicago, where a single government oversaw the entire city), New York confronted the basic challenges of metropolitan integration before many other urban areas in the United States. Linking New York and New Jersey, Manhattan and Long Island, downtown and suburbs, water users and watersheds, private wealth and public need, center and periphery—these tasks resolved what often appeared as strictly private or local problems from less comprehensive perspectives. Making a buck from commuters and improving the efficiency of corporate freight operations forced engineers to consider regional growth patterns; cleaning up the harbor and overcoming the water crisis encouraged similarly broad perspectives; the overwhelming need for public works fostered the notion that real estate values represented a collective resource; and addressing congestion yielded an understanding of the intimate connections between centralization and decentralization. Inherited distinctions of property rights and political jurisdictions did not make sense from this new perspective; such notions stood in the way of approaches to integration based on the reality of togetherness, rather than the fictions of individualism or autonomy.

In this milieu of problem solving, expertise took on a distinctly civic dimension. The application of technical skill and specialized knowledge to the challenges of living in the crowded metropolis did more than create remarkable structures linking different parts of the city and region. It generated new bonds of political obligation—in the form of expanded municipal claims on private wealth and regulatory powers over private property—and redrew the boundaries between citizens and among units of government. Embracing interdependence thus resulted in a more definite institutional articulation of what community entailed at this new level of aggregation, exceeding anything courts, parties, corporations, markets, and local governments had ever offered.

By no means did the civic culture of expertise exhaust the possibilities of community in the modern city. Even on its own terms, it could not always deliver on its promises. Neither could shared problems provide the only basis for social cohesion. By the time regional planners took up the challenge of linking center to periphery, they had fully embraced the notion that large-scale interconnections coexisted alongside even more compelling forms of local interaction based on other grounds. Their experiences confirm that livable, clean, efficient, successful, humane, interesting, diverse cities require multiple levels and systems of social integration with different standards of legitimacy.

Only by recognizing and organizing for the essential simultaneity of community can citizens and decision makers sustain those qualities over time.

All cities are experiments in the dealing with the implications of collective living. More so than in most other forms of organization, cities force us to come to grips with the possibility that we have a common interest in the consequences of social transactions, remote though many of them seem. To an even greater extent than in nation-states, in cities we negotiate the most palpable terms of togetherness and political obligation and relate multiplicity to unity in explicit institutional relationships. If, in the end, we do not find fully satisfactory responses to these issues in the civic culture of expertise created by engineers, lawyers, architects, and planners in New York City at the turn of the twentieth century, we can still acknowledge that their efforts provide one important part of the answer—for they posed the question correctly—as we attempt to reconcile autonomy (cultural and political) and interdependence in a world that seems to be demanding more of both.

Appendix

The graphs below and the discussion of New York City's finances in chapter 4 are based on an analysis of annual expenditures, tax rates, debt, and price indices drawn from Henry De Forest Baldwin, "The City's Purse," *Municipal Affairs* 1 (1897): 354; Edward Dana Durand, *The Finances of New York City* (New York: Macmillan, 1898), 372–76; Frederick L. Bird, *The Municipal Debt* (New York: Mayor's Committee on Management Survey of the City of New York, 1951), 28–30; *Historical Statistics of the United States, Colonial Times to 1970* (White Plains, N.Y.: Kraus International, 1989); and the *New York Times*.

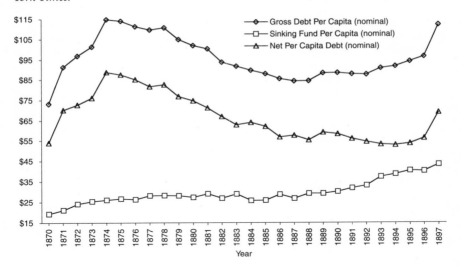

Graph 1. New York City Debt per Capita, 1870–1895

In the wake of the Tweed scandal, New York entered a twenty-year period of fiscal conservatism, setting the stage for a new era of taxing, spending, and borrowing for the consolidated city.

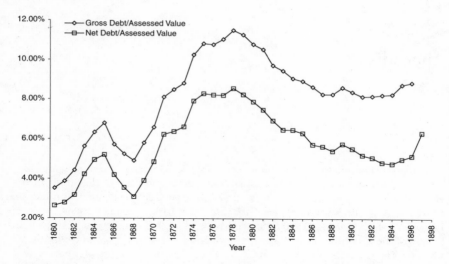

Graph 2. New York City Debt as a Percentage of Assessed Values, 1860–1898

Compared with the city's ability to incur debt—that is, compared with the value of its real estate—New York was progressively less burdened by debt between the end of the 1870s and the middle of the 1890s, thus setting the stage for the emergence of a new approach to fiscal management that made increasing claims on private wealth.

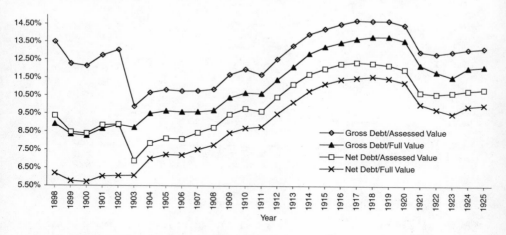

Graph 3. New York City Debt as a Percentage of the Value of Real Estate, 1898–1926

Beginning on the eve of consolidation, the constituent communities of Greater New York began a new round of borrowing. After a brief period of restraint, the city raised the ratio of the assessed value to full value of real estate (hence the sharp drop in gross debt/assessed value between 1902 and 1903) and then began borrowing feverishly until the war.

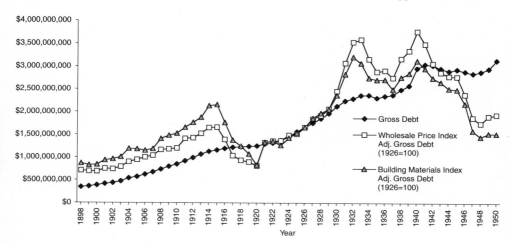

Graph 4. New York City Debt, 1898–1950

The bankers' agreement of 1914 eventually halted the city's borrowing spree until after World War I. Borrowing resumed its upward trend until the Great Depression, but New York never returned to this sort of self-financed investment in its physical plant.

Notes

Abbreviations

BCM	Benjamin C. Marsh Papers, Acc. No. 10,353, Library of Congress, Washington, D.C.
CFC	*Commercial and Financial Chronicle*
CWS	Catskill Water Supply for New York City, Division of Engineering and Industry, National Museum of American History, Smithsonian Institution, Washington, D.C.
EMB	Edward Murray Bassett Papers, Acc. No. 2708, Cornell University Library, Ithaca, N.Y.
EMS	Edward Morse Shepard Papers, Columbia University, NYC
ERAS	E. R. A. Seligman Papers, Columbia University Library, NYC
GBF	George B. Ford Papers, Francis Leob Library, Harvard University, Cambridge, Mass.
GBM	George B. McClellan Papers, Municipal Archives, NYC
GCW	George C. Whipple Papers, HUG 1876.3005, Harvard University Library, Cambridge, Mass.
GMC	George McAneny Papers, Rare Book and Manuscript Library, Columbia University, NYC
GMP	George McAneny Papers, Seeley G. Mudd Library, Princeton University, Princeton, N.J.
HBC	Heights of Buildings Commission Papers, Municipal Archives, NYC
ICCNA	Records of the Interstate Commerce Commission, National Archives, Washington, D.C.
JF	James Forgie Papers, Division of Engineering and Industry, National Museum of American History, Smithsonian Institution, Washington, D.C.
JFH	John Francis Hylan Papers, Municipal Archives, NYC
JFS	John F. Sullivan Papers, Manuscript and Archives Section, New York Public Library, NYC
JJW	James J. Walker Papers, Municipal Archives, NYC
JPM	John Purroy Mitchel Papers, Municipal Archives, NYC
JRF	John Ripley Freeman Papers, MC-51, Massachusetts Institute of Technology, Cambridge, Mass.

MES	Marc Eidlitz and Sons Papers, Manuscript and Archives Section, New York Public Library, NYC
ML	Metropolitan Life Insurance Archives, Metropolitan Life Building, NYC
NPL	Nelson Peter Lewis Papers, Acc. No. 2712, Cornell University Library, Ithaca, N.Y.
NYH	*New York Herald*
NYHC	*New York Harbor* case
NYT	*New York Times*
NYTr	*New York Tribune*
PBQD	Parsons, Brinckerhoff, Quade & Douglas Collection, Division of Engineering and Industry, National Museum of American History, Smithsonian Institution, Washington, D.C.
PNYA	*Port of New York Authority v. Atchison, Topeka & Santa Fe Railway*
PRR	Pennsylvania Railroad Co., department records, Executive Department, Acc. No. 1810, Hagley Museum and Library, Wilmington, Del.
RH	Rudolph Hering Collection, Division of Engineering and Industry, National Museum of American History, Smithsonian Institution, Washington, D.C.
RHW	Robert H. Whitten Papers, Francis Leob Library, Harvard University, Cambridge, Mass.
RPA	Regional Plan Association Papers, Acc. No. 2688, Cornell University Library, Ithaca, N.Y.
RVW	Robert Van Wyck Papers, Municipal Archives, NYC
SW	Schultze-Weaver Papers, Wolfsonian Museum, Florida International University, Miami Beach, Florida
TASCE	*Transactions of the American Society of Civil Engineers*
WGM	William Gibbs McAdoo Papers, Library of Congress, Washington, D.C.
WJG	William J. Gaynor Papers, Municipal Archives, NYC
WJW	William J. Wilgus Papers, Manuscript and Archives Section, New York Public Library, NYC
WWE	*Who's Who in Engineering*
WWNY	*Who's Who in New York*

INTRODUCTION: Conceiving the New Metropolis

The epigraph is from Edward H. H. Simmons, *Financing American Industry* (New York: n.p., 1930), 39–40.

1. Paul Bourget, *Outre-Mer: Impressions of America* (New York: Scribner's, 1895), 29.

2. Kenneth T. Jackson, "The Capital of Capitalism: The New York Metropolitan Region," in *Metropolis, 1890–1940*, ed. Anthony Sutcliffe (Chicago: University of Chicago Press, 1984), 319–53.

3. David C. Hammack, *Power and Society: Greater New York at the Turn of the Century* (New York: Russell Sage, 1982), 189–95; John Foord, *The Life and Public Services of Andrew Haswell Green* (New York: Doubleday, 1913), 178, 183.

4. "A Census of Skyscrapers," *American City* 41 (September 1929): 130; Marc A.

Weiss, "Density and Intervention: New York's Planning Traditions," in *The Landscape of Modernity: Essays on New York City, 1900–1940*, ed. David Ward and Olivier Zunz (New York: Russell Sage, 1992), 49.

5. Sam Bass Warner Jr., *The Private City: Philadelphia in Three Periods of Its Growth* (Philadelphia: University of Pennsylvania Press, 1968), 3–4.

6. Lucian W. Pye and Sidney Verba, eds., *Political Culture and Political Development* (Princeton, N.J.: Princeton University Press, 1965); Gabriel A. Almond and Sidney Verba, *The Civic Culture: Political Attitudes and Democracy in Five Nations* (Princeton, N.J.: Princeton University Press, 1963).

7. John Higham, "Hanging Together: Divergent Unities in American History," *Journal of American History* 61 (June 1974): 7.

8. Keith D. Revell, "The Road to *Euclid v. Ambler*: City Planning, State-Building, and the Changing Scope of the Police Power," *Studies in American Political Development* 13 (spring 1999): 53–54.

9. Compare L. Ray Gunn, *The Decline of Authority: Public Economic Policy and Political Development in New York, 1800–1860* (Ithaca, N.Y.: Cornell University Press, 1988).

10. Robert C. Wood, *Fourteen Hundred Governments: The Political Economy of the New York Metropolitan Region* (Cambridge: Harvard University Press, 1961), 1.

11. Martin J. Schiesl, *The Politics of Efficiency: Municipal Administration and Reform in America, 1880–1920* (Berkeley: University of California Press, 1977); Samuel Haber, *Efficiency and Uplift: Scientific Management in the Progressive Era, 1880–1920* (Chicago: University of Chicago Press, 1964); John M. Jordan, *Machine-age Ideology: Social Engineering and American Liberalism, 1911–1939* (Chapel Hill: University of North Carolina Press, 1994).

12. Samuel P. Hays, "The Politics of Reform in Municipal Government in the Progressive Era," *Pacific Northwest Quarterly* 55 (October 1964): 157–69; idem, *Conservation and the Gospel of Efficiency: The Progressive Conservation Movement, 1890–1920* (New York: Atheneum, 1969); John D. Fairfield, *The Mysteries of the Great City: The Politics of Urban Design, 1877–1937* (Columbus: Ohio State University Press, 1993); M. Christine Boyer, *Dreaming the Rational City: The Myth of American City Planning* (Cambridge: MIT Press, 1983).

13. Robin L. Einhorn, *Property Rules: Political Economy in Chicago, 1833–1872* (Chicago: University of Chicago Press, 1991); Philip J. Ethington, *The Public City: The Political Construction of Urban Life in San Francisco, 1850–1900* (Cambridge: Cambridge University Press, 1994).

14. *Autobiography of Lincoln Steffens* (New York: Harcourt, Brace, 1931), 237, 235, 232; Lincoln Steffens, *Shame of the Cities* (New York: McClure, Phillips, 1904); Charles E. Merriam, *American Political Ideas: Studies in the Development of American Political Thought, 1865–1917* (New York: Macmillan, 1920), 293, 296; Barry D. Karl, *Charles E. Merriam and the Study of Politics* (Chicago: University of Chicago Press, 1974), 103–4.

15. Jon C. Teaford, *The Unheralded Triumph: City Government in America, 1870–1900* (Chicago: University of Chicago Press, 1984), 307; Hendrik Hartog, *Public Property and Private Power: The Corporation of the City of New York in American Law, 1730–1870* (Ithaca, N.Y.: Cornell University Press, 1983).

16. James Bryce, *The American Commonwealth* (New York: Macmillan, 1911), 2: 593.

17. Jonathan R. T. Hughes, *The Governmental Habit Redux: Economic Controls from*

Colonial Times to the Present (Princeton, N.J.: Princeton University Press, 1991), 15; Benjamin Parke De Witt, *The Progressive Movement* (New York: Macmillan, 1915), 14; *Autobiography*, 492.

18. Thomas Adams, Harold M. Lewis, Theodore T. McCrosky, *Population, Land Values, and Government* [1929] (New York: Arno Press, 1974), 204.

19. Robert B. Fairbanks, *For the City as a Whole: Planning, Politics, and the Public Interest in Dallas, Texas, 1900–1965* (Columbus: Ohio State University Press, 1998).

20. *NYT*, January 22, 1929, 28:2.

21. Herbert Croly, *The Promise of American Life* (New York: 1909), 12, 22.

22. G. D. H. Cole, introduction to Jean Jacques Rousseau, *The Social Contract and Discourses* (New York: Dutton, 1950), xxxvi–xxxvii.

23. Steffens, *Autobiography*, 249; Steffens, *Shame of the Cities*, 199.

ONE: "The Public Be Pleased"

1. Walter Lippmann, *Drift and Mastery* [1914] (Madison: University of Wisconsin Press, 1985), 31–32, 49, 51; Nelson P. Lewis, "During my long journey from the rural district of Flatbush," 1903–4, NPL, box 1, file 6.

2. Contrast Thomas Bender, *Intellect and Public Life: Essays on the Social History of Academic Intellectuals in the United States* (Baltimore: Johns Hopkins University Press, 1993), xii–xiii, 3–15.

3. David A. Hollinger, "Historians and the Discourse of Intellectuals," in *In the American Province: Studies in the History and Historiography of Ideas* (Baltimore: Johns Hopkins University Press, 1985), 132.

4. A. J. County, "The Economic Necessity for the Pennsylvania Railroad Tunnel Extension into New York City," *Annals of the American Academy of Political and Social Science* (March 1907): 5, JF, box 46; Sharon Reier, *The Bridges of New York* (New York: Quadrant Press, 1977), 154–55.

5. Samuel Rea, "Pennsylvania Railroad New York Tunnel Extension," December 15, 1909, 58, JF, box 46; County, "Economic Necessity," 6; Walter Laidlaw, *Population of the City of New York, 1890–1930* (New York: Cities Census Committee, 1932), 25.

6. Rea, "Pennsylvania Railroad," 58; County, "Economic Necessity," 2; J. Vipond Davies, "Construction of a Rapid Transit Railroad in Relation to the Handling of Passengers," *Proceedings of the Engineers' Club of Philadelphia* 27 (1910): 318–19, JF, box 46.

7. Rea, "Pennsylvania Railroad," 57.

8. H. T. Hildage, "Underground Workings in New York City," 137, JF, box 1.

9. Ronald Steel, *Walter Lippmann and the American Century* (New York: Vintage, 1980), 169; John J. Broesamle, *William Gibbs McAdoo: A Passion for Change, 1863–1917* (Port Washington, N.Y.: Kennikat Press, 1973), 8–13; *WWNY* (1914), 470–71; Brian J. Cudahy, *Under the Sidewalks of New York* (Lexington, Mass.: Stephen Greene Press, 1988), 191; Carl W. Condit, *The Port of New York*, vol 2: *A History of the Rail and Terminal System from the Grand Central Electrification to the Present* (Chicago: University of Chicago Press, 1981), 115–18; William E. Leuchtenberg, *Franklin D. Roosevelt and the New Deal, 1932–1940* (New York: Harper & Row, 1963), 7.

10. Gilbert H. Gilbert, Lucius I. Wightman, and W. L. Saunders, *The Subways and Tunnels of New York* (New York: Wiley, 1912), 7–9, 155; King's Booklet, "Hudson Tunnel

System," *The Pennsylvania Railroad Tunnels and Terminals in New York City* (New York: Moses King, 1908), JF, box 46; *CFC* 54 (April 9, 1892): 597.

11. *CFC*, 61 (November 23, 1895): 925; 62 (June 27, 1896): 1177; 63 (July 25, 1896): 154. This financial odyssey can be traced in *CFC*, 61 (November 23, 1895): 925; 62 (June 27, 1896): 1177; 63 (July 25, 1896): 154; 68 (April 8, 1899): 671; 68 (June 17, 1899): 1182; 70 (May 19, 1900): 996; 70 (June 9, 1900): 1149.

12. William G. McAdoo, *Crowded Years* (Boston: Houghton Mifflin, 1931), 67–70; Davies, "Construction of a Rapid Transit Railroad," 319.

13. McAdoo, *Crowded Years*, 55–57, 71–78; *WWNY* (1914), 543; property inventory attached to letter from McAdoo to Young, October 5, 1892, WGM, box 91.

14. McAdoo to Watchorn, November 27, 1908, and memo beginning "The most forceful figure in the Hudson Tunnel enterprise," WGM, box 574; Vincent P. Carosso, *Investment Banking in America: A History* (Cambridge: Harvard University Press, 1970), 113, 127–30.

15. "Memoir of Charles Mattathias Jacobs," *TASCE* 83 (1919–20): 2236–38; *WWNY* (1914), 184; *WWE* (New York: John W. Leonard, 1922), 343; Gilbert, Wightman, and Saunders, *Subways and Tunnels*, v, 10–15.

16. John V. Davies, "The Tunnel Construction of the Hudson and Manhattan Railroad Company," *Proceedings of the American Philosophical Society* 49 (1910): 167–68, JF, box 1, file: H&M RR Co. construction papers; Gilbert, Wightman, and Saunders, *Subways and Tunnels*, 156–59.

17. McAdoo, *Crowded Years*, 90–93; Patricia C. Davis, *End of the Line: Alexander J. Cassatt and the Pennsylvania Railroad* (New York: Neale Watson Academic, 1978), 162–64.

18. McAdoo to Watchorn, November 27, 1908, WGM, box 574; memo in PRR, box 138, leading to "Award of Arbitrators," July 15, 1913, in file 21; McAdoo to Truesdale, January 19, 1903, and Truesdale to McAdoo, January 23, 1903, WGM, box 574.

19. Memo for McCrea, July 27, 1909, PRR, box 138, file 22.

20. Rea to Thayer, July 20, 1909, PRR, box 138, file 22; Rea to Sheppard, November 30, 1910, PRR, box 137, file 32; comparison, passengers carried in New York district on ferry steamers, 1907 to 1910, incl., WGM, box 574; Rea to Alexander, May 11, 1910, PRR, box 137, file 32; Michael Bezilla, *Electric Traction on the Pennsylvania Railroad, 1895–1968* (University Park: Pennsylvania State University Press, 1980), 26–27, 206.

21. Memo for McCrea, July 27, 1909, and Rea to Thayer, July 20, 1909.

22. Thomas K. McCraw, *Prophets of Regulation: Charles Francis Adams, Louis D. Brandeis, James M. Landis, Alfred E. Kahn* (Cambridge: Harvard University Press, 1984), 16; Lee Benson, *Merchants, Farmers, and Railroads: Railroad Regulation and New York Politics, 1850–1887* (Cambridge: Harvard University Press, 1955).

23. McAdoo, *Crowded Years*, 103–5.

24. *Poor's Manual of Public Utilities*, 1916, 1152; idem, 1918, 904; *Moody's Manual of Investments, Public Utility Securities*, 1928, 464; *Annual Report of the Hudson and Manhattan Railroad Company*, year ended March 31, 1911, 22, and idem, year ended December 31, 1915, 20, JF, box 1, file: H&M RR Co. annual reports; *Report to T. P. Shonts, Re Extension of Hudson-Manhattan Railroad, March 5th, 1909*, report #35, PBQD, box 3; *Hudson and Manhattan Railroad Company: Its Property, Finances, and Securities* (New York: Harvey Fisk, 1915), 5–6, WGM, box 574.

25. Rea, "Pennsylvania Railroad," 3, 6–8.

26. *WWNY* (1914), 597; Carl W. Condit, *The Port of New York*, vol. 1: *A History of the Rail and Terminal System from the Beginnings to Pennsylvania Station* (Chicago: University of Chicago Press, 1980), 263.

27. Rea, "Pennsylvania Railroad," 56–81, 17, 20.

28. Ibid., 8–21; Charles M. Jacobs, "The New York Tunnel Extension of the Pennsylvania Railroad: The North River Division," *Proceedings of the American Society of Civil Engineers* 36 (January 1910), 5, JF, box 46; Jameson W. Doig, "Politics and the Engineering Mind: O. H. Ammann and the Hidden Story of the George Washington Bridge," *Yearbook of German-American Studies* 25 (1990): 161–62.

29. Rea, "Pennsylvania Railroad," 13–17, 53, 20–22; Jacobs, "North River Division," 2–4, 10–11.

30. Jacobs, "North River Division," 22; Charles W. Raymond, "The New York Tunnel Extension of the Pennsylvania Railroad," *Proceedings of the American Society of Civil Engineers* 35 (September 1909): 21; pamphlet, *The New York Improvement and Tunnel Extension of the Pennsylvania Railroad* (July 1910), 29, 18, JF, box 46.

31. Condit, *Port of New York*, 1:333–34; "The New York Connecting Railroad," *Railroad Gazette* 41 (December 28, 1906): 570–71; *Putnam's Magazine* 6 (September 1909): 760, 764.

32. David P. Billington, *The Tower and the Bridge* (New York: Basic Books, 1983), 131–40; Condit, *Port of New York*, 1:340.

33. [Rea] to Schiff, October 11, 1910, PRR, box 141, file 23.

34. *New York Improvement and Tunnel Extension*, 3, 9, 12, 13.

35. Fiske Kimball, *American Architecture* [1928] (New York: AMS Press, 1970), 183.

36. Gail Fenske and Deryck Holdsworth, "Corporate Identity and the New York Office Building: 1895–1915," in *The Landscape of Modernity: Essays on New York City, 1900–1940*, ed. David Ward and Olivier Zunz (New York: Russell Sage, 1992), 143.

37. *Annual Report, Hudson and Manhattan Railroad*, 1911.

38. Richard Guy Wilson, *McKim, Mead, and White, Architects* (New York: Rizzoli, 1983), 211, 210–17; John V. Davies, "Discussion," *TASCE* 69 (October 1910): 408–9; James McCrea in *Evidence Taken by the Interstate Commerce Commission in the Matter of Proposed Advances in Freight Rates by Carriers (August to December, 1910)*, 61st cong., 3d sess., 1911, sen. doc. 725, vol. 4: 2296–97.

39. Davies, "Discussion," 401–2.

40. Second vice president to Berwind, November 19, 1910, PRR, box 142, file 1.

41. Albro Martin, *Enterprise Denied: Origins of the Decline of American Railroads, 1897–1917* (New York: Columbia University Press, 1971), 103–6.

42. Jacobs memorial, 2238; and Gubelman to Forgie, September 30, 1935, JF, box HRT-PRR-3, file: first automobile, PRR tunnel; William Worthington Jr., "A Few Words about This Picture," *Invention and Technology* 5 (fall 1989): 30–31.

43. Charles Jacobs, chief engineer's report to PT & TRR management, Pennsylvania Railroad N.Y. tunnel extension, North River div.: book 1 of vol. 1—chief engineer's review, 1–13, JF.

44. Terry Reynolds, "The Engineer in Twentieth-century America," in *The Engineer in America*, ed. Terry Reynolds (Chicago: University of Chicago Press, 1991), 169–73.

45. *Pennsylvania Tunnel Alumni Association of the North River Division* pamphlets, 1911 and 1912, JF, box 46; *WWE* (1922), 39, 798, 1380.

46. James Forgie, report to counsel Thomas F. Conway, readjustment claim, Bradley Contracting vs. Public Service Comm., JF, box 4.

47. *WWE* (1922), 319; "Memoir of Benjamin Franklin Cresson Jr.," *TASCE* 87 (1924): 1325–27; *NYT,* January 11, 1917, 5:1, January 12, 1917, 19:1, September 9, 1921, 4:3, January 8, 1922, II, 1:3, May 31, 1922, 29:2–3, July 23, 1922, II, 1:1, September 4, 1922, 13:4; B. F. Cresson Jr., "The Problem of the Lower West Side Manhattan Water-front of the Port of New York," *TASCE* 75 (1912): 226–46; Julius Henry Cohen, *They Builded Better than They Knew* [1946] (Freeport, N.Y.: Books for Libraries Press, 1971), 272, 281, 283, 286.

48. Davies, "Tunnel Construction," 164–68, 186–87; and Davies, "Discussion," 401–17.

49. John Nolen, ed., *City Planning* (New York: D. Appleton, 1915).

50. *WWE* (1922), 1381; Josef W. Konvitz, "William J. Wilgus and Engineering Projects to Improve the Port of New York, 1900–1930," *Technology and Culture* 30 (April 1989): 398–425; Kurt C. Schlichting, *Grand Central Terminal: Railroads, Engineering, and Architecture in New York City* (Baltimore: Johns Hopkins University Press, 2001).

51. *WWE* (1922), 1224; "Memoir of Francis Lee Stuart," *TASCE* 101 (1936): 1493–97.

52. Davies, "Tunnel Construction," 186; *Report on the Bank of New York Building,* report #80, PBQD, box 5; "The Vertical Growth of New York City," *Scientific American* 84 (March 2, 1901): 136–38.

53. Davies, "Tunnel Construction," 166–67.

54. Henry B. Lent, *The Waldorf-Astoria* (New York: Hotel Waldorf-Astoria, 1934), 20, SW.

55. Forgie, "Report to Counsel Thomas F. Conway."

56. Memo "The most forceful figure."

57. A. J. County, "What the Pennsylvania Railroad System Has Done for Greater New York and Long Island," Queens Chamber of Commerce, January 11, 1912, PRR, box 141, file 6.

58. Arthur D. Howden Smith, "'The Peak of the Load': What It Means to Light New York City and Transport Her Crowds," *Putnam's Magazine* 6 (August 1909): 520–21.

59. Samuel P. Hays, *Conservation and the Gospel of Efficiency: The Progressive Conservation Movement, 1890–1920* (Cambridge: Harvard University Press, 1969), xiii.

60. Richard L. McCormick, "The Party Period and Public Policy: An Explanatory Hypothesis," *Journal of American History* 66 (September 1979): 279–98.

61. *Evidence Taken before the Interstate Commerce Commission Relative to the Financial Transactions of the New York, New Haven & Hartford Railroad Company,* 63d cong., 2d sess., 1914, sen. doc. 543, 30.

62. Richard L. McCormick, *From Realignment to Reform: Political Change in New York State, 1893–1910* (Ithaca, N.Y.: Cornell University Press, 1981), 138–43.

63. Ibid., 142–43; *NYTr,* April 16, 1902, 2:1; June 17, 1902, 16:1; *NYT,* July 13, 1902, 2:2, and July 30, 1902, 1:7.

64. *NYT,* July 23, 1902, 2:2; *NYH,* November 1, 1902, in JF, book no. 1 of vol. 1, chief engineer's review.

65. *NYT,* July 12, 1902, and July 13, 1902, 2:2; *NYH,* November 1, 1902.

66. Jacobs, book no. 1 of vol. 1, chief engineer's review, 39, JF.

67. B. F. Cresson Jr., "Chief Engineer's Report to Management, P.T. & T.R.R. Co., Pennsylvania Railroad New York Tunnel Extension, North River Division: Book No. 3 of Volume 2—Construction of Terminal Station-West," 1–2, JF.

68. *NYT,* November 7, 1901, 1:7, 2:1–3; Harold Zink, *City Bosses in the United States* (Durham, N.C.: Duke University Press, 1930), 143–45.

69. *NYT,* December 24, 1939, 14:5; May 5, 1938, 23:3; September 28, 1919, 22:4.

70. Irwin Yellowitz, *Labor and the Progressive Movement in New York State, 1897–1916* (Ithaca, N.Y.: Cornell University Press, 1965), 158–66; Gerald Kurland, *Seth Low: The Reformer in an Urban and Industrial Age* (New York: Twayne, 1971), 126–27; *NYTr*, August 5, 1901, 10:3; December 1, 1902, 12:1; November 10, 1902, 4:1; September 29, 1902, 9:5.

71. *NYT*, July 13, 1902, 2:2; *NYH*, November 1, 1902, JF.

72. Yellowitz, *Labor and the Progressive Movement*; Ira Katznelson, *City Trenches: Urban Politics and the Patterning of Class in the United States* (New York: Pantheon, 1981).

73. *NYTr*, December 11, 1902, 1:2 and December 12, 1902, 6:1; *NYT*, November 27, 1902, 1:7 and December 17, 1902, 1:1.

74. *Tribune Almanac and Political Register* (1902), 398–99; *NYT*, November 7, 1901, 1:7, 2:1–3; *NYTr*, December 17, 1902, 1:1.

75. Zink, *City Bosses*, 147–63; Nancy Joan Weiss, *Charles Francis Murphy, 1858–1924: Respectability and Responsibility in Tammany Politics* (Northampton, Mass.: Smith College, 1968); *NYT Magazine* (October 28, 1917): 6; Gustavus Myers, *History of Tammany Hall*, 2d ed. (New York: Boni & Liveright, 1917), 311–12; Morris Robert Werner, *Tammany Hall* (New York: Doubleday, Doran, 1928), 486; *NYTr*, June 22, 1904, 12:2; *NYT*, February 16, 1905, 6:6.

76. *Tribune Almanac and Political Register* (1905), 267–68; *NYTr*, March 3, 1905, 11:1; *NYT*, February 16, 1905, 6:6, March 3, 1905, 6:3; interview with George McAneny, February 25, 1949, 59, GMC, box 1; *Financial Transactions of the New York, New Haven & Hartford*, 754–55.

77. Ferdinand Lundberg, *Imperial Hearst* [1936] (Westport, Conn.: Greenwood Press, 1970), 36–37, 101–4; W. A. Swanberg, *Citizen Hearst: A Biography of William Randolph Hearst* (New York: Scribner's, 1961), 49–50, 230–38; Kenneth Finegold, *Experts and Politicians: Reform Challenges to Machine Politics in New York, Cleveland, and Chicago* (Princeton, N.J.: Princeton University Press, 1995), 45–47.

78. *NYTr*, March 3, 1905, 11:1, March 14, 1905, 7:3.

79. *NYT*, March 23, 1905, 7:1; *Wilcox v. McClellan*, 185 NY 9 (1906) at 19–21 and 110 app. div. 378 (1905) at 388.

80. Milo Roy Maltbie, "A Century of Franchise History," *Municipal Affairs* 4 (March 1900): 197, 200; Edward M. Grout, "New York City Should Own the Gas Supply," *Municipal Affairs* 1 (June 1897): 225–44; R. R. Bowker, "The Piracy of Public Franchises," *Municipal Affairs* 5 (December 1901): 886–904.

81. George Foster Peabody, "Edward Morse Shepard," *American Monthly Review of Reviews* 24 (November 1901): 548–51; McAneny to Shepard, March 3, 1894, December 26, 1894, November 19, 1895, January 2, 1897, and Cassatt to Shepard, February 28, 1906, EMS; *NYT*, April 14, 1905, 5:4–5; Coler to Shepard, February 21, 1906, EMS.

82. Ervin Wardman, "Hearst versus McClellan: The New York Mayoralty Imbroglio," *Broadway Magazine* 18 (April 1907): 3–14; Lundberg, *Imperial Hearst*, 104.

83. *NYT*, April 7, 1905, 7:2.

84. Oscar Handlin, *Al Smith and His America* (Boston: Little, Brown, 1958); Thomas Kessner, *Fiorello H. La Guardia and the Making of Modern New York* (New York: Penguin, 1989); Michael P. Weber, *Don't Call Me Boss: David L. Lawrence, Pittsburgh's Renaissance Mayor* (Pittsburgh: University of Pittsburgh Press, 1988); Roger Biles, *Richard J. Daley: Politics, Race, and the Governing of Chicago* (DeKalb: Northern Illinois University Press, 1995).

85. Third vice president to McCrea, June 6, 1908, PRR, box 138, file 21.
86. William D. Middleton, *Grand Central* (San Marino, Calif.: Golden West Books, 1977), 29–30.
87. Petitioner's brief, Manufacturers' and Property Owners' Interests Association vs. PRR, 3, 8, JF, box 5, file: WSAS briefs.
88. *Jersey Journal,* July 1, 1922, JF, box 5, file: WSAS.
89. Petitioner's brief, 3–4, 28.
90. Petitioner's brief, 6–7, 12–13, and "West Side Avenue Station, Statement No. 2," JF, box 5, first file: WSAS 1915, 1916.
91. Forgie memo for Wall, April 5, 1915, JF, box 5, first file: WSAS 1915, 1916.
92. Forgie to Krick and Wall, April 29, 1916, JF, box 5, first file: WSAS 1915, 1916.
93. Brief for respondent, Manufacturers' and Property Owners' Interests Association vs. PRR, 7, 8–17, JF, box 5, file: WSAS briefs.
94. Reeves memo, June 8, 1916, JF, box 5, second file: WSAS 1915, 1916.
95. Memo as an argument against West Side Avenue Station, June 8, 1916 (from Mr. Reeves, I.R.T. Co.), JF, box 5, second file: WSAS 1915, 1916.
96. Petition for station at West Side Avenue—draft of possible line of examination, JF, box 5, second file: WSAS 1915, 1916.
97. Lippmann, *Drift and Mastery,* 155, 148; David Hollinger, "Science and Anarchy: Walter Lippmann's *Drift and Mastery,*" in *In the American Province,* 48.

TWO: Beyond Voluntarism

1. *NYHC,* 47 ICC 643, quotation at 742.
2. Calvin Tomkins, "A Comprehensive Plan and Policy for the Organization and Administration of the Inter-State Port of New York and New Jersey," address before the New Jersey Harbor Commission, February 19, 1912, 4; *NYHC,* 47 ICC 643 at 646, 651, 740, 741; Charles Jacobs and John V. Davies, "New York City West Side Freight Problem, April 9, 1912," PRR, box 141, file 9.
3. Freight and passenger transportation planning differed in four important respects. First, people could move themselves: once deposited on the Manhattan shoreline, passengers walked to other destinations, whereas freight had to be moved by hand and then by wagon or truck. This problem of freight mobility was resolved, at least partially, by the construction of automobile bridges, which allowed goods to be moved across the river on trucks: thus carried, the freight, like passengers, moved itself once on the Manhattan side. Second, the volume of goods was greater than the volume of people. Dozens of ferries moved passengers throughout the harbor, but on any given day thousands of freight cars waited to be unloaded and reloaded. Third, and most important, the problem of passenger transportation was taken up by other institutions once passengers moved beyond the waterfront. Railroad companies created new points of entry (like Penn Station), but that only produced a new problem—one that subway and traffic planners addressed. This can be seen clearly in the PRR's efforts to secure the Seventh Avenue subway line. Fourth, the problem of cost was more apparent in the case of freight transportation. The inefficiencies of moving goods across the Hudson translated into higher freight rates and thus impinged on the competitive position of New York vis-à-vis other ports. That problem would be confronted regarding passenger service most directly during the debate over subway fares.
4. Erwin W. Bard, *The Port of New York Authority* (New York: Columbia University

Press, 1942); Carl W. Condit, *The Port of New York* (Chicago: University of Chicago Press, 1980–81); Josef W. Konvitz, "William J. Wilgus and Engineering Projects to Improve the Port of New York, 1900–1930," *Technology and Culture* 30 (April 1989): 398–425; Jameson W. Doig, "Entrepreneurship in Government: Historical Roots in the Progressive Era," paper prepared for the annual meeting of the American Political Science Association, Washington, D.C., September 1988; idem, "Regional Planners and the New York Minefield in the 1920s: A Port Authority Tries to Light the Way," paper prepared for the annual conference of the American Historical Association, Chicago, Ill., December 1991.

5. Ann L. Buttenwieser, *Manhattan Water-Bound: Manhattan's Waterfront from the Seventeenth Century to the Present*, 2d ed. (Syracuse, N.Y.: Syracuse University Press, 1999), 81–196.

6. *NYHC*, 47 ICC 643 at 646, 651, 740, 741.

7. Jacobs and Davies, "West Side Freight," 2–4, 7; E. P. Goodrich and Harry P. Nichols, *Report upon the Elimination of Surface Freight Railroad Tracks of the New York Central and Hudson River Railroad (March 1911)*, 16–19.

8. Stenographer's minutes, 2886–89, 2893–95, "*NYHC*," formal docket no. 8994, ICCNA; Port of New York Authority, *Report with Plan for the Comprehensive Development of the Port of New York* (Albany: J. B. Lyons, 1921), 17.

9. Goodrich and Nichols, *Elimination*, 15–16, 44, plate no. 1; Jacobs and Davies, "West Side Freight," 3, 6–7; *NYHC*, 47 ICC 643 at 690.

10. *In the Matter of Proposed Advances in Freight Rates*, 9 ICC 382 (1903) at 383–84, 395, 413–14, 424–25.

11. *Advances in Freight Rates*, 9 ICC 382 at 401, 404–5.

12. *Five Percent* case, 31 ICC 351 at 359–60.

13. *Advances in Freight Rates*, 9 ICC 382 at 428–29.

14. *Central Yellow Pine Association v. Illinois Central*, 10 ICC 505 at 526, 543–44; *Illinois Central v. ICC*, 206 U.S. 441 (1907) at 461–63.

15. Louis D. Brandeis, *Other People's Money: And How the Bankers Use It* [1914] (Norwood Editions, 1980); *In Re Investigation of Advances in Rates by Carriers in Officials Classification Territory [Advances in Rates—Eastern Case]*, 20 ICC 243 at 280–81.

16. *Advances in Rates—Eastern Case*, 20 ICC 243 at 247, 249–61, 263, 266–69.

17. *Evidence Taken by the Interstate Commerce Commission in the Matter of Proposed Advances in Freight Rates*, 61st cong., 3d sess., 1910, sen. doc. 725, 2335, 2286–91, 2295–97; *Advances in Rates—Eastern Case*, 20 ICC 243 at 269, 270.

18. Elizabeth Sanders, *Roots of Reform: Farmers, Workers, and the American State, 1877–1917* (Chicago: University of Chicago Press, 1999); *Advances in Rates—Eastern Case*, 20 ICC 243 at 283; *In the Matter of Rates, Practices, Rules, and Regulations Governing the Transportation of Anthracite Coal*, 35 ICC 220 at 270–71.

19. *Five Percent* case, 31 ICC 351 at 384, 367, 359–60, 375–82.

20. Ibid., 31 ICC 351 at 384, 387–92.

21. Ibid., 32 ICC 325 at 340.

22. *Five Percent* case, *Letters from the Chairman*, 63d cong., 2d sess., 1915, sen. doc. 466, 4139, 4138–41; Samuel Rea, "Our Railroad Problem: How to Settle It Effectually in the Public Interest," address before the American Bankers Association, Saint Louis, October 1, 1919, 23–24.

23. Calvin Tomkins, *Report on Transportation Conditions at the Port of New York* (July 1910); idem, *Report on the Organization of the South Brooklyn Waterfront* (March 27, 1911); idem, *Supplementary Report on the Organization of the South Brooklyn Waterfront* (January 8, 1912); Tomkins address to Newark Board of Trade, December 21, 1911, WJG, box GWJ-26; preliminary statement by Tomkins at hearing on waterfront improvement plans for Borough of Brooklyn, July 10, 1912, and Tomkins to Gaynor, September 30, 1912, WJG, box GWJ-46.

24. *WWNY* (1914), 712; Calvin Tomkins, "A New Jersey Forecast," address at roads convention, Atlantic City, September 26, 1908.

25. "Comprehensive Plan and Policy," 23.

26. Tomkins, *Report on Transportation Conditions*, 5–6, 8–18; idem, *Report on the Organization of the South Brooklyn Waterfront*; idem, *Supplementary Report on the Organization of the South Brooklyn Waterfront*; Tomkins address to Newark Board of Trade, WJG, GWJ-26 (the emphasis is mine).

27. Peters to Browne, April 9, 1915, JPM, box MJP-219; summary of action affecting proposed South Brooklyn Marginal Railroad, JPM, box MJP-118; Tomkins, *Supplementary Report on the Organization of the South Brooklyn Waterfront*.

28. Franklin Escher, "Retrenchment Policy of the Railways," *Harper's Weekly* 55 (October 7, 1911): 22; idem, "The Delicate Question of Railway Credit," *North American Review* 197 (February 1913): 213; *NYT*, June 5, 1910, II, 7:1, March 9, 1911, 8:3, March 28, 1914, 1:4, March 17, 1911, 6:6, April 15, 1911, 15:4, August 27, 1911, 2:4, February 26, 1911, 5:3, April 16, 1911, 5:7, March 27, 1914, 10:1; "The Freight Rate Increase: A Crisis in Railroad Finance," *Review of Reviews* 50 (May 1914): 563.

29. *NYT*, February 21, 1914, 14:3, April 3, 1914, 12:2, February 25, 1912, II, 13:2, October 29, 1914, 10:1, July 29, 1912, 8:1, May 8, 1914, 14:3, November 2, 1914, 11:5 (emphasis in the original).

30. Franklin Escher, "When Railway Earnings Decline," *Harper's Weekly* 55 (April 29, 1911): 22.

31. Olivier Zunz, *Why the American Century?* (Chicago: University of Chicago Press, 1998), 52–55; David M. Kennedy, *Over Here: The First World War and American Society* (Oxford: Oxford University Press, 1980), 114, 118, 143; John M. Jordan, *Machine-age Ideology: Social Engineering and American Liberalism, 1911–1939* (Chapel Hill: University of North Carolina Press, 1994), 93–109, 110, 113–14.

32. Albro Martin, *Enterprise Denied: Origins of the Decline of American Railroads, 1897–1917* (New York: Columbia University Press, 1971), 335–40; Kennedy, *Over Here*, 5–6, 252–54; K. Austin Kerr, *American Railroad Politics, 1914–1920* (Pittsburgh: University of Pittsburgh Press, 1968), 40–41.

33. Kerr, *American Railroad Politics*, 128–30.

34. Thorstein Veblen, *The Engineers and the Price System* [1921] (New York: August M. Kelley, 1965), 1, 40–41, 52–54, 69–70; Samuel Haber, *Efficiency and Uplift: Scientific Management in the Progressive Era, 1880–1920* (Chicago: University of Chicago Press, 1964), 143, as quoted in Jordan, *Machine-age Ideology*, 104–5.

35. *NYHC*, 47 ICC 643 at 681–83; Tomkins, *Report on Transportation Conditions*, 3; Tomkins to Gaynor, April 10, 1911, Tomkins to Adamson, April 12, 1911, and "The Differential R.R. Freight Rate Against the Port of New York," WJG, box GWJ-25; *Chamber of Commerce v. New York Central* 24 ICC 55.

36. Bard, *Port of New York*, 16–24; Doig, "Entrepreneurship in Government"; Todd

to Wilgus, May 7, 1908, transfer case 2 (file 2), small car freight subway correspondence, March 29, 1908 to October 31, 1908, WJW, box 45; Jacobs and Davies, "West Side Freight," 9.

37. *NYHC*, 47 ICC 643 at 739.

38. Julius Henry Cohen, *They Builded Better than They Knew* [1946] (Freeport, N.Y.: Books for Libraries Press, 1971), 179–214, 276–77; idem, "Collective Bargaining and the Law as a Basis for Industrial Reorganization," *Annals of the American Academy of Political and Social Science* 90 (July 1920): 47–49.

39. *U.S. v. B&O*, 231 U.S. 274 (1913) at 288; Cohen, *Builded Better than They Knew*, 277–78, 280–81, 283.

40. *NYHC*, 47 ICC 643 at 739, 734, 733.

41. New York, New Jersey Port and Harbor Development Commission, *Joint Report With Comprehensive Plan and Recommendations* (Albany, N.Y.: J. B. Lyon, 1920), 2, 38.

42. Jameson W. Doig, *Empire on the Hudson: Entrepreneurial Vision and Political Power at the Port of New York Authority* (New York: Columbia University Press, 2001), 63–65.

43. Rea to Wilgus, February 26, 1924, WJW, box 105.

44. Wilgus to Cohen, August 20, 1925, and September 3, 1925, WJW, box 56.

45. Cohen, *Builded Better than They Knew*, 281 (emphasis in the original), 291–92, 296; John Francis Hylan, "To the Honorable, the Board of Estimate and Apportionment," May 6, 1924, JFH, box HJF-189; Doig, "Entrepreneurship in Government," 77–86.

46. Port of New York Authority, *Report, with Plan for the Comprehensive Development of the Port of New York* (Albany, N.Y.: J. B. Lyon, 1921), 11.

47. Bard, *Port of New York*, 41–45, 63–92; Cohen, *Builded Better than They Knew*, 279–80; Doig, "Regional Planners and the New York Minefield"; *Newark v. Central Railroad*, 267 U.S. 377 at 386.

48. Interstate Commerce Act, sec. 15A, para. 3, in Rogers Mac Veagh, *The Transportation Act, 1920* (New York: Henry Holt, 1923), 384; I. L. Sharfman, *The Interstate Commerce Commission* (New York: Commonwealth Fund, 1931), 1:177–244.

49. *Railroad Commission of Wisconsin v. C.B. & Q.*, 257 U.S. 563 at 583; *Dayton–Goose Creek Railway v. U.S.*, 263 U.S. 456 at 478 (the emphasis is mine).

50. Sec. 15A, paras. 2 and 3, in Mac Veagh, *Transportation Act*, 384–85.

51. *Dayton–Goose Creek*, 263 U.S. 456 at 478.

52. Sec. 15A, paras., 6, 7, 10, in Mac Veagh, *Transportation Act*, 386–87; Samuel Rea, "Objections to Government Guarantee of Return on Railroad Capital," *Journal of the National Institute of Social Sciences* 5 (June 1, 1919): 86–95; Kerr, *American Railroad Politics*, 153–55.

53. Kerr, *American Railroad Politics*, 111–19; Kennedy, *Over Here*, 254–58.

54. *New England Division cases*, 261 U.S. 184 at 190–91; *Dayton–Goose Creek*, 263 U.S. 456 at 480.

55. *Manufacturers Association of York v. PRR*, 73 ICC 40 at 49.

56. For an in-depth analysis of the joint-use controversy, see Keith D. Revell, "Cooperation, Capture, and Autonomy: The Interstate Commerce Commission and the Port of New York Authority in the 1920s," *Journal of Policy History* 12 (spring 2000): 177–214.

57. *33rd Annual Report of the ICC* (Washington, D.C.: Government Printing Office,

1919), 1–6; Claude Moore Fuess, *Joseph B. Eastman: Servant of the People* (Westport, Conn.: Greenwood Press, 1974), 98–100.

58. Homer Bews Vanderblue and Kenneth Farwell Burgess, *Railroads: Rates, Service, Management* (New York: Macmillan, 1923), 274; *Hearings before the Committee on Interstate and Foreign Commerce of the House of Representatives*, 66th cong., 1st sess., on H.R. 4378, vol. 1 (Washington, D.C.: Government Printing Office, 1919), 16–17.

59. *33rd Annual Report*, 6.

60. *Hearings before the Committee . . . on H.R. 4378*, 1: 73, 104.

61. Revell, "Cooperation, Capture, and Autonomy," 210–11.

62. Second *Hastings*, 107 ICC 208 at 212; *33rd Annual Report*, 6; Sharfman, *Interstate Commerce Commission*, 3A:416–21.

63. Mac Veagh, *Transportation Act*, 241.

64. *Jamestown Chamber of Commerce v. Jamestown, Westfield & Northwestern Railroad*, 195 ICC 289 at 292.

65. Second *Hastings*, 107 ICC 208 at 216; *PNYA*, 144 ICC 514 at 522.

66. Revell, "Cooperation, Capture, and Autonomy," 181–85.

67. Second *Hastings*, 107 ICC 208 at 215–17.

68. *PNYA*, 144 ICC 514.

69. Stenographer's minutes, 996, 1006, 1047–48, "Port of New York Authority v. Atchison, Topeka & Santa Fe Railway," docket no. 16923; LaRoe to Aitchison, March 9, 1926; Choate to Aitchison, March 13, 1926; LaRoe to Aitchison, March 17, 1929; LaRoe to Aitchison, March 20, 1926; all ICCNA.

70. Stenographer's minutes, 599, 1050–51; *PNYA*, 144 ICC 514 at 534.

71. Stenographer's minutes, 1052–53; R. Kent Newmyer, *Supreme Court Justice Joseph Story: Statesman of the Old Republic* (Chapel Hill: University of North Carolina Press, 1985), 224, 225, 226.

72. Stenographer's minutes, 1061, 1065.

73. Ibid., 1061–62, 1063.

74. *PNYA*, 144 ICC 514 at 522–24.

75. Ibid. at 532, 533, 527, 515, 534–35.

76. Ibid., at 523–24, 533–34.

77. Sharfman, *Interstate Commerce Commission*, 3A: 421.

78. Doig, "Regional Planners and the New York Minefield," 18–19; Bard, *Port of New York*, 111–28, 139, 186.

79. Cohen, *Builded Better than They Knew*, 272.

80. Pendleton Herring, *Public Administration and the Public Interest* [1936] (New York: McGraw-Hill, 1967), 8.

THREE: Buccaneer Bureaucrats, Physical Interdependence, and Free Riders

1. *NYTr*, October 8, 1909, 1; *NYT*, October 8, 1909, 1, 2, October 10, 1909, 6; A. A. Breneman, "Explosions of Sewer Gas in New York City," *Engineering News* 62 (December 2, 1909): 608–9.

2. Alfred D. Flinn, "Engineering Achievements and Activities of New York City," *TASCE* 72 (December 1913): 1814–15; Nelson P. Lewis, "During my long journey from the rural district," 1903–4, NPL, box 1, file 6.

3. Sharon Reier, *The Bridges of New York* (New York: Quadrant, 1977), 154–57; William A. Prendergast, "Financial Administration, Budget and Tax Rate," in *The Government of the City of New York* (New York: State Constitutional Convention Commission, 1915), 9–10, 157.

4. Josef W. Konvitz, *The Urban Millennium: The City-building Process from the Early Middle Ages to the Present* (Carbondale: South Illinois University Press, 1985), 132; Maureen Ogle, *All the Modern Conveniences: American Household Plumbing, 1840–1890* (Baltimore: Johns Hopkins University Press, 1996), 110, 159; Joanne Abel Goldman, *Building New York's Sewers: Developing Mechanisms of Urban Management* (West Lafayette, Ind.: Purdue University Press, 1997), 167.

5. H. Malcolm Pirnie, "Zoning and Water Supply," *TASCE* 85 (1925): 718.

6. Robin L. Einhorn, *Property Rules: Political Economy in Chicago, 1833–1872* (Chicago: University of Chicago Press, 1991), 104–9, 114, 133–43.

7. Einhorn, *Property Rules*, 240–41; Paul Boyer, *Urban Masses and Moral Order in America, 1820–1920* (Cambridge: Harvard University Press, 1978).

8. Charles A. Beard, "Some Aspects of Regional Planning," *American Political Science Review* 20 (May 1926): 278.

9. Clifton Hood, *Seven Hundred and Twenty-two Miles: The Building of the Subways and How They Transformed New York* (Baltimore: Johns Hopkins University Press, 1995), 12; Harold M. Lewis, *Transit and Transportation* (New York: Regional Plan of New York and Its Environs, 1928), 40, 36–37, 54; Clifton Hood, "The New York Subway, Transit Politics, and Metropolitan Expansion," in *The Landscape of Modernity: Essays on New York City, 1900–1940*, ed. David Ward and Olivier Zunz (New York: Russell Sage, 1992), 203–4.

10. Walter Laidlaw, *Population of the City of New York, 1890–1930* (New York: Cities Census Committee, 1932), 51.

11. Cf. Hood, *Seven Hundred and Twenty-two Miles*, 16, 190, 202, 230–31, 253.

12. Ibid., 21–28, 56–74, esp. 24 and 71, 59.

13. *NYT*, April 28, 1907, 3:1, October 15, 1909, 10:1; *Levy v. McClellan*, 196 N.Y. 178 (1909); second vice president to Shonts, May 25, 1910, PRR, box 141, file 22; *NYT*, July 18, 1907, 6:7.

14. *NYT*, October 23, 1909, 10:1, November 1, 1909, 11:6; Shonts to Rea, May 9, 1910, PRR, box 141, file 22; president to Willcox, July 5, 1910, PRR, box 141, file 23.

15. Digest of report on rapid transit conditions, n.d., PRR, box 141, file 21; Hood, *Seven Hundred and Twenty-two Miles*, 129.

16. *NYT*, October 15, 1909, 10:1; W. A. Swanberg, *Citizen Hearst: A Biography of William Randolph Hearst* (New York: Scribner's, 1961), 266–68; Hood, *Seven Hundred and Twenty-two Miles*, 220.

17. McAneny, Miller, and Cromwell to board of estimate, July 20, 1911, PRR, box 142, file 5; Hood, *Seven Hundred and Twenty-two Miles*, 123–24; Outerbridge to Rea, April 25, 1911, PRR, box 142, file 2.

18. Scott L. Bottles, *Los Angeles and the Automobile: The Making of the Modern City* (Berkeley: University of California Press, 1987), 33–42; Philip J. Ethington, *The Public City: The Political Construction of Urban Life in San Francisco, 1850–1900* (Cambridge: Cambridge University Press, 1994), 342–43, 395–96.

19. Second vice president to Hamilton, July 15, 1909, PRR, box 141, file 21.

20. *NYT*, August 8, 1911, 8:3; Parsons to Rea, August 8, 1911, PRR, box 142, file 5.

21. "New York Seventh Avenue Subway," May 24, 1910, PRR, box 141, file 22.
22. Hedley to Bassett, November 11, 1909, Shonts to Rea, May 9, 1910, PRR, box 141, file 22.
23. Willcox to Gaynor, July 7, 1910, PRR, box 141, file 23.
24. Boardman to Massey, October 16, 1911, first vice president to O'Brien, October 14, 1911, County to O'Brien, October 23, 1911, PRR, box 142, file 6.
25. *NYT,* December 9, 1910, 10:2; County to Rea, December 12, 1910, PRR, box 142, file 1.
26. Assistant to second vice president to McCrea, July 1, 1910, assistant to second vice president to Peters, PRR, box 141, file 23.
27. First vice president to McAneny, May 20, 1911, PRR, box 142, file 2; second vice president to Berwind, November 19, 1910, PRR, box 142, file 1; first vice president to Mitchel, October 27, 1911, PRR, box 142, file 6.
28. Shonts to Rea, July 12, 1909, PRR, box 141, file 21; *NYT,* November 18, 1910, 2:2; "Justice for the Pennsylvania," *Record and Guide,* September 16, 1911, PRR, box 142, file 5; first vice president to Massey, October 14, 1911, PRR, box 142, file 6.
29. Letter to O'Brien, October 26, 1911, PRR, box 142, file 6.
30. [Rea] to Schiff, October 11, 1910, PRR, box 141, file 23.
31. County to Rea, October 17, 1911, PRR, box 142, file 6.
32. *NYT,* November 18, 1910, 2:1.
33. Moschzisker to Rea, August 18, 1911, PRR, box 142, file 5; Rea to Shonts, July 28, 1909, PRR, box 141, file 22; Rea to Stern, September 2, 1909, Stern to Rea, September 13, 1909, PRR, box 141, file 22; Peters to County, July 12, 1910, PRR, box 141, file 23; Rea to Gaynor and Willcox, July 18, 1910, PRR, box 141, file 23; first vice president to Gimbel Brothers, May 23, 1911, April 21, 1911, PRR, box 142, file 2; Shepard to Rea, July 19, 1910, PRR, box 141, file 23; Outerbridge to Rea, July 21, 1910, PRR, box 141, file 23.
34. First vice president to McAneny, May 20, 1911, PRR, box 142, file 2; Outerbridge to Pratt, August 21, 1909, PRR, box 141, file 22; first vice president to Mitchel, October 27, 1911, PRR, box 142, file 6; County to McAneny, January 17, 1912, GMP, box 72.
35. Hood, *Seven Hundred and Twenty-two Miles,* 150–61.
36. "A Keeper of the City," *Outlook* 93 (November 27, 1909): 659–61; Henry F. Griffin, "A Reformer in Office," *Outlook* 98 (August 12, 1911): 829–32; George McAneny, "What I Am Trying to Do," *World's Work* 26 (June 1913): 177, 179; interview with McAneny, November 26, 1948, 8, GMC, box 1.
37. McAneny, "What I Am Trying To Do," 170; Tucker to McAneny, February 1, 1911, GMP, box 72; McAneny to Rea, May 29, 1911, PRR, box 142, file 2; McAneny, Miller, and Cromwell to board of estimate, July 20, 1911; Henry Farrand Griffin, "George McAneny," *Independent* 75 (July 31, 1913): 248; Hood, *Seven Hundred and Twenty-two Miles,* 152.
38. McAneny, "What I Am Trying To Do," 170; Hood, *Seven Hundred and Twenty-two Miles,* 156, 158.
39. Statement accompanying Pounds to McAneny, May 12, 1911, GMP, box 72.
40. Hood, *Seven Hundred and Twenty-two Miles,* 187–90, 193–97, 203–13, 221–22; Swanberg, *Citizen Hearst,* 307; Ferdinand Lundberg, *Imperial Hearst: A Social Biography* [1936] (Westport, Conn.: Greenwood Press, 1970), 248.
41. *Finances and Financial Administration of New York City* (New York: Columbia University Press, 1928), 222, 190–91.

42. Merchants' Association, *An Inquiry into the Conditions Relating to the Water Supply of the City of New York* (1900), 11; Charles H. Weidner, *Water for a City: A History of New York City's Problem from the Beginning to the Delaware River System* (New Brunswick, N.J.: Rutgers University Press, 1974), 132–33; Edward Hagaman Hall, *The Catskill Aqueduct and Earlier Water Supplies of the City of New York*, 2d ed. (New York: Mayor's Catskill Aqueduct Celebration Committee, 1917), 70–76; Nelson Manfred Blake, *Water for the Cities: A History of the Urban Water Supply Problem in the United States* (Syracuse, N.Y.: Syracuse University Press, 1956), 121–71, 277.

43. John R. Freeman, *Report upon New York's Water Supply* (New York: Martin B. Brown, 1900), 55; notes accompanying Taber to Freeman, September 24, 1899, JRF, box 101.

44. Freeman, *Report*, 50, 54; "Further Investigations of Water Waste in New York City," *Engineering News* 49 (April 9, 1903): 335–36.

45. Olivier Zunz, *The Changing Face of Inequality: Urbanization, Industrial Development, and Immigrants in Detroit, 1880–1920* (Chicago: University of Chicago Press, 1982), 116–17; report attached to Jensen to Moore, June 24, 1895, JRF, box 101.

46. "To the Commissioners of the Sinking Fund, New York City," May 8, 1895 and May 21, 1895, JRF, box 101.

47. *NYH*, February 1, 1901, 3; *NYT*, February 1, 1901, 1:1,3; Merchants' Association, *Inquiry into the Conditions*, 97–98, 389–90; Chambers to McClellan, January 20, 1905, GBM, box MGB-115.

48. Merchants, manufacturers, property owners, and residents of Brooklyn to the mayor and board of aldermen, December 2, 1896, and McLean to Coler, August 1899, JRF, box 101.

49. McLean to Coler, August 1899, and *Statement by the Commissioner of Water Supply, Gas and Electricity to the Board of Estimate Regarding Issue of Corporate Stock for Water Supply Purposes*, April 2, 1902, JRF, box 101; *NYT*, July 12, 1900, 12:1.

50. Freeman to Dougherty, July 15, 1902, JRF, box 102.

51. Merchants' Association, *An Inquiry into the Conditions*, 11–12; Lazarus White, *The Catskill Water Supply of New York City* (New York: John Wiley, 1913), 22–26; *Smith v. Brooklyn*, 46 N.Y.S. 141 (1897); Moon to McClellan, January 5, 1905, GBM, box MGB-115.

52. New York Board of Fire Underwriters to board of public improvements, June 10, 1898, JRF, box 101; McLean to Coler, August 1899.

53. Freeman to Chadwick, June 22, 1905, Freeman to Towne, October 31, 1904, JRF, box 103.

54. Freeman to Towne, October 12, 1905, JRF, box 103.

55. "A Review of the Brooklyn Water Supply System," November 19, 1901, 2, 3, 8, RVW, box WRV-11; "How Long Must We Suffer? Our Engineering Department at Present the Creature of Party Politics," *Scientific American* 91 (July 16, 1904): 38.

56. Towne to Freeman, May 31, 1904, JRF, box 103.

57. New York Board of Fire Underwriters to board of public improvements, June 10, 1898, and George T. Hope and Charles S. Smith, address of underwriters and merchants to the legislature, JRF, box 101.

58. Merchants' Association, *An Inquiry into the Conditions*, xxx, 17–18, 23, 597–604, esp. 603.

59. Bird Coler, "Municipal and Business Corruption," *Independent* 52 (March 15, 1900): 662–63; idem, "The Political Wrecking of Business Enterprises," *Munsey's Magazine* 23 (May 1900): 277–80; "Coler's Ramapo Platform," clipping dated August 2, 1900, "Coler a Candidate," *Boston Advertiser*, August 13, 1900, and Goodell to Freeman, May 23, 1900, JRF, box 102; *Ramapo v. NY*, 236 U.S. 579 (1915); Moore to Freeman, July 10, 1900, JRF, box 102; *NYT*, June 14, 1941, 17:1.

60. "Memoir of John Ripley Freeman," *TASCE* 98 (1933): 1471–76.

61. Freeman to Evans, February 9, 1905, JRF, box 103; Freeman, *Report*, 4–5.

62. "Memoir of Jonas Waldo Smith," *TASCE* 101 (1936): 1502–11; Freeman to Chadwick, June 22, 1905, "Civil Service Examinations for Assistant Engineer," January 24, 1906, and "Civil Service Examinations for Assistant Engineers, Continuation of Notes of January 24, 1906," February 14, 1906, all in JRF, box 103.

63. "Memoir of Alfred Douglas Flinn," *TASCE* 103 (1938): 1787–95; Deming to McClellan, April 4, 1905, GBM, box MGB-115; Weidner, *Water for a City*, 159–75; *Report of the Board of Water Supply of the City of New York to the Board of Estimate* (New York: Martin B. Brown, 1905), 5–8, 11–14, 15–18, Board of Water Supply, *Catskill Water Supply (January, 1913)*, 3, 5, both in CWS.

64. *Catskill Water Supply*, 32; Board of Water Supply, *The Water Supply of the City of New York* (September 1950), 7; Harold M. Lewis, *Physical Conditions and Public Services* (New York: Regional Plan of New York and Its Environs, 1929), 38.

65. Pirnie, "Zoning and Water Supply," 718–19; Glen E. Vogel, "City Tunnels No. 1 and 2: 'The Granddaddies of Tunneling,'" *One Hundred Twenty Years of Tunneling in New York City*, American Society of Civil Engineers, Metropolitan Section, February 14 and 15, 1994, 4, 5–8; Flinn, "Engineering Achievements," 1809; Hall, *Catskill Aqueduct*, 108–10, 113.

66. Freeman, *Report*, 6, 66.

67. James H. Fuertes, *Waste of Water in New York and Its Reduction by Meters and Inspection* (New York: Merchants' Association, 1906), 251–52.

68. Detroit had a similar problem; there, officials addressed it by more draconian means—they cut off water until the leaks were repaired; Zunz, *Changing Face*, 117.

69. Freeman to Towne, October 31, 1904.

70. Croes to Freeman, May 5, 1900 and April 28, 1900, JRF, box 102; "Memoir of James John Robertson Croes," *TASCE* 58 (1907): 524–31; Baker to Freeman, July 19, 1900, JRF, box 102.

71. De Varona to O'Brien, October 16, 1906, and Towne to Gaynor, June 14, 1911, WJG, box GWJ-95.

72. Gaynor to Crowell, May 31, 1911, WJG, box GWJ-95.

73. Gaynor to Thompson, November 7, 1919, Gaynor to Ettinger, June 1, 1911, and Towne to Gaynor, June 14, 1911, WJG, box GWJ-95; Fuertes, *Waste of Water*, 17; *NYT*, November 8, 1912, 3:2.

74. Freeman to Dougherty, July 15, 1902, JRF, box 102.

75. *NYT*, April 16, 1916, I, 15:1; Thompson to Gaynor, October 28, 1912, WJG, box GWJ-95.

76. *NYT*, April 19, 1916, 12:8, September 17, 1917, III, 2:2; Thompson to Gaynor, October 28, 1912, WJG, box GWJ-95; *NYT*, June 10, 1917, IV, 3:3–5, July 1, 1917, IV, 2:8.

77. Thompson to Gaynor, October 28, 1912.

78. "City Asks Advice on Water Meters," *NYT,* July 31, 1966, and "Wave of Meters to Hit," *News,* February 4, 1988, vertical file: New York City, water meters, Municipal Reference Library, New York City.

79. Weidner, *Water for a City,* 283; *New Jersey v. NY,* 283 U.S. 336 at 344 (1931); Lewis, *Physical Conditions,* 33, 40, 49.

80. Flinn, "Engineering Achievements," 1775.

81. Eugene P. Moehring, *Public Works and the Patterns of Urban Real Estate Growth in Manhattan, 1835–1894* (New York: Arno Press, 1981), 87–98; Flinn, "Engineering Achievements," 1776, 1785; Goldman, *Building New York's Sewers.*

82. Kenneth Allen, "The Pollution of Tidal Harbors by Sewage with Especial Reference to New York Harbor," *TASCE* 85 (1922): 435–36; Leonard P. Metcalf and Harrison P. Eddy, *Design of Sewers* (New York: McGraw-Hill, 1914), 22–23; Lewis, *Physical Conditions,* 54; William Schroeder, "New York Starts $300,000,000 Sewage Treatment Program," *American City* 45 (August 1931): 11.

83. New York differed significantly from Chicago, which disposed of its sewage in Lake Michigan—the source of its drinking water; Louis P. Cain, *Sanitation Strategy for a Lakefront Metropolis: The Case of Chicago* (DeKalb: Northern Illinois University Press, 1978), xi, xiii, 61, 68–76, 114, 117–18, 123, 126.

84. Leonard Metcalf and Harrison P. Eddy, *Construction of Sewers* (New York: McGraw-Hill, 1915), 494–95; MSC, *Sewage and Sewerage Disposal in the Metropolitan District* (April 30, 1910), 103, 235–36.

85. Louis P. Cain and Elyce J. Rotella, "Death and Spending: Urban Mortality and Municipal Expenditure on Sanitation," *Annales de démographie historique* (summer 2001): 139–47.

86. Gretchen A. Condran, "Changing Patterns of Epidemic Disease in New York City," in *Hives of Sickness: Public Health and Epidemics in New York City,* ed. David Rosner (New Brunswick, N.J.: Rutgers University Press, 1995), 28–35.

87. Flinn, "Engineering Achievements," 1653.

88. *NY v. New Jersey and Passaic Valley Sewerage Commissioners,* 256 U.S. 296 at 298–99.

89. *NY v. Passaic Valley,* 256 U.S. 296 at 298–300; Stuart Galishoff, "The Passaic Valley Trunk Sewer," *New Jersey History* 88 (winter 1970): 197–214, esp. 199, 201–4; idem, *Newark: The Nation's Unhealthiest City, 1832–1895* (New Brunswick, N.J.: Rutgers University Press, 1988), 117–30, esp. 128; "A Scientific Scheme For Sewage Disposal by Dilution in New York Harbor," *Engineering News* 60 (December 17, 1908): 678; Bronx Valley Sewer Commission, *Report* (New York: Martin H. Brown, 1986), 4, 6–7; Lewis, *Physical Conditions,* 68.

90. *Report of the New York Bay Pollution Commission,* State of New York, sen. doc. no. 39 (Albany: Brandow Printing, 1905), 26, 27–28, 31, 44, 107.

91. Lewis, *Physical Conditions,* 62; MSC, *Final Report* (April 30, 1914); George A. Soper, "Permissible Limits of Sewage Pollution as Related to New York Harbor," *Engineering Record* 66 (September 28, 1912): 354; *WWNY* (1929), 1594–95; Soper to Gaynor, March 1, 1910, WJG, box GWJ-12.

92. Soper to Gaynor, March 1, 1910.

93. Lewis, *Physical Conditions,* 55; MSC, *Supplementary Report on the Disposal of New York's Sewage* (June 30, 1914), 14–15, and *Report of Rudolph Hering to the MSC,* 17–21, both from RH, box 7.

94. *Supplementary Report*, 15–16; Lewis, *Physical Conditions*, 64.
95. "A Scientific Scheme for Sewage Disposal," 679; *Report of Rudolph Hering*, 33–38.
96. Consulting engineer's report, 9, JFS, box 1, file 1913.
97. Lewis, *Physical Conditions*, 64; *Supplementary Report*, 4, 16, 18, 34, 40, 42; Joel A. Tarr, Terry Yosie, and James McCurley III, "Disputes over Water Quality Policy: Professional Cultures in Conflict, 1900–1917," *American Journal of Public Health* 70 (April 1980): 427–35.
98. Lewis, *Physical Conditions*, 65–68, 61–62; *NYT*, May 7, 1933, VIII, 3:1.
99. *NYT*, October 20, 1929, X, 6:1; Lewis, *Physical Conditions*, 55; Allen, "Pollution of Tidal Harbors," 436; report from Allen to Tuttle, September 7, 1926, RPA, box 9–4.
100. *NYT*, July 1, 1927, 6:3; *New Jersey v. NY*, 290 U.S. 237 (1933).
101. *NYT*, March 25, 1926, 1:2.
102. *NYT*, July 21, 1926, 3:5, August 4, 1926, 21:4, July 10, 1927, I, 6:3, August 1, 1926, II, 1:2.
103. *WWNY* (1929), 139–40; *NYT*, July 24, 1928, 20:4, February 19, 1929, 21:4; comptroller to chairman, board of estimate, April 7, 1928, RPA, box 9–2; *NYT*, April 23, 1927, 16:7, March 17, 1929, XI, 14:1.
104. Flinn, "Engineering Achievements," 1786–87; *NYT*, January 31, 1928, 1:5, February 26, 1928, IX, 1:1, March 8, 1928, 14:2, June 9, 1928, 19:1, July 4, 1928, 1:1, September 28, 1928, 29:1, October 18, 1928, 2:3, 28:2; *People v. Connolly*, 253 N.Y. 330 (1930) and 227 N.Y. app. div. 167 (1929).
105. *NYT*, January 2, 1929, 3:5, February 29, 1929, 29:5, March 11, 1929, 46:1, March 17, 1929, XI, 14:1, March 29, 1929, 1:7, December 1, 1929, 20:1.
106. *NYT*, April 5, 1929, 27:3, May 13, 1929, 22:8, October 22, 1929, 31:8, October 27, 1929, XII, 2:7, November 6, 1929, 5:4, November 7, 1929, 20:7.
107. George W. Fuller, "Sewage Disposal Trends in the New York City Region," *Sewage Works Journal* 4 (July 1932): 642; Lewis, *Physical Conditions*, 65–68, 61–62.
108. *NYT*, November 9, 1928, 27:7; Schroeder, "$300,000,000 Program," 11; *NYT*, April 3, 1931, 24:4.
109. *NYT*, July 17, 1931, 1:3, September 14, 1937, 16:4, June 28, 1935, 5:1.
110. *NYT*, June 29, 1935, 14:3, June 23, 1934, 27:1, January 20, 1939, 39:7, May 7, 1933, VIII, 3:1, July 26, 1935, 6:5, September 22, 1935, II, 2:1; "Ten Years of Sewage Treatment Progress," *American City* 62 (June 1947): 96–97 and 62 (July 1947): 140–42.
111. "Metropolitan Sewerage Commission against the Discharge of Treated Passaic Valley Sewage into New York Bay," *Engineering News* 62 (June 9, 1910): 662; "Paying for Joint Outfall Sewers," *Engineering Record* 62 (October 15, 1910): 421–22; Lewis, *Physical Conditions*, 69; Galishoff, "Passaic Valley Trunk Sewer," 209–14; *Van Cleve v. Passaic Valley*, 71 N.J.L. 574 (1904); *NY v. Passaic Valley*, 256 U.S. 296 at 309–10, 311, 313; "Passaic Valley Trunk Sewer Completed," *American City* 31 (October 1924): 315–18.
112. *NYT*, May 26, 1929, X, 4:6.
113. *NYT*, July 19, 1931, 39:4, July 12, 1931, IX, 4:2, June 4, 1931, 29:8.
114. *NYT*, June 18, 1931, 50:1, September 16, 1930, 26:8, May 8, 1935, 10:5, June 11, 1935, 11:3, January 8, 1932, 23:4, July 7, 1935, V, 10:4, August 16, 1935, 17:1, March 1, 1936, IV, 10:8; Fuller, "Sewage Disposal Trends," 643–45; "Connecticut Joins Interstate Commission," *American City* 57 (March 1942):11.
115. President to Day, March 7, 1933, RPA, box 69.
116. *NYT*, March 1, 1936, IV, 10:8, August 13, 1947, 25:1; *L. L. F. Realty v. Fuchs*, 273 N.Y.

app. div. 111 (1947); *ISC v. Weehawken,* 141 N.J. eq. 536 (1948); *Middlesex Concrete v. Carteret Industrial,* 37 N.J. 507 (1962); *Westchester v. Mamaroneck,* 41 N.Y. misc. 2d 811 (1964).

117. *NYT,* July 3, 1948, 17:8; "Bigger and Better Post-War Sewers for New York City," *American City* 60 (February 1945): 67–68; *NYT,* August 28, 1950, 1:2, 18:3; *Tursellino v. Paduano,* 202 N.Y. misc. 74 (1951).

118. *NYT,* August 18, 1991, VI, 16; Thomas M. Brosnan and Marie L. O'Shea, "Long-Term Improvements in Water Quality Due to Sewage Abatement in the Lower Hudson River," *Estuaries* 19 (December 1996): 890–900, esp. 895; *NYT,* April 12, 1991, B1, B4.

119. *NYT,* June 3, 1948, 27:1, August 28, 1950, 1:2, 18:3.

120. *NYT Magazine,* July 5, 1936, 8, 15; *NYT,* March 3, 1934, 12:2, July 3, 1948, 17:8, June 28, 1931, XI, 2:1, June 4, 1931, 29:8, January 20, 1943, 21:5.

121. Michael J. Ganas, Michael P. Hunnemann, and Danni R. Goulet, "Marine Borer Activity on the Rise in New York Harbor," *Public Works* 124 (January 1993): 32–35, John Waldman, *Heartbeats in the Muck: The History, Sea Life, and Environment of New York Harbor* (New York: Lyons, 1999).

FOUR: Taxing, Spending, and Borrowing

1. Nathan Matthews, *The City Government of Boston* (Boston: Rockwell & Churchill, 1895), 174–75; Pendleton Herring, *Public Administration and the Public Interest* [1936] (New York: Russell & Russell, 1967), 3.

2. Martin Shefter, *Political Crisis/Fiscal Crisis: The Collapse and Revival of New York City* (New York: Basic Books, 1985), xii–xiii, 3, 13, 24, 26–29, 34–37, 41–42, 219–25; Ester R. Fuchs, *Mayors and Money: Fiscal Policy in New York and Chicago* (Chicago: University of Chicago Press, 1992), 13–15, 17–19, 40–41, 72–74, 78–86, 94–96, 100.

3. Terrence J. McDonald, *The Parameters of Urban Fiscal Policy: Socioeconomic Change and Political Culture in San Francisco, 1860–1906* (Berkeley: University of California Press, 1986); Jon C. Teaford, *The Unheralded Triumph: City Government in America, 1870–1900* (Chicago: University of Chicago Press, 1984), 283–306.

4. Eugene P. Moehring, *Public Works and the Patterns of Urban Real Estate Growth in Manhattan, 1835–1894* (New York: Arno Press, 1981), 291–324; Seymour J. Mandelbaum, *Boss Tweed's New York* [1965] (Chicago: Ivan R. Dee, 1990), 59–80; Edward Dana Durand, *The Finances of New York City* (New York: Macmillan, 1898), 375.

5. Durand, *Finances,* 325–32, 296–97.

6. McDonald, *Parameters,* 40–41; *NYT,* August 27, 1896, 12:7; Teaford, *Unheralded Triumph,* 283–90.

7. Thomas Bender, *New York Intellect: A History of Intellectual Life in New York City, from 1750 to the Beginnings of Our Own Time* (Baltimore: Johns Hopkins University Press, 1987), 181–91.

8. David C. Hammack, *Power and Society: Greater New York at the Turn of the Century* (New York: Russell Sage, 1982), 130–31.

9. Durand, *Finances,* 331, 332.

10. *In the Matter of the Application of the Board of Rapid Transit Railroad Commissioners,* 5 N.Y. app. div. 290 (1896) at 296–97.

11. *NYT,* March 11, 1895, 10:4, December 12, 1895, 4:3, January 26, 1896, 16:3, January 3, 1897, 17:3, January 19, 1897, 3:6.

12. Bird Coler, "The Government of Greater New York," *North American Review* 169 (July 1899): 91; *NYT,* January 3, 1898, 6:4, January 1, 1898, 5:5, July 14, 1898, 4:5, November 24 1898, 12:3, December 15, 1898, 12:1; January 4, 1898, 8:2, March 18, 1898, 12:1, April 10, 1898, 12:4; Durand, *Finances,* 81–82; *Matter of the Application . . . Rapid Transit Railroad Commissioners,* 23 N.Y. app. div. 472 (1897) at 494–95.

13. McDonald, *Parameters; Finances and Financial Administration of New York City* (New York: Columbia University Press, 1928), 144–45.

14. Henry Bruère, *Reorganization of the Office of Chamberlain, 1914* (New York: Clarence S. Nathan, 1915), 24; *Finances and Financial Administration,* 337–38, 191; Henry Bruère, "New York's Nine-Hundred Million Debt," *Century Magazine* 77 (March 1909): 792; *Historical Statistics of the United States, Colonial Times to 1970* (White Plains, New York: Kraus International, 1989), 1118; F. Dodd McHugh, *Financing New York City's Future Permanent Improvements* (study by Works Progress Administration, sponsored by the mayor's committee on city planning, New York City, February 1938), fig. 22; William A. Prendergast, "Financial Administration, Budget and Tax Rate," in *The Government of the City of New York* (New York: State Constitutional Convention Commission, 1915), 9–10, 157.

15. Yin-ch'u Ma, "The Finances of the City of New York," *Studies in History, Economics, and Public Law,* vol. 61, whole no. 149 (New York: Columbia University, 1914), 187–93; Franklin Escher, "The Recent New York City Bond Sale," *Harper's Weekly* 55 (February 11, 1911): 22; Arthur M. Anderson, "The History of the Pay-as-You-Go Policy," *Proceedings of the Academy of Political Science of the City of New York* 8 (July 1918): 107.

16. Edgar J. Levey, "Sinking Funds of the City of New York," *Municipal Affairs* 4 (1900): 680; Delos F. Wilcox, "Finances of New York City," *Municipal Affairs* 2 (1898): 308.

17. Milo Roy Maltbie, "Cost of Government in City and State," *Municipal Affairs* 4 (1900): 686, 697; Durand, *Finances,* 293.

18. Henry De Forest Baldwin, "The City's Purse," *Municipal Affairs* 1 (1897): 338, 342, 351; Heydecker to McAneny, September 22, 1932, RPA, box 44.

19. Baldwin, "City's Purse," 331, 337; Levey, "Sinking Funds," 665–66; Bird Coler, "Financial Problems of a Great City," *North American Review* 173 (October 1901): 466–69.

20. Escher, "Bond Sale," 22; Ma, "Finances, City of New York," 188–90, 194–97; Gaynor to Menken, February 24, 1913, Gaynor to Beyer & Co., October 1, 1912, WJG, box GWJ-83.

21. Jonathan Kahn, *Budgeting Democracy: State Building and Citizenship in America, 1890–1928* (Ithaca, N.Y.: Cornell University Press, 1997), 39–45, 29–58, 88–91; Tilden Adamson, "The Preparation of Estimates and the Formulation of the Budget—The New York City Method," *Annals of the American Academy of Political and Social Sciences* 62 (November 1915): 250–51, 260–61.

22. *Reports of Division of Expert Accounting to the Comptroller, 1911–1912,* vols. 1 and 2, WJG, box GWJ-48; William Prendergast, "New York City Finances," *National Municipal Review* 2 (April 1913): 223–29.

23. Kahn, *Budgeting Democracy,* 41, 42, 44; *NYT,* November 18, 1917, VIII, 5:1.

24. Martin J. Schiesl, *The Politics of Efficiency: Municipal Administration and Reform in America, 1880–1920* (Berkeley: University of California Press, 1977), 88–110.

25. Matthews, *City Government of Boston*, 179–80.
26. *Municipal Yearbook of the City of New York* (1913), 44, 50, 51; *Finances and Financial Administration*, 137.
27. Nelson P. Lewis, "Paying the Bills for City Planning," *Fourth National Conference on City Planning* (1912), RHW; Advisory Commission on Taxation and Finance, *Final Report* (New York: Martin B. Brown, 1908), 81; Lawson Purdy, "Condemnation, Assessments, and Taxation in Relation to City Planning," *Third National Conference on City Planning* (1911): 118–25; Frank J. Goodnow, in "Discussion," *Third National Conference on City Planning* (1911): 129.
28. *Special Studies and Other Memoranda Relative to Municipal Revenues, February 1911 to January 1913*, nos. 18, 37, 42, 54, ERAS, box 89; *Finances and Financial Administration*, 326–30; "New Sources of City Revenue Recommended for New York," *American City* 8 (February 1913): 209, 211.
29. Marling to Mitchel, January 18, 1915, JPM, box MJP-227.
30. *Dictionary of American Biography*, vol. 11, supp. 2 (1958), s.v., "Edwin Robert Anderson Seligman," by Joseph Dorfman; Thomas Bender, "E. R. A. Seligman and the Vocation of Social Science," in *Intellect and Public Life: Essays on the Social History of Academic Intellectuals in the United States* (Baltimore: Johns Hopkins University Press, 1993), 49–77; Edwin R. A. Seligman, *Essays in Economics* (New York: Augustus M. Kelley, 1967), 18–19.
31. Seligman, *Essays in Economics*, 311, 302, 323.
32. Edwin R. A. Seligman, *Essays in Taxation*, 3d ed. (New York: Macmillan, 1905), 21; W. Hastings Lyon, *Investment Bankers Association of American Bulletin* 2 (August 31, 1914): 15, ERAS, box 53.
33. Edwin R. A. Seligman, "The New York Income Tax," *Political Science Quarterly* 34 (December 1919): 526 (the emphasis is mine).
34. Seligman, "Income Tax," 524.
35. Seligman to Mitchel, January 27, 1915, JPM, box MJP-227.
36. Marling to Mitchel, January 18, 1915, JPM, box MJP-228.
37. Frank Goodnow, in "Discussion," *Third National Conference on City Planning*, 129; Robert Whitten, "Incidence of Tax on Real Estate," 17, RHW; conference: board of estimate and mayor's commission on taxation, February 1, 1915, JPM, box MJP-227, 4, 21.
38. Biographical note, 2, *Guide*, BCM; Benjamin C. Marsh, *Lobbyist for the People: A Record of Fifty Years* (Washington, D.C.: Public Affairs Press, 1953), 17.
39. John D. Fairfield, *Mysteries of the Great City: The Politics of Urban Design, 1877–1937* (Columbus: Ohio State University Press, 1993), 27; Henry George, *Progress and Poverty* [1879] (New York: Robert Schalkenbach, 1979), 454–55.
40. Seligman to Simone, August 14, 1915, Marling to Seligman, June 17, 1914, ERAS, box 53.
41. Frank LeRoy Spangler, *Operation of Debt and Tax Rate Limits in the State of New York* (Albany: J. B. Lyon, 1932), 21, 105; *Finances and Financial Administration*, 348, 141–47; Charles L. Craig, "The Pay-as-You-Go Policy in New York City," *Proceedings of the Academy of Political Science of the City of New York* 8 (July 1918): 97. Compare McDonald, *Parameters*, 243.
42. *Finances and Financial Administration*, 203–8, although experts in the field of public finance focus on overlapping debt (gross debt accumulated by all governing

bodies with jurisdiction over a geographic area); Lennox L. Moak, *Administration of Local Government Debt* (Chicago: Municipal Finance Officers Association, 1970), 158.

43. Craig, "Pay-as-You-Go Policy," 94–96; Levey, "Sinking Funds," 682–83; *NYT,* January 12, 1916, 7:1.

44. Johnson to Prendergast, February 10, 1913, Prendergast to Gaynor, February 11, 1913, Gaynor to Prendergast, February 13, 1913, WJG, box GWJ-83 (the emphasis is mine).

45. Bruère, "New York's Nine-Hundred Million Debt," 793, 794, 795 (emphasis in the original).

46. McAneny, "The Physical Problem of the City," Yale Lectures, II, 1914, GMC, box 1.

47. Ma, "Finances, City of New York," 163; Anderson, "Pay-as-You-Go Policy," 108.

48. *NYT,* August 13, 1907, 11:3, August 15, 1907, 9:1, August 17, 1907, 11:4, August 28, 1907, 1:1; Herbert L. Satterlee, *J. Pierpont Morgan: An Intimate Portrait* (New York: Macmillan, 1939), 454–93; Edwin P. Hoyt Jr., *The House of Morgan* (New York: Dodd, Mead, 1966), 279–307; Andrew Sinclair, *Corsair: The Life of J. Pierpont Morgan* (Boston: Little, Brown, 1981), 174–91; Frederick Lewis Allen, *The Great Pierpont Morgan* (New York: Harper & Brothers, 1949), 239–66; Lewis Corey, *The House of Morgan: A Social Biography of the Masters of Money* [1930] (New York: AMS Press, 1969), 338–48; Ron Chernow, *The House of Morgan: An American Banking Dynasty and the Rise of Modern Finance* (New York: American Monthly Press, 1990), 121–30; Vincent P. Carosso, *Investment Banking in America: A History* (Cambridge: Harvard University Press, 1970), 128–31; Thomas W. Lamont, *Henry P. Davison: The Record of a Useful Life* [1933] (New York: Arno Press, 1975), 72–87; "A Review of the World," *Current Literature* 43 (December 1907): 585–94.

49. *NYT,* September 11, 1907, 1:7, October 31, 1907, 1:7.

50. *NYT,* October 31, 1907, 6:2, October 24, 1907, 10:7; Chernow, *House of Morgan,* 126.

51. *NYT,* January 1, 1908, 16:5, February 15, 1908, 1:3, February 15, 1908, 6:2.

52. *NYT,* September 6, 1914, 12:2; Harold Nicolson, *Dwight Morrow* [1935] (New York: Arno Press, 1975), 168, 164–69; Lamont, *Davison,* 175, 172–76.

53. Nicolson, *Dwight Morrow,* 169–70; Lamont, *Davison,* 179–84; *NYT,* September 14, 1914, 12:1, September 15, 1914, 12:4, September 5, 1914, 10:4; Craig, "Pay-as-You-Go Policy," 94, 96; Anderson, "Pay-as-You-Go Policy," 106.

54. Morgan to McAdoo, August 21, 1914, as quoted in Lamont, *Davison,* 177.

55. Henry Bruère, *New York City's Administrative Progress, 1914–1916* (New York: M. B. Brown, 1916), 55; Craig, "Pay-as-You-Go Policy," 94; "Cash Basis for New York," *Outlook* 108 (October 28, 1914): 444–45.

56. Anderson, "Pay-as-You-Go Policy," 107–9, 111.

57. Lolabel Hall, "Community Civics: Paying the City's Bills," *Outlook* 124 (March 17, 1920): 475; Anderson, "Pay-as-You-Go Policy," 111.

58. *NYT,* September 8, 1914, 10:2, May 27, 1915, 12:2, January 12, 1916, 7:1, January 14, 1916, 19:1.

59. *NYT,* November 9, 1917, 1:1, 3:3, November 11, 1917, VIII, 4:4.

60. *NYT,* March 5, 1918, 22:3.

61. *NYT,* August 4, 1918, IX, 12:3.

62. *NYT,* October 16, 1919, 19:1, November 1, 1919, 4:5.

63. *NYT,* August 4, 1918, IX, 12:3; *Schieffelin v. Hylan,* 106 N.Y. misc. 347 (1919), and 188 N.Y. app. div. 192 (1919).

64. Craig, "Pay-as-You-Go Policy," 94, 98–101; *Finances and Financial Administration,* 211; *NYT,* March 7, 1922, 26:3, August 17, 1923, 1:5, October 27, 1924, 12:2.

65. *NYT,* October 11, 1925, 1:4; *Finances and Financial Administration,* 141–47, xxxiv–lxviii.

66. Mills to Seligman, August 17, 1915, Seligman to Mitchel, September 5, 1915, ERAS, box 53; Seligman, "Income Tax," 527–28, 545; *Finances and Financial Administration,* 162–65, 136–37; *NYT,* January 17, 1932, IX, 4:1; Fuchs, *Mayors and Money,* 61.

67. Benjamin M. Anderson Jr., "State and Municipal Borrowing in Relation to the Business Cycle," *Chase Economic Bulletin* 5 (June 10, 1925): 2–18; Guy Alchon, *The Invisible Hand of Planning: Capitalism, Social Science, and the State in the 1920s* (Princeton, N.J.: Princeton University Press, 1985); Paul Studensky, *Public Borrowing* (New York: National Municipal League, 1930), 83.

68. Lindsay Rogers, "Fiscal Crisis of the World's Largest City," *Outlook* 161 (January 1933): 24, 25; Fuchs, *Mayors and Money,* 56–72; Thomas Kessner, *Fiorello H. La Guardia and the Making of Modern New York* (New York: Penguin, 1989), 218–20; *NYT,* September 28, 1933, 1:8, September 29, 1933, 1:6–8; Henry J. Rosner, "The Bankers Rule New York," *World Tomorrow* 15 (December 28, 1932): 613. Before the September agreement was struck, New York was not as subject to the control of financial institutions as Chicago and Detroit; see William C. Beyer, "Financial Dictators Replace Political Boss," *National Municipal Review* 22 (April 1933): 162–67, and Mark I. Gelfand, *A Nation of Cities: The Federal Government and Urban America, 1933–1935* (New York: Oxford University Press, 1975), 30–35.

69. Robert M. Haig and Carl S. Shoup, *The Financial Problem of the City of New York* (New York: Mayor's Committee on Management Survey, 1952), 9, 5, 6, 7, 15; Roy W. Bahl, Alan K. Campbell, and David Greytak, *Taxes, Expenditures, and the Economic Base: Case Study of New York City* (New York: Praeger, 1974), 163–74.

70. Haig and Shoup, *Financial Problem,* 20, 21, 24, 25.

71. Ibid., 489–90.

72. Ibid., 537, 523, 544, 545.

73. Frederick L. Bird, *The Municipal Debt* (New York: Mayor's Committee on Management Survey of the City of New York, 1951), 50.

74. Haig and Shoup, *Financial Problem,* 523, 547.

FIVE: City Planning versus the Law

1. *NYT,* November 28, 1912, 1:3, December 2, 1912, 10:2, April 30, 1914, 10:8, May 3, 1914, IX, 1:3.

2. Appended report, zoning and districting, minutes, heights of buildings commission, June 9, 1913, HBC, box 2507.

3. *Development and Present Status of City Planning in New York City* (New York: Board of Estimate, 1914), 7; *NYT,* November 23, 1914, 5:6.

4. Report to the officers and executive committee, zoning committee, December 12, 1929, EMB, box 10, file 17; *NYT,* December 15, 1915, 7:1.

5. Walter Laidlaw, *Population of the City of New York, 1890–1930* (New York: Cities Census Committee, 1932), 208–9.

6. George B. Ford, paper read before the mayor's commission on congestion, New York, October 28, 1910; idem, "The City Plan as a Basis For Housing Betterment," New Jersey State Housing Conference, May 28, 1915, GBF.

7. Robert H. Whitten, preliminary report to the heights of buildings commission on the police power in relation to building-height restrictions, 43, HBC, box 2507.

8. Edward M. Bassett, "Distribution of Population in Cities," *American City* 13 (July 1915): 7–8.

9. *Statement of the Fifth Avenue Association on the Limitations of Building Heights to the New York City Commission* (June 19, 1913), 3–4, 39–44; Bruce Falconer, memo on behalf of the Fifth Avenue Association, and minutes, heights of buildings commission, June 19, 1913, HBC, box 2507; *NYT,* January 12, 1913, V, 12:1, January 12, 1913, VII, 17:4, December 14, 1913, IV, 6:1.

10. *Report of the Heights of Buildings Commission* (December 23, 1913), 15; president to McAdoo, January 21, 1908, WGM, box 91; Vincent W. Lanfear, "The Metal Industry," in *Chemical, Metal, Wood, Tobacco, and Printing Industries* [1928] (New York: Arno Press, 1974), 33.

11. Keith D. Revell, "Law Makes Order: The Search for Ensemble in the Skyscraper City, 1890–1930," in *The American Skyscraper,* ed. Roberta Moudry (Cambridge: Cambridge University Press, 2003).

12. Lawson Purdy, *Zoning: As an Element in City Planning, and for Protection of Property Values, Public Safety, and Public Health,* American Civic Association pamphlet, series 2, no. 15 (June 30, 1920), 3; "Reminiscences of Lawson Purdy," Columbia University Oral History Collection, 2; CBDR, *Final Report* (New York: Board of Estimate, 1916), 142, 166; S. Adolphus Knopf, *Tuberculosis: A Preventable and Curable Disease* (New York: Moffat, Yard, 1909), 84, 5, 122–23; Knopf to Bassett, July 29, 1913, HBC, box 2507.

13. McAneny to Eidlitz, April 12, 1913, MES, box 2.

14. CBDR, *Final Report,* 75–76; *Final Report of the Committee on Taxation of the City of New York* (New York: O'Connell Press, 1916), 334; Undated memo on "litigation now pending," EMB, box 10, file 16; Robert A. Dahl, *Who Governs? Democracy and Power in an American City* (New Haven: Yale University Press, 1961), 246–48; S. J. Makielski Jr., *The Politics of Zoning: The New York Experience* (New York: Columbia University Press, 1966).

15. Garrett Power, "Apartheid Baltimore Style: The Residential Segregation Ordinances of 1910–1913," *Maryland Law Review* 42 (1983): 289–328.

16. George B. Ford, "Future Negro Population" (September 11, 1923), GBF.

17. Whitten to Lasker, April 28, 1922, RHW.

18. Zoning committee to Harman, August 18, 1926, EMB, box 10, file 15; Bassett, *Zoning: The Laws, Administration, and Court Decisions During the First Twenty Years* (New York: Russell Sage, 1936), 49–50.

19. Bassett to McAneny, September 28, 1914, EMB, box 4, file 65.

20. "The Lawyer as a City Planning Advisor," *American City* 13 (September 1915): 172.

21. Duncan Kennedy, "Toward an Historical Understanding of Legal Consciousness: The Case of Classical Legal Thought in America, 1850–1940," *Research in Law and Sociology* 3 (1980): 5, 12; Thomas C. Grey, "Langdell's Orthodoxy," *University of Pittsburgh Law Review* 45 (1983): 1–53; T. Alexander Aleinikoff, "Constitutional Law in the Age of Balancing," *Yale Law Journal* 96 (April 1987): 949–52.

22. *Welch v. Swasey*, 214 U.S. (1909) 91 at 105.
23. Morton J. Horwitz, *The Transformation of American Law, 1870–1960* (New York: Oxford University Press, 1992), 27–30.
24. Kennedy, "Toward an Historical Understanding," 9–14
25. Aleinikoff, "Constitutional Law," 952–58; Roscoe Pound, "Mechanical Jurisprudence," *Columbia Law Review* 8 (December 1908): 609–10.
26. *Cochran v. Preston*, 70 atl. rep. 113 at 114, 115; Garrett Power, "High Society: The Building Height Limitation on Baltimore's Mt. Vernon Place," *Maryland Historical Magazine* 79 (fall 1984): 202–5.
27. *Welch v. Swasey*, 214 U.S. 91 at 92–93, 107, 105–6; Statement of the Fifth Avenue Association, 22–24; John F. Dillon, *Commentaries on the Law of Municipal Corporations*, 5th ed. (Boston: Little, Brown, 1911), 2:1064.
28. *Ex parte Quong Wo*, 118 Pac. Rep. 714 at 715–18.
29. *People ex rel. Lincoln Ice Company v. Chicago*, 260 Ill. 150 at 153.
30. *People ex rel. Friend v. Chicago*, 261 Ill. 16 at 20, 21.
31. *Willison v. Cooke*, 54 Colo. 320 at 326, 327, 328.
32. *Calvo v. New Orleans*, 136 La. 480 at 482.
33. *People ex rel. Lankton v. Roberts*, 90 misc. rep. 439 at 442, 444.
34. *Calvo*, 136 La. 480 at 483; *Willison*, 54 Colo. 320 at 329; *Friend*, 261 Ill. 16 at 21; *Lankton*, 90 misc. rep. 439 at 443.
35. *Passaic v. Patterson Bill Posting*, 72 N.J.L. 285 (1905) at 287.
36. Andrew Crawford Wright, "Discussion," *Proceedings of the Fifth National Conference on City Planning* (1913): 65; Frederick Law Olmsted, "Discussion," *Proceedings of the Sixth National Conference on City Planning* (1914): 14.
37. Lawrence Veiller, "Protecting Residential Districts," *Proceedings of the Sixth National Conference on City Planning* (1914): 110, 95, 96, 111.
38. John J. Walsh, "Discussion," *Proceedings of the Sixth National Conference on City Planning* (1914): 120.
39. Ernst Freund, "Discussion," *Proceedings of the Third National Conference on City Planning* (1911): 246.
40. Phillip Kates, "Discussion," *Proceedings of the Third National Conference on City Planning* (1911): 250–52.
41. Remarks of Lawson Purdy at meeting of zoning division, October 28, 1926, JJW, box WJJ-231.
42. Digest of legal conference, July 7, 1913, 3, HBC, box 2507.
43. *WWNY* (1914), 489; Man to Purdy, July 3, 1913, HBC, box 2507.
44. Miller to Haag, December 30, 1913, HBC, box 2507; Timmis to McAneny, February 10, 1915, JPM, box MJP-208.
45. "*Districting*: Regulations Governing the Heights of Buildings in Districts C to H, Proposed by the Committee on Districting" and minutes, heights of buildings commission, June 4, 1913, and June 9, 1913, HBC, box 2507.
46. Bassett to Bard, May 18, 1915, EMB, box 1, file 183.
47. Edward M. Bassett, "A Survey of the Legal Status of a Specific City in Relation to City Planning," *Proceedings of the Fifth National Conference on City Planning* (1913): 46, 57; Alfred Bettman, "Discussion," *Proceedings of the Sixth National Conference on City Planning* (1914): 111.

48. Leonard W. Levy, *The Law of the Commonwealth and Chief Justice Shaw* (New York: Oxford University Press, 1957), 247–54 (the emphasis is mine).
49. *C.B. & Q. Railway v. Drainage Commissioners*, 200 U.S. 561 at 592.
50. *Noble State Bank v. Haskell*, 219 U.S. 104 at 111.
51. *Reinman v. Little Rock*, 237 U.S. 171 at 172, 174.
52. Misc. notes on law, EMB, box 10, file 18.
53. *Reinman v. Little Rock*, 237 U.S. 171 at 176.
54. *Ex parte Hadacheck*, 132 Pac. Rep. 584 at 585; *Hadacheck v. Sebastian*, 239 U.S. 394 at 405, 412.
55. *Hadacheck v. Sebastian*, 239 U.S. 394 at 410 (the emphasis is mine).
56. Misc. notes, EMB, box 10, file 18.
57. Whitten, *Preliminary Report*, 3.
58. CBDR, *Final Report*, 233–35; George Ford, *New York City Building Zone Resolution* (New York: New York Title & Mortgage, 1917), 5; idem, "How New York City Now Controls the Development of Private Property," *City Plan* 2 (October 1916): 3.
59. CBDR, *Final Report*, 236–37, figs. 129–31.
60. CBDR, *Final Report*, 238–42, figs. 132–35; Ford, *Building Zone Resolution*, 12; Ford, "How New York City Controls," 3.
61. CBDR, *Tentative Report* (March 1916), 9; *Report of the Heights of Buildings Commission*, 26; CBDR, *Final Report*, 235.
62. Herbert Swan, "Making the New York Zoning Ordinance Better," *Architectural Forum* 35 (October 1921): 126.
63. CBDR, *Final Report*, 53 (the emphasis is mine).
64. George C. Whipple and Melvin C. Whipple, "Air Washing as a Means of Obtaining Clean Air in Buildings," *Transactions of the Fourth International Congress on School Hygiene* 2 (1914): 228–29.
65. George C. Whipple, "Zoning and Health," *TASCE* 88 (1925): 603; idem, "The Philosophy of Sanitation-Lecture No. 3: Urban Problems in Sanitation," 27, 4–6, GCW, box 3; idem, "Municipal Economy and Sanitation," March 20, 1913, GCW, box 1.
66. CBDR, *Final Report*, 86–157, passim.
67. *Report of the Heights of Buildings Commission*, diagram 6.
68. CBDR, *Final Report*, 55, 54.
69. Ibid., 56.
70. Raphaël Fischler, "Health, Safety, and the General Welfare: Markets, Politics, and Social Science in Early Land-Use Regulation and Community Design," *Journal of Urban History* 24 (September 1998): 692–94.
71. Alfred Bettman, "Constitutionality of Zoning," *Harvard Law Review* 37 (May 1924): 845, 839, 841.
72. CBDR, *Final Report*, 19, 40–41; Bassett to Hornstein, January 3, 1922, EMB, box 3, file 133.
73. Minutes, heights of buildings commission, May 26, 1913, 4, HBC, box 2507.
74. Keith D. Revell, "The Road to *Euclid v. Ambler*: City Planning, State-Building, and the Changing Scope of the Police Power," *Studies in American Political Development* 13 (spring 1999): 50–145.
75. Report to the officers and executive committee of the zoning committee, November 20, 1922, EMB, box 10, file 16.

76. Henry Campbell Black, *Black's Law Dictionary* (Saint Paul, Minn.: West Publishing, 1933), 301, 1152; W. F. Bailey, *A Treatise on the Law of Habeas Corpus and Special Remedies* (Chicago: T. H. Flood, 1913), 1:621–22, 2:773; Edith G. Henderson, *Foundations of English Administrative Law: Certiorari and Mandamus in the Seventeenth Century* (Cambridge: Harvard University Press, 1963), 1–2.

77. Report to the officers and executive committee, zoning committee, December 3, 1925, EMB, box 10, file 16; Frank D. Stringham, "Memorandum for the City Planning Commission," 1924, EMB, box 9, file 2.

78. Bassett to Comey, February 6, 1924, EMB, box 2, file 33.

79. *People ex rel. Sheldon v. Board of Appeals*, 115 misc. 449 (1921), 200 app. div. 907 (1922), and 234 N.Y. 484 (1923).

80. Bassett to Comey, January 24, 1923, EMB, box 1, file 30; *NYT*, October 20, 1919, 15:1.

81. Bassett to Comey, February 6, 1924, EMB, box 2, file 33.

82. Committee to Protect the Zoning Resolution, December 7, 1916, EMB, box 10, file 13.

83. Midyear report to the officers of the zoning committee, August 7, 1922, EMB, box 10, file 16.

84. Report on work of zoning committee during 1921, December 5, 1921, EMB, box 10, file 16; report to the officers and executive committee, zoning committee, November 23, 1931, EMB, box 10, file 17; memo of meeting, zoning committee, December 6, 1927, EMB, box 10, file 13; supplemental report of counsel, November 26, 1928, and report to officers and executive committee, zoning committee, December 2, 1942, EMB, box 10, file 17.

85. Bassett to Adams, December 27, 1929, RPA, box 44.

86. Minutes of annual meeting, executive committee of the zoning committee, December 2, 1920, EMB, box 10, file 13.

87. Order of business, annual meeting of the zoning committee, January 17, 1918, EMB, box 10, file 13; midyear report to officers of the zoning committee, August 7, 1922, EMB, box 10, file 16.

88. Report on work of zoning committee during 1921, December 5, 1921, EMB, box 10, file 16.

89. Ibid.

90. Report to the officers and executive committee, zoning committee, November 20, 1922.

91. Lawrence Veiller, "Mistakes to be avoided in Zoning," 4, EMB, box 12.

92. Report to the officers and executive committee, zoning committee, December 3, 1925, EMB, box 10, file 16.

93. Idem, November 25, 1924, and December 4, 1923, EMB, box 10, file 16.

94. *Ignaciunas v. Risley*, 98 N.J.L 712 (1923); Bassett to Olmsted, January 5, 1923, EMB, box 11; Newman F. Baker, "Zoning Legislation," *Cornell Law Quarterly* 11 (February 1926): 176–77; *Handy v. South Orange*, 118 atl. rep. 838 (1922); *Vernon v. Westfield*, 98 N.J.L 600 (1923); *Handy v. East Orange*, 2 N.J. misc. 884 (1924); *Krumgold v. Jersey City*, 102 N.J.L 170 (1925); *Steinberg v. Bigelow*, 3 N.J. misc. 1228 (1925).

95. Report to the officers and executive committee, zoning committee, November 25, 1924, EMB, box 10, file 16.

96. *People ex rel. Rosevale Realty v. Kleinert*, 268 U.S. 646.

97. *Euclid v. Ambler,* 272 U.S. 365 at 394, 395; Bettman to Underwood, September 29, 1924, as cited in Daniel R. Mandelker and Roger A. Cunningham, *Planning and Control of Land Development* (Indianapolis: Bobbs-Merrill, 1979), 213, note 5; argument of Edward M. Bassett, in *Steerman vs. District of Columbia,* April 29, 1925, 163, EMB, box 11.

98. Thomas Reed Powell, "The Supreme Court and State Police Power, 1922–1930-IV," *Virginia Law Review* 18 (November 1931): 33.

99. *Euclid v. Ambler,* 272 U.S. 365 at 394, 395; Revell, "Road to *Euclid v. Ambler,*" 112–20.

100. Report to the officers and executive committee, zoning committee, December 1, 1927, EMB, box 10, file 17.

101. *Women's Kansas City St. Andrew Society v. Kansas City,* 58 F. 2d 593 at 605, 606 (the emphasis is mine).

102. Report to the officers and executive committee, zoning committee, December 3, 1925; remarks of Mr. Walsh at meeting of the zoning division, October 28, 1926, JJW, box WJJ-231.

103. Bassett to Comey, February 6, 1924.

104. Report to the officers and executive committee, zoning committee, December 3, 1925.

105. Minutes, board of standards and appeals, December 21, 1926, and Bassett to Purdy, December 17, 1926, EMB, box 11, file 20.

106. William B. Northrup and John B. Northrup, *The Insolence of Office: The Story of the Seabury Investigations* (New York: G. P. Putman, 1932), 127–36, 193–203; report to the officers and executive committee, zoning committee, November 24, 1930, EMB, box 10, file 17; *NYT,* August 10, 1930, 15:1, December 12, 1930, 27:8, December 19, 1930, 24:7.

107. Report to the officers and executive committee, zoning committee, November 23, 1931.

108. *NYT,* November 5, 1928, 22:3; Thomas Adams, "The Character, Bulk, and Surroundings of Buildings," in *Buildings: Their Uses and the Spaces about Them* (New York: Regional Plan of New York and Its Environs, 1931), 158–59; Edward Bassett, confidential report of the zoning division on administrative and general features, December 29, 1927, JJW, box WJJ-231; report to the officers and executive committee, zoning committee, December 1, 1932, EMB, box 10, file 17.

109. Midyear report to officers of the zoning committee, August 7, 1922, EMB, box 10, file 10.

110. Report on work of zoning committee during 1921.

111. Report to the officers and executive committee, zoning committee, November 20, 1922.

112. Walter Stabler, "Some of the Effects of City Planning on Real Estate Values," Cleveland Chamber of Commerce, December 18, 1919, ML, 6.

113. Report to the officers and executive committee, zoning committee, November 20, 1922.

114. *People ex rel. Rosevale Realty v. Kleinert,* 268 U.S. 646 at 648–51; brief for plaintiff in error, *People ex rel. Rosevale Realty v. Kleinert,* 30, 33, 35–36, 45, 52.

115. Report to the officers and executive committee, zoning committee, December 3, 1925.

116. *Ward's Appeal,* 289 Pa. 458 at 471.

117. Chester C. Maxey, *An Outline of Municipal Government* (New York: Doubleday, 1924), 9, as quoted in Baker, "Zoning Legislation," 165.
118. Brief for plaintiff in error, 70.
119. Report to the officers and executive committee, zoning committee, December 14, 1933, EMB, box 10, file 17.
120. Idem, December 3, 1925.

SIX: "They shall splash at a ten-league canvas with brushes of comets' hair"

1. "Why the City of New York Was Originally Unplanned—and Remains So," *American City* 40 (May 1929): 98, 96.
2. Robert Caro, *The Power Broker: Robert Moses and the Fall of New York* [1974] (New York: Vintage, 1975), 20, 779–86.
3. Robert Moses, "Mr. Moses Dissects the 'Long-Haired Planners,'" *NYT Magazine*, June 25, 1944, 39.
4. Wallace S. Sayre and Herbert Kaufman, *Governing New York City: Politics in the Metropolis* (New York: Russell Sage, 1960), 372.
5. *The Graphic Regional Plan* [1929] (New York: Arno Press, 1974), 127.
6. Robert C. Wood, "The New Metropolis: Green Belts, Grass Roots, or Gargantua?" *American Political Science Review* 52 (March 1958): 111–12.
7. *Graphic Regional Plan*, 127.
8. Thomas Adams, *The Building of the City* [1931] (New York: Arno Press, 1974), 122, 126.
9. Edwin R. Lewinson, *John Purroy Mitchel: The Boy Mayor of New York* (New York: Astra Books, 1965), 245.
10. Interview with George McAneny, 26 November 1948, 7, GMC, box 1; *NYT*, February 2, 1918, 18:1.
11. Report to the officers and executive committee, zoning committee, December 3, 1941, EMB, box 10, file 17.
12. *Graphic Regional Plan*, 127; Irving T. Bush, "The Congestion Problem," *North American Review* 226 (September 1928): 289–95; Calvin Coolidge, "A Plea for Fundamental Consideration of Congestion in Cities," *American City Magazine* 33 (July 1925): 1–2.
13. David A. Johnson, *Planning the Great Metropolis: The 1929 Regional Plan of New York and Its Environs* (London: Chapman & Hall, 1996), 48–67.
14. Walter L. Fisher, "Legal Aspects of the Plan of Chicago," in Daniel H. Burnham and Edward H. Bennett, *Plan of Chicago* [1909], ed. Charles Moore (New York: Princeton Architectural Press, 1993), 127, 140–41.
15. Johnson, *Planning the Great Metropolis*, 63.
16. Frederick Delano, foreword to Robert Murray Haig and Roswell C. McCrea, *Major Economic Factors in Metropolitan Growth and Arrangement* (New York: Regional Plan of New York and Its Environs, 1927), v; Olivier Zunz, *Why the American Century?* (Chicago: University of Chicago Press, 1998), 34–35.
17. Haig and McCrea, *Major Economic Factors*, 13.
18. Interview with McAneny, 7, 48.
19. Harrison to James, May 1, 1924, RPA, box 44.
20. Harrison to Keppel, "Notes on the Relation of City and Regional Planning to Health," December 13, 1923, RPA, box 44.

21. Adams to James, May 8, 1924, RPA, box 44.
22. Delano to James, June 25, 1924, RPA, box 44.
23. Delano to Wilgus, August 19, 1924, WJW, box 56.
24. Heydecker to James, June 8, 1924, 2, 3, 4, RPA, box 44 (the emphasis is mine).
25. Heydecker to James, June 8, 1924, 6, 7.
26. Robert Murray Haig, "Toward an Understanding of the Metropolis, I," *Quarterly Journal of Economics* 40 (February 1926): 182; Haig and McCrea, *Major Economic Factors*, 14.
27. Haig and McCrea, *Major Economic Factors*, 33, 34 (the emphasis is mine).
28. Vincent W. Lanfear, "The Metal Industry," in *Chemical, Metal, Wood, Tobacco and Printing Industries* [1928] (New York: Arno Press, 1974), 37.
29. Lucy Winsor Killough, "The Tobacco Products Industry," in *Chemical, Metal, Wood*, 11, 12, 55; Haig and McCrea, *Major Economic Factors*, 10, 34.
30. Mark Carter Mills, "The Wood Industries," in *Chemical, Metal, Wood*, 9, 10, 51, 52.
31. B. M. Selekman, Henriette R. Walter, and W. J. Cooper, "The Clothing and Textile Industries," in *Food, Clothing, and Textile Industries, Wholesale Markets and Retail Shopping and Financial Districts* [1928] (New York: Arno Press, 1974), 15; Haig and McCrea, *Major Economic Factors*, 77–80; Lanfear, "Metal Industry," 10, 33, 35, 36.
32. Haig and McCrea, *Major Economic Factors*, 81, 87, 94; Selekman, Walter, and Cooper "Clothing and Textile Industries," 21–26, 37, 44–48.
33. Haig and McCrea, *Major Economic Factors*, 35; Donald H. Davenport, Lawrence M. Orton, and Ralph W. Roby, "The Retail Shopping and Financial Districts," in *Food, Clothing, and Textile Industries*, 32, 54.
34. Haig and McCrea, *Major Economic Factors*, 104–5; "Basic General Assumptions Underlying the Regional Plan," December 30, 1926, 4, RPA, box 43; Robert Murray Haig, "Toward an Understanding of the Metropolis, II," *Quarterly Journal of Economics* 40 (May 1926): 431–32.
35. Haig and McCrea, *Major Economic Factors*, 105–6.
36. Haig, "Toward an Understanding, II," 433; Haig and McCrea, *Major Economic Factors*, 67, 70; Haig, "Toward an Understanding, I," 186; Faith M. Williams, "The Food Manufacturing Industries," in *Food, Clothing, and Textile Industries*, 16–17, 20, 24.
37. Haig, "Toward an Understanding, II," 420–23; Haig, "Toward an Understanding, I," 184–85.
38. Haig, "Toward an Understanding, II," 406–7, 419.
39. Haig and McCrea, *Major Economic Factors*, 18.
40. Thomas Adams, foreword to Thomas Adams, Harold M. Lewis, and Theodore T. McCrosky, *Population, Land Values, and Government* [1929] (New York: Arno Press, 1974), 3.
41. Adams, Lewis, and McCrosky, *Population, Land Values*, 71, 93, 99–100.
42. Ibid., 28–29, 61, 63; Deborah Dash Moore, "On the Fringes of the City: Jewish Neighborhoods," in *The Landscape of Modernity: Essays on New York City, 1900–1940*, ed. David Ward and Olivier Zunz (New York: Russell Sage, 1992), 252–72.
43. Adams, Lewis, and McCrosky, *Population, Land Values*, 33, 25, 59, 58; *NYT*, January 29, 1929, XI, 1:3.
44. *Graphic Regional Plan*, 151, 143, 149; Adams, *Building of the City*, 34, 37.
45. Adams, Lewis, and McCrosky, *Population, Land Values*, 33, 175–76; *NYT*, October 31, 1928, 33:8.

46. Adams, Lewis, and McCrosky, *Population, Land Values*, 63, 169, 58, 27.

47. *Graphic Regional Plan*, 125.

48. Adams, Lewis, and McCrosky, *Population, Land Values*, 201, 36.

49. Ibid., 27.

50. *Graphic Regional Plan*, 149–50, 152; Nelson P. Lewis, "Regional Planning," *TASCE* 86 (1923): 1327; "Basic General Assumptions," December 30, 1926, 3; Adams, *Building of the City*, 75, 112, 114, 116, 275–321, 364–67, 400, 529–30, 540–42.

51. Adams, Lewis, and McCrosky, *Population, Land Values*, 8, 62, 31.

52. *Graphic Regional Plan*, 143–49, 173; Adams, *Building of the City*, 321; Bassett to Adams, February 28, 1925, 5, RPA, box 43.

53. Contrast Robert Fishman, "The Regional Plan and the Transformation of the Industrial Metropolis," in *Landscape of Modernity*, 106–25, esp. 109.

54. Adams to Delano, April 8, 1927, RPA, box 44; *Graphic Regional Plan*, 162, 176, 183; Lewis to Adams, 10/24/25, and Adams to Wilgus, November 3, 1925, WJW, box 56.

55. *Graphic Regional Plan*, 210, 137.

56. Delano to Adams, March 23, 1925, 1–2, RPA, box 43; "Basic General Assumptions on which the Regional Plan Is Being Prepared," December 13, 1926, 4, RPA, box 43; *Graphic Regional Plan*, 139, 215, 233, 213, 136.

57. Lewis Mumford, "The Plan of New York," in *Planning the Fourth Migration: The Neglected Vision of the Regional Planning Association of America*, ed. Carl Sussman (Cambridge: MIT Press, 1976), 227, 239–43, 252.

58. Ibid., 248, 249 (the emphasis is mine).

59. Thomas Adams, "A Communication: In Defense of the Regional Plan," in *Planning the Fourth Migration*, 262, 263.

60. Johnson, *Planning the Great Metropolis*, 193; memo regarding proposals of Colonel Wilgus, August 28, 1924, Adams to Wilgus, November 3, 1925, and Desmond memo re. conversation with Swales, September 15, 1927, WJW, box 56.

61. Adams, *Building of the City*, 371, 376, 578; *Graphic Regional Plan*, 408.

62. Lewis Mumford, *Technics and Civilization* (New York: Harcourt, Brace, 1934), 366.

63. *Graphic Regional Plan*, 160–61.

64. "Basic General Assumptions," December 30, 1926, 7; *Graphic Regional Plan*, 8; Adams to Wilgus, November 3, 1925.

65. *Graphic Regional Plan*, 160, 163.

66. Adams to Glenn, May 10, 1930, RPA, box 44.

67. Adams to Ford, May 10, 1930, RPA, box 44.

68. "Basic General Assumptions," December 13, 1926, 1.

69. Adams to Ford, May 10, 1930, RPA, box 44.

70. *Graphic Regional Plan*, 409.

71. Adams, *Building of the City*, 122.

72. Press release from the Regional Plan Association, July 9, 1930, WJW, box 57.

73. *Graphic Regional Plan*, 163.

74. Adams, Lewis, and McCrosky, *Population, Land Values*, 256.

75. *NYT*, September 20, 1900, 12:3, January 17, 1901, 6:3.

76. Adams, Lewis, and McCrosky, *Population, Land Values*, 258–63, esp. 261, 263; *NYT*, March 23, 1902, 12:3.

77. *NYT*, March 31, 1909, 3:2; Adams, *Population, Land Values*, 265.

78. Gene Fowler, *Beau James: The Life and Times of Jimmy Walker* (New York: Viking, 1949), 141–55, 11; *NYT,* April 26, 1926, 20:1.

79. Fowler, *Beau James,* 11; "Remarks of the Mayor on Organization of the Committee on Planning and Survey," June 21, 1926, JJW, box WJJ-232.

80. O'Brien to Walker, June 5, 1928, in *Report of the City Committee on Plan and Survey* (1928); statement by Mayor James J. Walker, June 13, 1926, JJW, box WJJ-232; *NYT,* June 16, 1926, 26:2.

81. Wilgus to Adams, February 20, 1927, WJW, box 56; memo on submission of proposals to the mayor's city plan and survey committee, December 21, RPA, box 43.

82. *Report of the City Committee on Plan and Survey,* 3, 4, 8, 10, 12, 15.

83. *NYT,* November 14, 1928, 16:4, November 15, 1928, 28:5.

84. Edward M. Bassett, "Enlarged Usefulness of City Planning Commissions in New York State," *Twenty-first National Conference on City Planning* (1929): 68–71; *Laws of New York,* 1926, chapters 690 and 719.

85. *Report of the City Committee on Plan and Survey,* 34; Board of Estimate, *Statutory Set-up of a Planning Board for Greater New York* (prepared by Edward M. Bassett, December 15, 1928, rev., January 10, 1929), JJW, box WJJ-232; *NYT,* January 22, 1929, 28:2.

86. *NYT,* March 12, 1929, 31:5, March 16, 1929, 18:4, March 23, 1929, 21:2, March 27, 1929, 3:4, March 29, 1929, 2:5; "Transition," *Newsweek* 3 (January 27, 1934): 29.

87. Memo on Mayor Walker's city planning bill, n.d., 1–2, JJW, box WJJ-232.

88. *NYT,* June 26, 1930, 12:2, July 8, 1930, 5:5, July 18, 1930, 21:8, July 7, 1930, 1:1, 4:3, June 13, 1931, 5:7; McAneny to Wilgus, July 29, 1930, WJW, box 57; Childs to board of aldermen, June 26, 1930, JJW, box WJJ-232.

89. *NYT,* December 5, 1932, 1:1, 12:3, January 21, 1933, 1:4.

90. Edward J. Flynn, *You're the Boss: The Practice of American Politics* [1947] (New York: Collier, 1962), 62–68; *NYT,* March 23, 1929, 21:2.

91. *NYT,* April 1, 1909, 8:1.

92. Joseph McGoldrick, "The Board of Estimate and Apportionment of New York City," *National Municipal Review Supplement* 18 (February 1929): 128; *The City of the Future: A Permanent City Planning Commission—What It Means to New York* (New York: City Club, n.d.), 6, JJW, box WJJ-232; remarks of Edward M. Bassett, zoning division of the subcommittee on housing, zoning, and distribution of population, October 21, 1926, JJW, box WJJ-231; Heydecker to McAneny, September 22, 1932, RPA, box 44.

93. Memo on city planning bill, 3; Pedrick to O'Brien, December 6, 1926, 5, 6, JFS, box 3.

94. Press release from Regional Plan Association.

95. George H. McCaffrey, "Proportional Representation in New York City," *American Political Science Review* 33 (October 1939): 841–42; Joseph McGoldrick, "Proposals for Reorganization of the Government of the City of New York," *American Political Science Review* 27 (April 1933): 336; Howard Lee McBain, "Proportional Representation in American Cities," *Political Science Quarterly* 37 (June 1922): 281–98.

96. *NYT,* June 14, 1934, 1:1, August 7, 1934, 1:1, 8:1,3, August 3, 1934, 1:1, February 17, 1935, IV, 10:7.

97. *NYT,* September 23, 1929, 26:1, November 28, 1933, 1:4, November 29, 1933, 18:2, December 17, 1933, VIII, 4:1, April 13, 1934, 22:5.

98. *Graphic Regional Plan*, 167.

99. *NYT*, January 13, 1935, 1:4, April 27, 1936, 26:3; McCaffrey, "Proportional Representation," 843.

100. New York Charter Revision Commission, *Preliminary Report and Draft of Proposed Charter for the City of New York* (April 27, 1936), 19–22; *NYT*, December 16, 1935, 1:8, 2:3; Laurence Tanzer, *The New York City Charter* (New York: Clark-Boardman, 1937), 75–82; "Reminiscences of Cleveland Rodgers," Columbia University Oral History Collection, 237, 238; *NYT*, May 3, 1936, 20:1, May 14, 1936, 36:5; "The Colossus Moves," *Saturday Evening Post* 209 (December 12, 1936): 22.

101. *NYT*, March 10, 1935, IV, 11:4, December 17, 1935, 11:5.

102. *NYT*, May 27, 1936, 24:1, July 16, 1936, 10:2, October 29, 1936, 1:4, 11:2.

103. *NYT*, October 28, 1936, 1:8; Tanzer, *New York City Charter*, 7–8; *Mooney v. Cohen*, 272 N.Y. 33 (1936).

104. *NYT*, November 5, 1936, 6:3, November 25, 1936, 15:3.

105. Thomas Kessner, *Fiorello H. La Guardia and the Making of Modern New York* (New York: Penguin, 1989), 404–5; Mark I. Gelfand, "Rexford G. Tugwell and the Frustration of Planning in New York City," *Journal of the American Planning Association* 51 (spring 1985): 153, 160; *NYT*, May 5, 1936, 25:5, July 30, 1937, 33:2, November 4, 1936, 1:3.

106. McCaffrey, "Proportional Representation," 844–45.

107. Tanzer, *New York City Charter*, 484, 483.

108. *NYT*, December 10, 1937, 10:2.

109. *NYT*, December 11, 1937, 18:3.

110. Bennett to Brennan, December 9, 1926, JJW, box WJJ-232.

111. Caro, *Power Broker*, 84–85; Cleveland Rodgers, *Robert Moses: Builder for Democracy* (New York: Henry Holt, 1952), 18, 19.

112. *NYT*, December 17, 1931, 2:5.

113. *NYT*, December 14, 1940, 16:7; Moses, "Mr. Moses Dissects,"16, 39.

114. Caro, *Power Broker*, 17–18.

115. *NYT*, February 10, 1935, IV, 11:3; *NYT World's Fair Section*, March 5, 1939, 47.

116. *NYT*, December 12, 1940, 56:1.

117. Cleveland Rodgers, *New York Plans for the Future* (New York: Harper, 1943), 261–62.

118. Caro, *Power Broker*, 318.

119. *NYT*, December 31, 1937, 17:5, April 10, 1938, 1:3, August 19, 1938, 5:4.

120. Kenneth T. Jackson, *Crabgrass Frontier: The Suburbanization of America* (Oxford: Oxford University Press, 1987), 195; *NYT*, December 14, 1940, 16:7.

121. Rexford G. Tugwell, "Implementing the General Interest," *Public Administration Review* 1 (autumn 1940): 34.

122. Ibid., 43; Gelfand, "Tugwell," 151–60.

123. *NYT*, July 28, 1940, XI, 3:8, July 30, 1940, 15:8, July 31, 1940, 19:1, January 5, 1942, 19:5; Citizen's Budget Commission, *Report on the Master Plan of Land Use Proposed by the City Planning Commission* (December 1941), 9, 6, 29, 42–43; *NYT*, July 26, 1941, 13:3, November 23, 1941, 47:8.

124. Cleveland Rodgers, "Robert Moses," *Atlantic Monthly* 163 (February 1939): 234.

125. *NYT*, November 12, 1933, X, 1:3, February 20, 1934, 23:5.

126. Mayor's Committee on City Planning, *Progress Report: A Preliminary Report*

upon *Planning Surveys and Planning Studies* (June 1936), 36; *NYT,* March 29, 1936, II, 1:7, September 20, 1936, IV, 10:1.

127. *NYT,* November 10, 1938, 29:2.

128. *NYT,* March 25, 1940, 17:1; Citizens' Housing Council of New York, *Ailing City Areas: Economic Study of Thirteen Depressed Districts in Manhattan* (May 1941), 6; *NYT,* April 21, 1941, 21:6; Jackson, *Crabgrass Frontier,* 195–203.

129. Robert M. Haig and Carl S. Shoup, *The Financial Problem of the City of New York* (New York: Mayor's Committee on Management Survey, 1952), 476–77.

130. Pendleton Herring, *Public Administration and the Public Interest* [1936] (New York: Russell & Russell, 1967), 378.

131. Harold Lewis, "Basic Information Needed for a Regional Plan," October 6, 1926, WJW, box 56 (the emphasis is mine).

CONCLUSION: "An almost mystical unity"

The epigraph is from John Dewey, *The Public and Its Problems,* 136–37.

1. Cleveland Rodgers, *New York Plans for the Future* (New York: Harper, 1943), xi.
2. Robin L. Einhorn, *Property Rules: Political Economy in Chicago, 1833–1872* (Chicago: University of Chicago Press, 1991), 144, 68, 240–41.
3. Harold L. Platt, *City Building in the New South: The Growth of Public Services in Houston, Texas, 1830–1910* (Philadelphia: Temple University Press, 1983).
4. Edward C. Banfield and James Q. Wilson, *City Politics* (Cambridge: Harvard University Press, 1967), 40–41, 138–50, 330.
5. John Dewey, *The Public and Its Problems* (New York: Henry Holt, 1927), 96, 141, 130, 131, 128, 126.
6. Ibid., 91, 151.
7. Ibid., 191, 193, 194, 211, 188, 208, 184.
8. Pendleton Herring, *Public Administration and the Public Interest* [1936] (New York: Russell & Russell, 1967), 9, 5.
9. Ibid., 397, 24, 16, 5, 24, 378, 16.
10. E. E. Schattschneider, *Party Government* [1942] (Westport, Conn.: Greenwood Press, 1977), 32, 31, 87; idem, *The Semisovereign People: A Realist's View of Democracy in America* [1960] (Hindsdale, Ill.: Dryden Press, 1975), 23.
11. E. E. Schattschneider, "Political Parties and the Public Interest," *Annals of the American Academy of Political and Social Science* 280 (March 1952): 23–24; idem, *Party Government,* 209.
12. Jane Jacobs, *The Death and Life of Great American Cities* [1961] (New York: Vintage, 1992), 117, 118–19, 418.

Index

Adams, Charles Francis, 23, 72
Adams, Henry Carter, 158
Adams, Thomas, 234, 238, 239, 240, 243, 244, 248, 264, 265, 266
Adler, Felix, 77, 158
Aitchison, Clyde Bruce, 90
Allan, Eugene, 32
Allen, William, 156
American Museum of Natural History, 175
American Radiator Building, 176
Ammann, O. H., 26
Anderson, Arthur, 170, 171, 172
Astor, William Waldorf, 211–12

Baker, Benjamin, 26
Baker, George F., 168
Baldwin, Henry De Forest, 154
Baltimore and Ohio (B&O) Railroad, 35
Banfield, Edward C., 273
Barney, Charles T., 168, 169
Bassett, Edward, 112, 270, 275; and planning, 227, 248, 249, 250, 254, 255; and zoning, 188–226, 258
Beard, Charles, 105
Belmont, August, 106, 108, 109
Bennett, Edward, 34
Berle, Adolphe A., Jr., 261
Berry, Charles, 134
Bettman, Alfred, 207, 216
board of aldermen, 40–49
board of estimate: and franchises, 45, 46, 47, 49, 108, 110, 112; and municipal finances, 165, 173, 178; Nelson P. Lewis, chief engineer of, 102, 158; and planning, 231, 245–47, 248, 249, 250, 251–52, 254–55, 256; and sewage, 135; and zoning, 189, 196, 201, 210, 212, 218, 220, 222
borough government, 5, 41, 45; and municipal finances, 148–49; and planning, 230–31, 245–57, 261, 262, 264, 266; and sewage, 133, 134–35; and zoning, 219–20, 222
Boston, 63, 88, 105, 143, 151, 157, 181, 194, 202
Boston Water Works, 130
Bourget, Paul, 1
Brady, Anthony, 21
Brandeis, Louis D., 66, 67, 77
bridges, 116; Bayonne, 93; Brooklyn, 3, 9, 102; George Washington, 26, 93; Goethals, 93; Hell Gate, 28, 77, 86–92, 269; Manhattan, 102; Outerbridge Crossing, 93; Queensboro, 102; Triborough, 259–60; Williamsburg, 102
Bronx, 1, 18, 19, 25, 38, 43, 44, 45, 61, 105, 176, 197; borough president James Lyons, 253, 255; and sewage, 131, 135; and subways, 106, 109, 113; and water, 115, 116; and zoning, 202, 219
Brooklyn, 18, 112, 118; and consolidation, 1, 246; and freight movement, 61, 63, 69–71, 75, 90; industry, 236, 237, 238; and municipal finances, 148–49, 175, 176; and planning, 250, 251, 252, 255; population, 3, 17, 105, 190, 226, 239, 270; and PRR franchise battle, 38, 43, 45, 46, 47, 49; and railroads, 25, 27, 28, 32, 34, 50; and sewage, 126–27, 131; and subways, 106, 108, 109, 110, 113; and water, 4, 114, 115–16, 117, 121, 125; and zoning, 188, 197, 202, 219–20, 221, 231
Brooklyn Marginal Railroad (BMR), 69–72, 173
Brooklyn Museum of Arts and Sciences, 175–76

Brooklyn Rapid Transit (BRT), 21, 109
Brown, Elon, 173
Bruère, Henry, 156, 164–65, 166, 181, 248
Bryce, James, 9
Byrne, James, 250
Bureau of Municipal Research, 150, 156, 158, 162, 178
Burnham, Daniel, 188, 232–33
Bush, Irving, 77
Bush Terminal Company, 63, 69, 70

Caro, Robert, 228
Cassatt, Alexander J., 23, 26, 29, 55, 56; and PRR franchise battle, 39–40, 44, 49
Cassidy, "Curly Joe," 134
Catskill water system. *See* water
Central Federated Union, 42
Central Park, 9
centralization, 38–39, 47, 246–47, 253
chamber of commerce, 42, 71, 77, 110, 137
Charles River Bridge case, 87, 89, 90, 92
charter, 5, 266; and franchises, 39, 40, 45, 46, 49, 70; and municipal finance, 162, 173, 179; and planning, 246, 249, 252–57, 261, 263; and public works, 133, 246–47. *See also* borough government; consolidation; Green, Andrew
Chicago, 272; finances, 308 n. 68; sewage, 302 n. 83; vs. New York, 1, 3, 279
Chrysler Building, 176
city as a system. *See* system ideal
city as a whole, 6, 11, 45, 48, 49, 95, 104, 111, 271, 272, 273, 278; and municipal finances, 140, 141, 165–66, 182; and planning, 228–29, 230, 246, 247–49, 250, 252, 253, 255, 256–57, 259, 260, 262, 266; and zoning, 189, 198, 218, 219, 221
City Beautiful movement, 191, 194, 209, 231
City Island, 134
civic culture: and aldermen, 41, 45, 47–48; and engineers, 31–37; of expertise, 5, 7–14, 18, 55–57, 92–97, 269, 271, 279–80; and municipal finances, 181–82; and planning, 228–29; and railroads, 31, 47–48, 55–57
Clark, Edgar, 83
Clean Water Act (1972), 138
Cleveland, Frederick, 156
Cohen, Julius Henry, 76–80, 128, 189, 234, 270
Coke, Edward, 188

Coler, Bird S., 118, 119, 149
Columbia University, 150
Committee on the Regional Plan, 35, 229–30, 232–45, 270
commuting, 3, 18–19
comptrollers, 156–57
Coney Island, 131, 136
congestion, 3, 11, 109, 186–88, 223, 227–28, 232
Connecticut, 137, 237, 239, 264, 271
Connolly, Maurice, 134–35
consolidation, 1, 2, 5, 6, 49, 117, 118, 246, 266, 279; and municipal finance, 107, 144, 149, 151
Converse, E. C., 21
Corbin, Austin, 21
Corbett, Harvey Wiley, 248
corruption, 10, 46, 134–35, 218–19
County, Albert J., 87, 92, 109, 110
Craig, Charles, 79, 174–76
Cresson, Benjamin Franklin, Jr., 33–34
Croker, Richard, 41, 42, 43

Daley, Richard J., 48
Davies, John Vipond, 21, 26, 29, 33, 34, 54, 269
Davis Bournonville Company, 51
Davison, Henry, 170, 171
Day, Joseph, 137, 139
Delano, Frederic A., 232–34, 240, 248, 264
Delaware, Lackawanna & Western (DL&W) Railroad, 23
Detroit, 301 n. 68, 308 n. 68
Dewey, John, 268, 273–75
De Witt, Benjamin Park, 9
Dietz, John J., 41
diffuse centralization, 7, 272
distributive politics, 11, 12, 253, 267, 276; and franchises, 39–40, 48–49; and municipal finances, 144–45, 146, 147, 148, 149, 164–66, 175–76, 181–82; and planning, 251, 252, 253, 267; and public works, 104–5, 113, 123–24, 125–26, 135, 141–42; and zoning, 188, 192, 220, 222
docks department, 58–59, 60, 70, 163
Doig, Jameson W., 79
Dowling, Frank, 41
Doyle, William F., 219
Durand, Edward Dana, 154
Dutchess County, 116, 124

East River Gas tunnel, 21
Edge, Walter, 79
efficiency, 7–8, 191–92, 229; and municipal finances, 144, 157, 164, 170, 172, 173; and public works, 105, 114, 119, 124, 125–26, 140; and railroads, 63, 66, 76, 78, 80, 86
Eidlitz, Otto, 209
Einhorn, Robin L., 272–73
electric trains, 26
Elsberg Bill (1906), 107–8
Emerson, Haven, 205
engineers, 7, 10, 12, 18, 72, 76, 145, 147, 212, 222, 244, 250; and public works, 103, 115, 116, 117, 121, 123, 124, 125, 133, 134; and railroads, 31–37, 47–48, 50, 55–56, 59, 60, 70, 78, 79, 93. *See also* Allan, Eugene; Ammann, O. H.; Baker, Benjamin; Cassatt, Alexander J.; County, Albert J; Davies, John Vipond; Delano, Frederic A.; Delano; Flinn, Alfred D.; Forgie, James; Freeman, John Ripley; Fuller, W. B.; Greathead, James; Jacobs, Charles M.; Lewis, Nelson P.; Lindenthal, Gustav; McAllister, Daniel B.; McCrea, James; Rea, Samuel; Ropes, Horace; Seely, Frederick; Smith, J. Waldo; Soper, George A.; Stuart, Francis Lee; Sullivan, John F.; Taylor, Frederick Winslow; Tuttle, Arthur; Wild, Herbert; Wilgus, William
Equitable Building, 185–87, 192
Erie Railroad, 23, 24, 29, 35, 60
Euclid v. Ambler, 216–17

Fenner, Burt, 209
ferries, 19, 23
Fifth Avenue, 192, 194, 236
fiscal crises, 144, 167–73, 178, 252–53
Fisk, Pliny, 21
Flagg, Ernest, 3
Flinn, Alfred D., 120, 126
Flushing Bay, 134, 136
Ford, George B., 189, 190, 191, 192–93, 208, 209, 223, 245, 270
Forgie, James, 33, 36, 53–54
franchises, 37–49, 95
free riders, 104–5, 106, 113, 114, 126
Freeman, John Ripley, 119–21, 123, 124, 140, 141, 270
Freund, Ernst, 196
friction of space, 238, 240, 241, 242, 246, 261

Fuller, W. B., 120
Fusion, 6, 41, 42, 152, 170, 174

Gaffney, James, 44
garment industry, 191, 236–37
Gary, Elbridge, 21
Gaynor, William J., 108, 124, 157, 158, 163
general interests, 5, 59, 113, 256–57, 258, 260, 264, 272–73, 275, 277; aggregative and disaggregative approaches to, 12–14, 94, 131, 166, 261–63, 276
George, Henry, 158, 159, 161
Gleason, "Battle Axe," 134
Godkin, E. L., 147
Grand Central Station, 35, 67, 68, 71
Gravesend Bay, 134
Greathead, James, 26, 33
Green, Andrew, 5, 229, 246, 247, 256

Haber, Samuel, 76
Haig, Robert Murray, 178–80, 235–38, 242
harbor. *See* port
Harvey, George U., 135, 250, 251, 253
Harvey Fisk & Sons, 21
Haskins, De Witt, 20, 25
Hastings Commercial Club, 82–83
Hawley, Edwin, 108
Hayek, Friedrich, 259
Hays, Samuel, 38
Hearst, William Randolph, 44, 47, 108, 110, 113
Heinze, F. Augustus, 168
Hepburn Act (1906), 66
Hepburn Committee, 24
Herring, Pendleton, 96, 143, 266, 275–76
Heydecker, Wayne, 235
Higham, John, 5
Hillquit, Morris, 77
Hofstadter, Richard, 273
Holland Tunnel, 32, 33
Hudson and Manhattan (H&M) Railroad, 19–24, 29, 50, 60
Hughes Commission, 45
Hylan, John Francis, 79, 113–14, 174–76, 197, 231, 247, 249, 258, 266

institutions of collective decision making, 2, 4, 5, 13, 188, 192, 223, 229, 231, 238, 269, 277

Interborough Rapid Transit (IRT) Company, 106–14
interdependence, 10–12, 103–4, 273–75, 279; financial, 145; and planning, 234–43, 267; and water, 123, 126; and zoning, 198
Interstate Commerce Commission (ICC), 60, 64–72, 74–78; joint use of terminals, 81–92; recaptured earnings, 81; retained earnings, 31, 64–69; weak railroad problem, 68, 81, 84
Interstate Commerce Commission cases: *Advance Rate*, 64–65, 66; *Anthracite Coal*, 71; *Central Yellow Pine Association*, 64, 65–66; *Eastern Rate*, 66, 71; *Five Percent*, 67; *Hastings Commercial Club*, 82–83, 85–86, 90; *Hell Gate Bridge*, 86–93; *New York Harbor*, 76–77, 80–86, 89, 90; *Port Arthur Chamber of Commerce*, 85, 90; *York Manufacturers Association*, 85, 90. *See also* Transportation Act (1920)
Interstate Sanitation Commission. *See* sewage
invisible government, 8, 49, 63, 64, 65, 70

Jacobs, Charles M., 21, 26, 29, 31, 33, 54, 61, 269
Jacobs, Jane, 277–78
John Wanamaker Store argument, 87
Johnson, Joseph, 163

Kates, Phillip, 196
Kelly, Frank V., 255
Kennedy, David M., 73
Ketten, H. G., 51
Kimball, Fiske, 29
Kipling, Rudyard, 228, 231
Kleinfeld, Phillip, 250
Knickerbocker Trust, 168
Knopf, Adolphus, 205
Koch, Ed, 125

La Farge, C. Grant, 208
La Guardia, Fiorello, 48, 175; and planning, 253–57, 261, 264–65; and sewage, 136, 137, 139, 140
laissez faire, 9, 147
Lamont, Thomas, 171
Lawrence, David, 48
Lehman, Herbert, 178

L'Enfant, Pierre Charles, 232
Levy, Samuel, 255
Lewis, Harold, 267
Lewis, Nelson P.,. 102, 158, 189, 223
Lindenthal, Gustav, 26, 28, 59
Lindner, Walter, 197, 221
Lippmann, Walter, 17, 55
Littleton, Martin, 47
localism, 5, 38; and municipal finances, 164–66, 175–76, 181–82; and planning, 229, 230, 244, 246, 257, 262, 267; and public works, 104, 126–28, 130–42; and zoning, 218–23
London, 2
Long Island City, 86, 134
Long Island Railroad (LIRR), 21, 26, 29, 32, 34, 50, 71, 86, 90, 222
Lorillard, P., & Company, 51, 54
Low, Seth, 42, 163
Lyons, James J., 253, 255

Madison, James, 13, 276
Maloney, Thomas, 51
Maltbie, Milo, 154
Man, Alrich, 197
Mangel Box factory, 51, 54
Manhattan: alderman, 41–42; assessed values, 148, 162; central business district, 3, 17, 18, 105; and consolidation, 1, 246; industry, 115, 235–38, 241; Lower East Side, 3, 190; and planning, 241, 252, 254, 256, 258, 260, 261, 262, 264, 266; population, 105–6, 229, 239, 251, 253, 264; and PRR franchise battle, 38, 43, 45, 48; and railroads, 21, 23, 24, 25, 26, 27, 28, 29, 35, 37, 50, 53, 54, 269; and sewage, 127, 131, 135; subterranean structures, 36, 121; and subways, 106, 108, 109, 111; tax delinquencies, 265; tax rate, 174; and water, 115, 116, 117, 119, 121; waterfront, 2, 18–19, 61–63, 69, 70, 78, 93; and zoning, 202. *See also* congestion; skyscrapers
Manhattan Elevated, 108
Manhattan Transfer, 50
Mann-Elkins Act (1910), 66
Marion, New Jersey, 51–54, 95
Marsh, Benjamin, 161
Matthews, Nathan, Jr., 143, 157

Mazet Committee, 41
McAdoo, William Gibbs, 17, 19–24, 56, 81, 109, 171, 269
McAllister, Daniel B., 32–33
McAneny, George, 46, 140, 141, 270; and municipal finance, 165–66; and planning, 233, 248, 249; and subways, 111–14; and zoning, 192, 193, 223
McCarren, "Long Pat," 46
McClellan, George B., 45, 47, 108, 157, 163, 168
McCooey, John, 250
McCrea, James, 109, 110
McKim, Mead & White, 28–29
Mellen, Charles, 39, 49
Merchants' Association, 42, 76, 119, 125, 135
Merriam, Charles, 156
metropolitan dilemma, 229–30
Metropolitan Life Insurance Company, 220–21; Tower, 3, 185, 205
Metropolitan Museum of Art, 175
Metz, Herman, 156, 168, 169, 270
Meyer, Charles, 254
Milef Realty Corporation, 219
Miller, Cyrus, 197
Mills, Ogden, 177
Mitchel, John Purroy, 157, 158, 163, 170, 173, 178, 186, 231, 258
Mitchell, Wesley, 177
Moore, F. C., 119
Morgan, J. P., 168–169, 170
Morgan, J. P., Jr., 170, 171–72, 176, 211–12
Morrow, Dwight, 170, 171
Moses, Robert, 4, 136, 228, 248, 257–64, 265, 266, 272
Moskowitz, Belle, 248, 258
muckraking, 9, 14
Mumford, Lewis, 243, 259
municipal finances, 281–83; assessed values, 162, 174, 176; budget, 150–51, 176, 179; debt, 146–49, 151–52, 155, 162, 163–65, 166–67, 174, 176, 179, 180; municipal bonds, 152; philosophy of, 153–55, 159–61, 164–66, 172–73; and subways, 107; tax rate, 150, 161, 174; and water, 117. *See also* Bruère, Henry; Bureau of Municipal Research; comptrollers; fiscal crises; Marsh, Benjamin; Metz, Herman; pay-as-you-go policy; Prendergast, William; Seligman, E. R. A.
Municipal Ownership League, 44
Murphy, Charles Francis, 43–45, 47, 48, 113, 247

New Jersey, 2, 3, 4, 125; commuters, 17, 20, 269; Danbury, 235; Elizabeth, 235; Harrison, 50; Hoboken, 21, 23, 63; industry, 235, 236, 237, 238; Jersey City, 23, 26, 35, 50, 51, 63, 127, 137, 214, 235, 241; Newark, 27, 69, 127, 137, 214, 235, 236; New Brunswick, 236; Paterson, 235; Perth Amboy, 235; population, 18–19, 239; and railroads, 21, 23, 25, 26, 27, 35; regional planning, 240, 242, 264, 271; Roselle, 235; and sewage, 127–28, 129, 137, 138; Weehawken, 35, 60; and zoning, 212, 214–15, 221. *See also* Marion; Passaic River; West Side Avenue Station case; zoning
New Jersey Board of Public Utility Commissioners, 51
Newmeyer, R. Kent, 88
New York Bay Pollution Commission. *See* sewage
New York Central Railroad, 19, 61, 77, 86, 110
New York Connecting Railroad, 28, 38, 44, 45, 47, 49, 86
New York Contracting and Trucking Company, 43–44
New York Dock Company, 63, 69, 70
New York, New Haven & Hartford Railroad, 27–8, 39, 44, 71
New York, New Jersey Port and Harbor Development Commission, 78
Nixon, S. Fred, 118
Nolen, John, 34
Norton, Charles Dyer, 232–33, 240, 254, 264

Oakman, Walter G., 21
O'Brien, Morgan J., 256
O'Dell, Benjamin, 118
Olmsted, Frederick Law, Jr., 34, 233
Olvany, George, 248
Outerbridge, Eugenius, 77, 95

Passaic River, 127–28, 129
Patton, Simon, 156

pay-as-you-go policy, 171–76, 177, 180
Penn Station, 28–29, 67, 68, 109, 269, location, 29–30, 40
Pennsylvania policy, 66–67
Pennsylvania Railroad (PRR), 21, 24–57, 60, 65, 71, 95
Perkins, Frances, 260, 262
Peters, Ralph, 70–71, 110
Phillips, John, 134
planning, 11, 12, 25, 36, 48, 103, 266; Committee on the City Plan, 189, 197, 250, 256, 258; and engineers, 33–35, 37; entrepreneurial, 19–24; freight vs. passenger, 293n. 3; legal aspects of, 193, 196, 197, 201, 203, 207, 208, 217, 218; Mayor's Committee on Plan and Survey, 248–49; and municipal finances, 144, 158, 161, 164–66, 180; port, 58–60, 64, 69, 70, 74, 75, 77, 78, 82, 86, 90, 92, 93, 94, 95, 97; railroad, 24–31, 38, 39–40, 46, 49, 50, 68; regional, 227–67; and sewage, 135, 136; and subways, 106, 108, 109, 110, 112–13; Veblenian, 60, 75–76, 87, 90, 96; and zoning, 188, 189, 191, 192, 205, 207, 209, 220, 222–23, 226. See also *Regional Plan of New York and Its Environs*
Platt, Harold L., 273
Platt, Thomas C., 39
police power. *See* zoning
port, 1, 2, 34, 35, 146, 234, 242, 257, 275; and railroads, 58–63, 73, 74, 75–80, 82, 86–87, 90, 92–93, 95, 265, 270
Port of New York Authority, 26, 34, 51, 55, 59, 74–80, 86–97, 126, 128, 180, 230, 242, 243, 248, 270, 271
Powell, Thomas Reed, 216
Prendergast, William, 156–57, 163, 170, 181, 270
privatism, 4–5, 104, 257
Public Service Commission (New York State), 108, 110
Pulitzer, Joseph, 108
Purdy, Lawson, 189, 193, 196, 208, 223, 248

Queens, 1, 3, 9, 17, 61, 247, 251; and industry, 236, 238; and municipal finances, 149, 176; and planning, 234, 252; population growth in, 239, 251, 253, 270; and railroads, 25, 26, 27, 28, 38, 63, 75, 86, 90; and sewage, 134–35, 136; 250; and subways, 105, 109, 110, 176; and water, 114, 116, 125; and zoning, 190, 202, 219, 220, 222, 226

Ramapo Water Company. *See* water

Rapid Transit Commission, 40, 108
Rea, Samuel, 19, 23, 25–26, 54, 56, 68–69, 78, 93–94, 109, 110–11, 269
Real Estate Board of New York, 125, 263
Regional Plan of New York and Its Environs, 11, 112, 135, 178, 212, 232–45, 259, 263, 264, 265–66, 267, 270, 271–72, 275, 278. *See also* planning
Richmond. *See* Staten Island
Richter, Frederick, 41
Riis, Jacob, 190
Roberts, George B., 25
Robinson, Charles Mulford, 34
Rockaways, 136
Rodgers, Cleveland, 268, 260
Rogers, Lindsay, 178
Roosevelt, Franklin, 137, 232, 256
Ropes, Horace, 120
Rousseau, Jean Jacques, 13

Sandy Hook, 131
San Francisco, 44, 63, 108, 119, 152, 168
Savarese, John, 248
Schattschneider, E. E., 276–77
Schlicher, Frank, 51
Schuyler, Montgomery, 191
Seabury, Samuel, 44, 253, 255
Seely, Frederick, 134
segmented system, 104
Seligman, E. R. A., 158–61, 177, 178, 181, 189, 235
sewage, 4, 126–39, 140–41, 142; Interstate Sanitation Commission, 137–39, 141, 179, 243, 270; Metropolitan Sewage Commission, 128–33, 137, 173; New York Bay Pollution Commission, 128; Passaic Valley Sewerage Commission, 127–28, 129, 137; sewer explosions, 101–2; Tri-State Anti-Pollution Commission, 137
Shepard, Edward Morse, 46, 112
Simkovich, Mary, 248
Simmons, Edward H. H., 1
Singer Building, 3

skyscrapers, 3, 115, 176, 185–88, 191, 225–26
Smith, Al, 48, 251, 253, 254, 258, 260
Smith, J. Waldo, 120
Soper, George A., 130, 133, 135, 136, 137, 140, 141, 189, 233, 270
Southern Pacific Railroad, 44, 108
Stabler, Walter, 220–21
Staten Island, 1, 43, 105, 114, 121, 149, 174, 190, 202, 247, 251, 256
Steffens, Lincoln, 9, 14
Stuart, Francis Lee, 35
Stillman, James, 168
Story, Joseph, 87
Swanstrom, J. Edward, 45
subways, 105–14; Seventh Avenue, 29–30, 37, 109–10, 293n. 3
Suffolk County, 116, 124
Sullivan, John F., 250–51
Sun Tzu, 188, 208
system ideal, 11, 269–70; and municipal finances, 145; and planning, 235, 240, 244, 267; and public works, 103, 104, 111, 139–42; and railroads, 38, 55–56, 57, 58–59, 60, 70, 75–76, 78, 79, 80–81, 87, 89–90, 94–97, 271

Taft, William Howard, 80–81
Tallman Island, 136
Tammany Hall, 6, 250, 251, 252, 259; and charter revision, 253, 254, 255, 256; and municipal finances, 152, 154, 157, 163, 174, 175; and PRR franchise battle, 38–49; and public works, 106, 107, 113, 117, 118, 135
Taney, Roger B., 87
Taylor, Frederick Winslow, 66
Thatcher, Thomas D., 254–55
Tobin, Austin, 93
Tomkins, Calvin, 34, 58–59, 69–72, 74, 92, 234
Tracy, Benjamin F., 118
Transportation Act (1920), 74, 79–86, 89–90, 95, 97, 158
Transportation Reform League of Brooklyn, 45
Tugwell, Rexford, 261–63, 265, 266
Tuttle, Arthur, 222
Tweed, William Marcy, 10, 42, 145–47, 149, 150, 281

Union freight terminal, 93
Union Station (Washington, D.C.), 65, 67

Vanderbilt, Cornelius, 17, 24
Van Kleeck, Mary, 177
Van Wyck, Robert, 247
Veblen, Thorstein, 60, 72, 75–76, 86, 87, 89, 90, 94, 96, 156, 262
Veiller, Lawrence, 196, 214, 248
voluntarism, 5, 24, 37, 55–57, 59, 60, 87, 94, 96–97, 245, 257

Waldorf-Astoria Hotel, 36
Walker, Jimmy, 11, 134, 135, 177–78, 254; and city planning, 227, 239, 247–53
Walsh, John, 196
Walsh, William E., 218, 219
Ward's Island treatment plant, 131, 134, 135–36
water, 3–4, 114–26; and Brooklyn, 114, 115–17; Catskill system, 120–23, 125; Croton Aqueduct, 3, 9, 114, 116, 117, 120; leakage, 121–26; and Manhattan, 115, 116, 117, 119, 121; Ramapo, 117–18, 119
West Side Avenue Station case, 49–54, 57
west side freight problem, 61–63
Whipple, George, 205
Whitman, Charles, 79, 174
Whitten, Robert, 189, 190, 193, 199–201, 202, 203, 208, 223, 270
Wilcox, Delos, 153
Wild, Herbert, 33
Wilgus, William, 34–35, 59, 78, 234, 248
Williamson, Frederick, 77
Wilson, James Q., 273
Wilson, Woodrow, 20, 73, 81, 193
Wood, Fernando, 46
Woolworth Building, 3, 185, 191
World War I: and municipal finances, 144, 151, 153, 166, 170; and railroads, 2, 59, 64, 67, 68, 72–74

Yonkers, 121

zoning, 185–226, 231; and aesthetics, 195–96, 208–10; board of appeals, 210–12, 214, 218–19; in New Jersey, 214–15; and police power, 188, 194–201, 202, 204, 206–26